GLOBAL OPTIMIZATION

From Theory to Implementation

Nonconvex Optimization and Its Applications

VOLUME 84

GLOBAL OPTIMIZATION

From Theory to Implementation

Edited by

LEO LIBERTI
Politecnico di Milano, Italia

NELSON MACULAN
COPPE, Universidade Federal do Rio de Janiero, Brasil

Springer

e-ISBN: 0-387-30528-9
ISBN: 978-1-4419-3930-2 e-ISBN: 978-0-387-30528-8

Printed on acid-free paper.

AMS Subject Classifications: 90C26, 90C30, 90C11, 65K05

9 8 7 6 5 4 3 2 1

springeronline.com

To Anne-Marie and Kja

Contents

Preface

The idea for this book was born on the coast of Serbia-Montenegro, in October 2003, when we were invited to the thirtieth Serbian Conference on Operations Research (SYM-OP-IS 2003). During those days we talked about many optimization problems, going from discussion to implementation in a matter of minutes, reaping good profits from the whole "hands-on" process, and having a lot of fun in the meanwhile. All the wrong ideas were weeded out almost immediately by failed computational experiments, so we wasted little time on those. Unfortunately, translating ideas into programs is not always fast and easy, and moreover the amount of literature about the implementation of global optimization algorithm is scarce.

The scope of this book is that of moving a few steps towards the systematization of the path that goes from the invention to the implementation and testing of a global optimization algorithm. The works contained in this book have been written by various researchers working at academic or industrial institutions; some very well known, some less famous but expert nonetheless in the discipline of *actually getting global optimization to work.*

The papers in this book underline two main developments in the implementation side of global optimization: firstly, the introduction of symbolic manipulation algorithms and automatic techniques for carrying out algebraic transformations; and secondly, the relatively wide availability of extremely efficient global optimization heuristics and metaheuristics that target large-scale nonconvex constrained optimization problems directly.

The book is divided in three parts. The first part is about new global optimization methods. The chapters in the first part are rather theoretical in nature, although a computational experiments section is always present. The second part is oriented towards the implementation, focusing on description of existing solvers and guidelines about building new global optimization software. This part follows two main trends: the first four chapters deal with continuous methods, the last three with combinatorial ones. The third (and last) part presents two applications of global optimization in Data Mining and Molecular Conformation.

More specifically, a lot of work has been carried out on the application of Variable Neighbourhood Search to global optimization (Chapters 6, 8, 10 and 11). A MultiStart-type algorithm based on low-discrepancy sequences generated deterministically has also been thoroughly explored (Chapters 5, 8). A full description of an API for interfacing to metaheuristic codes is given in Chapter 11. Deterministic algorithms can be found in Chapters 1 (Branch-and-Bound algorithms), 3 (a Cutting Plane algorithm), 4 (a Branch-and-Bound based method for stochastic mixed-integer nonlinear problems) and 8 (where the implementation of a spatial Branch-and-Bound algorithm is described).

Chapter 1 and 2 are more theoretical than most other chapters. Chapter 1 considers global optimization problems where the objective functions and constraints are difference of monotonic functions, and proposes some deterministic solution methods; Chapter 2 reports on a special local search method for reverse convex problems. In both chapters, a section on computational results is presented, discussing the efficiency of different solution approaches.

Chapter 4 describes one of the very few existing implementations of a deterministic global optimization software targeting robust nonconvex programming. In order to face the huge computational resources needed to solve multi-scenario nonconvex problems, the author proposes a Branch-and-Bound approach where the lower bounds are computed by solving a nonconvex Lagrangian relaxation through a standard global optimization algorithm. This multi-level solution method requires careful software design to obtain a working implementation.

As has been mentioned, a particularly important development is the introduction of symbolic manipulation algorithms in optimization. Chapter 7 describes a modelling language by which it is possible to keep track of the convexity property of the optimization problem being described. Although Chapter 7 is about convex programming, the role of convexity is so important in Branch-and-Bound based algorithms for global optimization that it was decided to include it in this book. In Chapter 8 the reader can find the description of some symbolic algorithms for differentiation, algebraic simplification and generation of convex relaxations. Chapter 3 introduces some effective convexity transformations for a large class of multilinear problems, as well as discussing some nonlinear cuts. Chapter 10 employs even more sophisticated symbolic techniques about automated theorem proving.

Chapters 9 and 12 describe working implementations of commercial-grade software. In particular, Chapter 9 is about the Lipschitz Global Optimization (LGO) solver suite, and its embedding within the Mathematica software framework; Chapter 12 describes a solver for Mixed-Integer Linear Programming problems (commercialized by Process Systems Enterprise, Ltd.): this software relies on CORBA techniques to automate the parallelization and distributed running of the solver.

As far as the applications are concerned, Chapter 13 describes an extremely interesting class of problems arising in Data Mining and Nonlinear

Classification. Chapter 14 describes a new way to generate instances for the Molecular Distance Geometry Problem, which is one of the hardest problems in Molecular Conformation.

Some of these papers have inter-relations and cross-references, due both to collaborations among the authors and to emergence of new trends in global optimization. Most of these inter-relations have been emphasized by means of footnotes, which have all been added by the editors.

We hope that the reader will find this book interesting and enlightening, and that it will serve as a source of ideas as well as a desktop companion for people who need to implement global optimization software.

Milano, Rio de Janeiro
June 2005

Leo Liberti
Nelson Maculan

Part I: Methods

Optimization under Composite Monotonic Constraints and Constrained Optimization over the Efficient Set

Hoang Tuy and N.T. Hoai-Phuong

Institute of Mathematics, VAST, 18 Hoang Quoc Viet, 10307 Hanoi, Vietnam
{htuy,htphuong}@math.ac.vn

Summary. We present a unified approach to a class of nonconvex global optimization problems with composite monotonic constraints. (By composite monotonic function is meant a function which is the composition of a monotonic function on $\mathbb{R}^n$ with a mapping from $\mathbb{R}^n \to \mathbb{R}^m$ with $m \leq n$.) This class includes problems with constraints involving products of linear functions, sums of ratio functions, etc., and also problems of constrained optimization over efficient/weakly efficient points. The approach is based on transforming the problem into a monotonic optimization problem in the space $\mathbb{R}^p$, which can then be efficiently solved by recently developed techniques. Nontrivial numerical examples are presented to illustrate the practicability of the approach.

Key words: Global optimization. Monotonic optimization, difference of monotonic (d.m.) optimization. Composite monotonic constraint. Nonconvex optimization. Branch-reduce-and-bound method. Constrained optimization over the efficient/weakly efficient set. Multiplicative constraint. Sum-of-ratio constraint.
Mathematics Subject Classification 90C26, 65K05, 90C20, 90C30, 90C56, 78M50

1 Introduction

Convexity is essential to modern optimization theory. However, it is not always the natural property to be expected from many nonlinear phenomena. Another property, perhaps at least as pervasive in the real world as convexity, is monotonicity.

A function $f : \mathbb{R}^n \to \mathbb{R}$ is said to be *increasing* on a box $[a, b] = \{x \in \mathbb{R}^n |\ a \leq x \leq b\}$ if $f(x) \leq f(x')$ whenever $a \leq x \leq x' \leq b$ (throughout this paper, inequalities between vectors are understood in the componentwise sense); it is said to be decreasing if $-f(x)$ is increasing, monotonic if it is either increasing or decreasing. A function which can be represented as a difference of two monotonic functions is referred to as a *d.m. function*. For instance, a polynomial in $x \in \mathbb{R}^n$ with positive coefficients is increasing on $\mathbb{R}^n_+$, whereas an arbitrary polynomial in $x \in \mathbb{R}^n$ is a d.m. function on $\mathbb{R}^n_+$. An optimization problem which can be described by means of d.m. functions is termed a d.m. optimization problem. In particular polynomial programming problems or, more generally, synomial programming problems are d.m. optimization problems.

Obviously, d.m. optimization problems form a very broad class of highly nonconvex problems encountered in different fields of mathematical, physical, engineering and economical sciences. In the last few years a theory of monotonic optimization has emerged which provides a general mathematical framework for the study of these problems. The basic concepts and problems of monotonic optimization, together with the basic methods of polyblock and reverse polyblock (copolyblock) approximation were introduced in [26], while their relation to and their analogy with d.c. optimization were discussed in [29], [32]. Applications to special problems were considered in [16], [27], [12], [30], [32], while improved outer approximation and branch and cut methods were developed in [31], and extended to discrete problems in [34].

Computational experience with numerical methods of monotonic optimization has demonstrated their efficiency for problems of limited dimension ([27], [16], [12], [30]). On the other hand, a host of large scale nonconvex optimization problems can be reduced to monotonic problems of much smaller dimension tractable by conventional monotonic optimization methods. An important class of such problems includes d.m. optimization problems with an additional composite monotonic constraint, i.e. a constraint of the form $h(C(x)) \geq 0$, where $C : \mathbb{R}^n \to \mathbb{R}^m$ is some continuous mapping with $m < n$, and $h : \mathbb{R}^m \to \mathbb{R}$ is an increasing function.

The aim of the present paper is to investigate this class of problems using the approach earlier developed in [26] and recently refined in [31].

In the next section we will first review some basic concepts and results from monotonic optimization. In Section 3 the class of optimization problems involving a composite monotonic constraint will be introduced. From a numerical point of view, our interest in this class stems from the important property that by passing to the image space of the above mentioned mapping C they can be converted into a prototype monotonic optimization problem (Q) of much reduced dimension. This property allows them to be tractable by a unified approach. In Section 4 it will be shown that this class includes as special case the problem of constrained optimization over the weakly efficient set. Although the problem of optimization over the efficient set does not satisfy all requirements for this class, it can be solved via solving a problem (Q), subject to some additional inexpensive manipulations. In Section 5, a method is developed for solving the prototype problem (Q). On the basis of a branch and bound scheme, this method involves, aside from branching and bounding, also a basic operation called reduction aimed at exploiting monotonicity to reduce the size of the current partition sets before computing the bounds. The corresponding algorithm is formulated and its convergence established under mild conditions. In Section 6 several substantial improvements of the algorithm for the problem of optimization over the efficient set are discussed in the most usual case when C is concave. In Section 7, the approach is extended to problems with a composite monotonic objective function, such as multiplicative programs and the like. Finally, Section 8 closes the paper with some nontrivial numerical examples illustrating the practicability of the approach.

2 Some basic concepts and results of monotonic optimization

In this section we review some basic concepts and results from the theory of monotonic optimization as developed in [26], [31] (see also [29], [34]).

In its general form a d.m. optimization problem can be formulated as

$$\min\{f(x)|\ g_i(x) \leq 0\ (i = 1, \ldots, m), x \in [a, b]\}, \qquad \text{(DM)}$$

where $f,\ g_i, i = 1, \ldots, m$, are d.m. functions.

Let $g_i(x) = u_i(x) - v_i(x)$ with $u_i(x), v_i(x)$ being increasing functions. Since $g_i(x) \leq 0 (i = 1, \ldots, m) \Leftrightarrow g(x) := \max_{i=1,\ldots,m} g_i(x) \leq 0$, while

$$g(x) = \max_i [u_i(x) + \sum_{j \neq i} v_j(x)] - \sum_j v_j(x) = u(x) - v(x)$$

with $u(x) = \max_i [u_i(x) + \sum_{j \neq i} v_j(x)],\ v(x) = \sum_j v_j(x)$ being increasing, the set of d.m. constraints $g_i(x) \leq 0,\ i = 1, \ldots, m$ can always be replaced by a single d.m. constraint $g(x) = u(x) - v(x) \leq 0$.

Furthermore, it can be shown (see e.g. [26]) that by simple manipulations and using additional variables, any d.m. optimization problem can be converted into the canonical form

$$\min\{f(x)|\ g(x) \leq 0 \leq h(x),\ x \in [a, b]\} \qquad \text{(MO)}$$

where f, g, h are increasing functions.

In this paper we will assume that the functions $f(x), g(x)$ are l.s.c., and $h(x)$ is u.s.c., so that the sets

$$G := \{x \in [a, b]|\ g(x) \leq 0\},\ H := \{x \in [a, b]|\ h(x) \geq 0\} \qquad (1)$$

are closed. It can be easily verified that these sets have the following properties:

$$x \in G \Rightarrow [a, x] \subset G; \quad x \in H \Rightarrow [x, b] \subset H.$$

These properties are expressed by saying that G is a *normal set*, and H is a *conormal set* (or a *reverse normal set*).

Obviously, $[a,b]$ is the largest normal set in $[a,b]$ and the intersection or the union of an arbitrary collection of normal sets is normal. For any given set $A \subset [a,b]$, the intersection of all normal sets containing A, i.e. the smallest normal set containing A, is called the *normal hull* of A, and is denoted by $A^{\rceil}$. Analogously, the smallest conormal set containing A is called the *conormal hull* of A and is denoted by $\lfloor A$.

Proposition 1. $A^{\rceil} = \cup_{x\in A}[a,x]$, $\lfloor A = \cup_{x\in A}[x,b]$. *If A is compact then so are $A^{\rceil}$ and $\lfloor A$.*

Proof. The set $\cup_{x\in A}[a,x]$ is obviously normal, so $A^{\rceil} \subset \cup_{x\in A}[a,x]$. Conversely, if $y \in [a,x]$ for some $x \in A \subset A^{\rceil}$ then $y \in A^{\rceil}$ because $A^{\rceil}$ is normal. This proves that $A^{\rceil} = \cup_{x\in A}[a,x]$. For any sequence $\{z^r\} \subset A^{\rceil}$ there exists a subsequence $\{z^{r_\nu}\}$ converging to some $\bar{z} \in [a,b]$. Let $z^{r_\nu} \in [a, x^{r_\nu}] \subset A$. By taking a subsequence if necessary, we can assume $x^{r_\nu} \to \bar{x}$ and since A is closed, $\bar{x} \in A$. Thus, $\bar{z} \in [a,\bar{x}]$ with $\bar{x} \in A$, proving that $A^{\rceil}$ is compact. The proof is analogous for $\lfloor A$.

A point $z \in G$ is called an *upper boundary point* of a closed normal set $G \subset [a,b]$ if $(z,b] \subset [a,b] \setminus G$. (here and throughout $(a,b] = \{x|\ a < x \le b\}$, while $[a,b) = \{x|\ a \le x < b\}$). The set of upper boundary points of G is called its *upper boundary* and is denoted by $\partial^+ G$.

Similarly, a point $z \in H$ is called a *lower boundary point* of a closed conormal set H if $[a,z) \subset [a,b] \setminus H$. The set of lower boundary points of G is called its *lower boundary* and is denoted by $\partial^- H$.

Proposition 2. *A closed nonempty normal set G (conormal set H, resp.) has a nonempty upper boundary $\partial^+ G$ (lower boundary $\partial^- H$, resp.) and is just equal to the normal (conormal, resp.) hull of this upper (lower, resp.) boundary. For the normal set G and conormal set H in (1) we have*

$$\partial^+ G \subset \{x \in [a,b] \mid g(x) = 0\}, \quad \partial^- H \subset \{x \in [a,b] \mid h(x) = 0\}.$$

Proof. It suffices to prove the Proposition for a normal set G. Take an arbitrary point $x \in G$ and let $z = a + \lambda(x-a)$ with $\lambda = \max\{\alpha|\ a + \alpha(x-a) \in G\}$. Then $z \in G$ because G is closed and if $u > z$, i.e., $u_i > z_i$, $i = 1,\dots,n$ then $u_i = a_i + \lambda_i(x_i - a_i)$ with $\lambda_i > \lambda$, so that if $\alpha = \min_{i=1,\dots,m} \lambda_i > \lambda$ and $\tilde{u}_i = a_i + \alpha(x_i - a_i)$ then $\tilde{u} = a + \alpha(x-a) \notin G$. Thus, $(z,b] \subset [a,b] \setminus G$, and so $z \in \partial^+ G$. Therefore $\partial^+ G \neq \emptyset$. Since $\partial^+ G \subset G$, we have $\partial^+ G \subset G^{\rceil}$. Conversely, since any $x \in G$ belongs to the line segment joining a to a point $z = a + \lambda(x-a) \in \partial^+ G$, it follows that $G \subset (\partial^+ G)^{\rceil}$, hence $G = (\partial^+ G)^{\rceil}$. The last assertion is obvious in view of the l.s.c. of $g(x)$.

A point $z \in \partial^+ G$ is called an *upper extreme point* of G if $x \in G$, $x \ge z$ always implies that $x = z$. A point $z \in \partial^- H$ is called a *lower extreme point* of H if $x \in H$, $x \le z$ always implies that $x = z$.

Proposition 3. *A closed normal set $G \subset [a,b]$ (conormal set $H \subset [a,b]$, resp.) has at least one upper (lower, resp.) extreme point and is equal to the normal (conormal, resp.) hull of the set V of its upper (lower, resp.) extreme points.*

Proof. In view of Proposition 2 $V^{\rceil} \subset (\partial^+ G)^{\rceil} = G$, so it suffices to show that $\partial^+ G \subset V^{\rceil}$. Let $z \in \partial^+ G$. Define $x^1 \in \mathrm{argmax}\{x_1 | \ x \in G, \ x \geq z\}$, and $x^i \in \mathrm{argmax}\{x_i | \ x \in G, \ x \geq x^{i-1}\}$ for $i = 2, \dots, n$. Then $v := x^n \in G$ and $v \geq x$ for all $x \in G$ satisfying $x \geq z$. Therefore, $x \in G, \ x \geq v$ implies that $x = v$. This means that $z \leq v \in V$, hence $z \in V^{\rceil}$, as was to be proved.

3 Problems with composite monotonic constraints

Let there be given a compact convex set $D \subset \mathbb{R}^n$, a continuous mapping $C : \mathbb{R}^n \to \mathbb{R}^m$, an u.s.c. increasing function $h : \mathbb{R}^m_+ \to \mathbb{R}$, together with a continuous d.m. function $f : D \to \mathbb{R}$. Consider the problem

$$\min\{f(x) | \ x \in D, \ h(C(x)) \geq 0\}. \tag{P}$$

This problem includes as special case the problem earlier studied in [27] where $f(x)$ is linear and $C_i(x), \ i = 1, \dots, m$ are concave. Since $C(D) = \{y = C(x), x \in D\}$ is compact we can assume that it is contained in some box $[a,b] \subset \mathbb{R}^m_+$. Define the set

$$\Omega = \{y \in [a,b] | \ h(y) \geq 0\} \tag{2}$$

and for each fixed $y \in [a,b]$, let

$$\varphi(y) := \min\{f(x) | \ x \in D, \ C(x) \geq y\}. \tag{R(y)}$$

Proposition 4. *Ω is a conormal and compact set, while $\varphi(y)$ is an increasing function, convex if $f(x)$ is convex and $C_i : \mathbb{R}^n \to \mathbb{R}, \ i = 1, \dots, m$, are concave functions.*

Proof. The set Ω is conormal and compact because $h(y)$ is increasing and u.s.c.. The function $\varphi(y)$ is increasing because every x feasible to R(y) is feasible to R(y') whenever $y' \leq y$. Let $y^1, \ y^2 \in [a,b], \ \alpha \in [0,1]$ and $\varphi(y^i) = f(x^i), \ i = 1,2$. Then $x = \alpha x^1 + (1-\alpha)x^2 \in D$, and if $C_i, \ i = 1, \dots, m$, are concave then $C(x) \geq \alpha C(x^1) + (1-\alpha)C(x^2) \geq \alpha y^1 + (1-\alpha)y^2 = y$, so that $\varphi(y) \leq f(x) \leq \alpha f(x^1) + (1-\alpha)f(x^2) = \alpha\varphi(y^1) + (1-\alpha)\varphi(y^2)$, proving the convexity of $\varphi(y)$.

Proposition 5. *Problem* (P) *is equivalent to the problem*

$$\min\{\varphi(y) | \ y \in \Omega\}. \tag{Q}$$

Specifically, if y^* solves (Q) *then any optimal solution x^* of* R(y^*) *solves* (P) *and conversely, if x^* solves* (P), *then $y^* = C(x^*)$ solves (Q).*

Proof. This can be easily seen by writing the problem as

$$\begin{aligned}
&\min_{x,y}\{f(x)|\ x \in D,\ C(x) \geq y,\ y \in [a,b],\ h(y) \geq 0\} \\
&= \min_{y\in[a,b],h(y)\geq 0}\ \min_{x}\{f(x)|\ x \in D,\ C(x) \geq y\} \\
&= \min\{\varphi(y)|\ y \in [a,b],\ h(y) \geq 0\}.
\end{aligned}$$

In more detail, let y^* be an optimal solution of (Q) and let x^* be an optimal solution of $(\mathrm{R}(y^*))$, so that $\varphi(y^*) = f(x^*)$, $x^* \in D$, $C(x^*) \geq y^*$. For any $x \in D$ satisfying $h(C(x)) \geq 0$, we have $h(y) \geq 0$ for $y = C(x)$, so $f(x) \geq \varphi(y) \geq \varphi(y^*) = f(x^*)$, proving that x^* solves (P).

Conversely, let x^* be an optimal solution of (P), so that $x^* \in D$, $h(y^*) \geq 0$ with $y^* = C(x^*)$. For any $y \in \Omega$ satisfying $h(y) \geq 0$, we have $\varphi(y) = f(x)$ for some $x \in D$ with $C(x) \geq y$, hence $h(C(x)) \geq h(y) \geq 0$, so that x is feasible to (P), and hence $\varphi(y) = f(x) \geq f(x^*) \geq \varphi(y^*)$, i.e. y^* is an optimal solution of (Q).

Remark 3.1 Define the normal compact set

$$\Delta = \{y \in [a,b] \mid y \leq C(x),\ x \in D\}. \tag{3}$$

Since $\varphi(y) = +\infty$ for $y \notin \Delta$ the problem (Q) can also be written as

$$\min\{\varphi(y)|\ y \in \Delta \cap \Omega\}. \tag{4}$$

Thus, (P) is equivalent to (Q) which is a conventional monotonic optimization problem in $\mathbb{R}^m$ (with m usually much smaller than n).

Note, however, that the objective function $\varphi(y)$ in (Q) is given implicitly as the optimal value of a subproblem $\mathrm{R}(y)$. In the cases of interest to be considered in the sequel, $\mathrm{R}(y)$ is solvable by standard methods, while the problem (Q) itself can be solved by monotonic optimization techniques previously developed in [26] and recently refined in [31].

Below are some of the most important examples of problems of the class (P).

Example 3.1

$$\min\{f(x)|\ x \in D,\ \sum_{i=1}^{m} \frac{u_i(x)}{v_i(x)} \geq 1\}, \tag{5}$$

where D is a polytope, $f(x),\ u_i(x), v_i(x),\ i = 1,\ldots,m$ are continuous d.m. functions such that $v_i(x) > 0\ \forall x \in D$ and the set $\{y|\ y_i = \frac{u_i(x)}{v_i(x)}\ (i = 1,\ldots,m),\ x \in D\}$ is contained in a box $[a,b] \subset \mathbb{R}^m_+$.

Here $C_i(x) = u_i(x)/v_i(x)$, $h(y) = \sum_{i=1}^m y_i - 1$, so this problem can be rewritten as a problem (Q) with

$$\varphi(y) = \min\{f(x)|\ x \in D,\ y_i v_i(x) \le u_i(x)\ (i = 1, \dots, m)\}, \tag{6}$$

$$\Omega = \{y \in [a,b]|\ \sum_{i=1}^{m} y_i \ge 1\}. \tag{7}$$

Example 3.2

$$\min\{f(x)|\ x \in D,\ \max_{i=1,\dots,m} \frac{u_i(x)}{v_i(x)} \ge 1\} \tag{8}$$

with $D,\ f(x),\ u_i(x),\ v_i(x)$ as in Example 3.1. Here $C_i(x) = u_i(x)/v_i(x)$, $h(y) = \max\{y_1, \dots, y_m\}$, so the problem is equivalent to (Q) with $\varphi(y)$ as in (6) and

$$\Omega = \{y \in [a,b]|\ \max_{i=1,\dots,m} y_i \ge 1\}.$$

Example 3.3

$$\min\{f(x)|\ x \in D,\ \prod_{i=1}^{m} \frac{u_i(x)}{v_i(x)} \ge 1\} \tag{9}$$

with $D,\ f(x),\ u_i(x),\ v_i(x)$, as in Example 3.1. Here $C_i(x) = u_i(x)/v_i(x)$, $h(y) = \prod_{i=1}^{m} y_i - 1$, so the problem can be rewritten as a problem (Q) with $\varphi(y)$ as in (6) and

$$\Omega = \{y \in [a,b)|\ \prod_{i=1}^{m} y_i \ge 1\}.$$

A special case of this problem when $v_i(x) \equiv 1$ has been studied earlier in [27].

Example 3.4

$$\min\{f(x)|\ x \in D,\ \sum_{j=1}^{r} c_j \prod_{i=1}^{m} \left[\frac{u_i(x)}{v_i(x)}\right]^{\alpha_j} \ge 1\} \tag{10}$$

with r being a natural number, c_j, α_{ij} being real positive numbers and $D,\ f(x),\ u_i(x),\ v_i(x)$ as in the previous examples. Here again $C_i(x) = u_i(x)/v_i(x)$, $h(y) = \sum_{j=1}^{r} c_j \prod_{i=1}^{m} [y_i]^{\alpha_{ij}}$, so the problem is equivalent to (Q) with $\varphi(y)$ as in (6) and

$$\Omega = \{y \in [a,b]|\ \sum_{j=1}^{r} c_j \prod_{i=1}^{m} [y_i]^{\alpha_{ij}} \ge 1\}.$$

Example 3.5 (Constrained optimization over the weakly efficient set)

$$\min\{f(x)|\ x \in X,\ g(x) \le 0,\ x \in X_{WE}\} \tag{OWE}$$

where $f(x),\ g(x)$ are continuous d.m. functions, X is a compact convex set in $\mathbb{R}^n$, and X_{WE} denotes the set of weakly efficient points of X w.r.t. a given continuous mapping $C : X \to \mathbb{R}^m$.

Recall that a point $x^0 \in X$ is said to be *weakly efficient* w.r.t. C if there is no point $x \in X$ such that $C(x) > C(x^0)$.

Up to now the problem (OWE), a weak variant of the problem (OE) to be discussed in the next Section, has been studied mostly in the special case when $f(x), C_i(x)$ are linear, $g(x) \equiv 0$, and X is a polytope (see e.g. [25], [15] and references therein). Although multi-ratio measures of efficiency such as return/investment, return/risk, cost/time, output/input occur in many practical applications, the case when $C_i(x) = u_i(x)/v_i(x)$ with $u_i(x), v_i(x)$ affine, while $f(x)$ is linear and X is a polytope, has received attention only recently, and to a much lesser degree ([17], [18], [19]).

Assume that $C(X) := \{C(x)|\ x \in X\} \subset [a,b] \subset \mathbb{R}^m_+$ and let

$$\Gamma = \{y \in [a,b] \mid y \leq C(x),\ x \in X\}. \tag{11}$$

Proposition 6. *Γ is a normal compact set such that $\Gamma = C(X)^{\rceil}$ and*

$$x \in X_{WE} \Leftrightarrow C(x) \in \partial^+\Gamma. \tag{12}$$

Proof. Clearly Γ is normal; it is compact because of the continuity of $C(x)$ and the compactness of X. Since $C(X)^{\rceil} = [a,b] \cap (C(X) - \mathbb{R}^m_+) = \{y \in [a,b]|\ y \leq z,\ z = C(x),\ x \in X\}$, the equality $\Gamma = C(X)^{\rceil}$ follows. To prove (12) observe that if $x^0 \in X_{WE}$ then $y^0 := C(x^0) \in \Gamma$ and there is no $x \in X$ such that $C(x) > C(x^0)$ i.e. no $y = C(x), x \in X$, such that $y > C(x^0) = y^0$, hence $y^0 \in \partial^+\Gamma$. Conversely, if $x^0 \in X$ satisfies $C(x^0) \in \partial^+\Gamma$, then there is no $y \in \Gamma$ such that $y > C(x^0)$, i.e. no $x \in X$ with $C(x) > C(x^0)$, hence $x^0 \in X_{WE}$.

Proposition 7. $\Gamma = \{y \in [a,b] \mid h(y) \leq 0\}$, *where*

$$h(y) = \min\{t|\ y_i - C_i(x) \leq t, i = 1,\dots,m,\ x \in X\} \tag{13}$$

is an increasing, continuous function.

Proof. By definition, $y \in \Gamma$ if and only if there exists $x \in X$ such that $\max_{i=1,\dots,m}(y_i - C_i(x)) \leq 0$. Since X is compact and the function $x \mapsto \max_i(y_i - C_i(x))$ is continuous, this is equivalent to saying that $\min_{x\in X}\max_i(y_i - C_i(x)) \leq 0$, i.e. $h(y) \leq 0$, with $h(y)$ being defined by (13). Clearly the function $h(y)$ is increasing. Suppose $y^\nu \to y^0$, and let $l(y^\nu) = y^\nu_{i_\nu} - C_{i_\nu}(x^\nu) \geq y^\nu_i - C_i(x^\nu)\ \forall i = 1,\dots,m$. By taking a subsequence if necessary we can assume $x^\nu \to x^0$, $i_\nu = i_0\ \forall\nu$. Then $l(y^\nu) \to y^0_{i_0} - C_{i_0}(x^0) \geq y^0_i - C_i(x^0)\ \forall i$, i.e. $l(y^\nu) \to \max_i(y^0_i - C_i(x^0)) = l(y^0)$. Therefore, $h(y)$ is continuous, and obviously $\Gamma = \{y|\ h(y) \leq 0\}$.

Corollary 1.

$$x \in X_{WE} \Leftrightarrow h(C(x)) \geq 0$$

Proof. By Proposition 6, $x \in X_{WE}$ if and only if $C(x) \in \partial^+ \Gamma$, hence, by Proposition 7, if and only if $h(C(x)) = 0$. Since for $x \in X$ we have $C(x) \in \Gamma$, i.e., by Proposition 7, $h(C(x)) \leq 0$, the equality $h(C(x)) = 0$ is equivalent to $h(C(x)) \geq 0$.

Setting $D = \{x \in X |\ g(x) \leq 0\}$, the problem (OWE) can thus be written as

$$\min\{f(x)|\ x \in D,\ h(C(x)) \geq 0\}.$$

In this form it provides a further important example of optimization under composite monotonic constraints. Here $h(y)$ is given by (13) and the equivalent problem (Q) is

$$\min\{\varphi(y)|\ y \in \Omega\}, \tag{14}$$

with

$$\Omega = \{y \in [a, b] \mid h(y) \geq 0\},$$
$$\varphi(y) = \min\{f(x) \mid x \in X,\ g(x) \leq 0,\ C(x) \geq y\}.$$

Setting $\Delta = \{y \in [a, b] \mid y \leq C(x),\ g(x) \leq 0,\ x \in X\}$, the problem can also be written as

$$\min\{\varphi(y)|\ y \in \Delta \cap \Omega\}. \tag{15}$$

In the simplest case when $f(x)$, $C(x)$ are linear with $C(X) \subset \mathbb{R}^m_+$, and the constraint $g(x) \leq 0$ is absent, the above formulation of (OWE) was earlier established in [16].

4 Constrained optimization over the efficient set

A general problem in multi-criteria decision making, of which (OWE) is only a relaxed variant, is the following constrained optimization problem over the efficient set:

$$\alpha := \min\{f(x)|\ x \in X,\ g(x) \leq 0,\ x \in X_E\}, \tag{OE}$$

where X_E denotes the set of efficient points of X w.r.t. C.

Recall that a point $x^0 \in X$ is said to be *efficient* w.r.t. C if there is no $x \in X$ such that $C(x) \geq C(x^0)$ and $C(x) \neq C(x^0)$.

The concept of efficiency and the properties of the efficient set, as well as the problem (OE), have been extensively discussed in the literature (see e.g. [20], [35], [5] and references therein, [13], [11], [8], [9], [22], [15] and references therein).

Although efficiency is a much stronger property than weak efficiency, it turns out that the problem (OE) can be solved, roughly speaking, by the same method as (OWE). To see this, consider, as previously, the set $\Gamma = \{y \in [a, b]|\ y \leq C(x),\ x \in X\}$ (see (11)). Recall that $\text{ext}^+\Gamma$ denotes the set of upper extreme points of the normal set Γ.

Proposition 8. $x \in X_E \Leftrightarrow C(x) \in \mathrm{ext}^+\Gamma$.

Proof. If $x^0 \in X_E$ then $y^0 = C(x^0) \in \Gamma$ and there is no $x \in X$ such that $C(x) \geq C(x^0)$, $C(x) \neq C(x^0)$ i.e. no $y = C(x)$, $x \in X$, such that $y \geq y^0$, $y \neq y^0$, hence $y^0 \in \mathrm{ext}^+\Gamma$. Conversely, if $y^0 = C(x^0) \in \mathrm{ext}^+\Gamma$ with $x^0 \in X$, then there is no $y = C(x), x \in X$, such that $y \geq y^0$, $y \neq y^0$, i.e. no $x \in X$ with $C(x) \geq C(x^0)$ $C(x) \neq C(x^0)$, hence $x^0 \in X_E$.

For every $y \in \mathbb{R}^m_+$ define

$$\rho(y) = \min\{\sum_{i=1}^{m}(y_i - z_i)|\ z \geq y, z \in \Gamma\}. \tag{16}$$

Proposition 9. *The function* $\rho(y) : \mathbb{R}^m_+ \to \mathbb{R}_+ \cup \{+\infty\}$ *is increasing and satisfies*

$$\rho(y) \begin{cases} \leq 0 & \text{if } y \in \Gamma \\ = 0 & \text{if } y \in \mathrm{ext}^+\Gamma \\ = +\infty & \text{if } y \notin \Gamma. \end{cases} \tag{17}$$

Proof. If $y' \geq y$ then $z \geq y'$ implies that $z \geq y$, hence $\rho(y') \geq \rho(y)$. Suppose $\rho(y) = 0$. Then $y \in \Gamma$ since there exists $z \in \Gamma$ such that $y \leq z$. There cannot be any $z \geq y, z \neq y$, for this would imply that $z_i > y_i$ for at least one i, hence $\rho(y) \leq \sum_{i=1}^m (y_i - z_i) < 0$, a contradiction. Conversely, if $y \in \mathrm{ext}^+\Gamma$ then for every $z \in \Gamma$ such that $z \geq y$ one must have $z = y$, hence $\sum_{i=1}^m (y_i - z_i) = 0$, i.e. $\rho(y) = 0$. That $\rho(y) \leq 0\ \forall y \in \Gamma$, $\rho(y) = +\infty\ \forall y \notin \Gamma$ is obvious.

Setting $D = \{x \in X \mid g(x) \leq 0\}$, $\tilde{h}(y) = \min\{h(y), \rho(y)\}$, the problem (OE) can thus be formulated as

$$\min\{f(x)|\ x \in D,\ \tilde{h}(C(x)) \geq 0\}, \tag{18}$$

which has the same form as (P) with the difference, however, that $\tilde{h}(y)$, though increasing, is not u.s.c. as required. Upon the same transformations as previously, this problem can be rewritten as

$$\min\{\varphi(y)|\ y \in \Omega,\ \rho(y) = 0\}, \tag{$\tilde{\mathrm{Q}}$}$$

where $\varphi(y), \Omega$ are defined as in (14).

Clearly ($\tilde{\mathrm{Q}}$) differs from problem (Q) only by the additional constraint $\rho(y) = 0$, whose presence is what makes (OE) a bit more complicated than (OWE).

There are two possible approaches for solving (OE). In the first approach, proposed in [22], the problem (OE) is approximated by an (OWE) differing from (OE) only in that the criterion mapping C is slightly perturbed and replaced by C^ε with

$$C^\varepsilon(x)_j = C(x)_j + \varepsilon \sum_{i=1}^{m} C_i(x), \qquad j = 1, \ldots, m,$$

where $\varepsilon > 0$ is sufficiently small. Denote by X^ε_{WE} and X^ε_E resp. the weakly efficient set and the efficient set of X w.r.t. the mapping $C^\varepsilon(x)$ and consider the problem

$$\alpha(\varepsilon) := \min\{f(x)|\ g(x) \leq 0,\ x \in X^\varepsilon_{WE}\} \qquad (\text{OWE}^\varepsilon).$$

Lemma 1. ([22]) (i) *For any* $\varepsilon > 0$ *we have* $X^\varepsilon_{WE} \subset X_E$.

(ii) $\alpha(\varepsilon) \to \alpha$ *as* $\varepsilon \searrow 0$.

(iii) *If* X *is a polytope, and* C *is linear, then there is* $\varepsilon_0 > 0$ *such that* $X^\varepsilon_{WE} = X_E$ *for all* ε *satisfying* $0 \leq \varepsilon \leq \varepsilon_0$.

Proof. (i) Let $x^* \in X \setminus X_E$. Then there is $x \in X$ satisfying $C(x) \geq C(x^*)$, $C_i(x) > C_i(x^*)$ for at least one i. This implies that $\sum_{j=1}^m C_j(x) > \sum_{j=1}^m C_j(x^*)$, hence $C^\varepsilon(x) > C^\varepsilon(x^*)$, so that $x^* \notin X^\varepsilon_{WE}$. Thus if $x^* \in X \setminus X_E$, then $x^* \in X \setminus X^\varepsilon_{WE}$, i.e. $X^\varepsilon_{WE} \subset X_E$.

(ii) The just established property (i) implies that $\alpha(\varepsilon) \geq \alpha$. Define

$$\Gamma^\varepsilon = \{y \in [a,b]|\ y \leq C^\varepsilon(x), x \in X\}, \qquad (19)$$

$$\varphi_\varepsilon(y) = \min\{f(x)|\ x \in X,\ g(x) \leq 0,\ C^\varepsilon(x) \geq y\}, \qquad (20)$$

so that the problem (OWE$^\varepsilon$) reduces to

$$\min\{\varphi_\varepsilon(y)|\ y \in \text{cl}([a,b] \setminus \Gamma^\varepsilon\}. \qquad (21)$$

Next consider an arbitrary point $\bar{x} \in X_E \setminus X^\varepsilon_{WE}$. Since $\bar{x} \notin X^\varepsilon_{WE}$, there exists $x^\varepsilon \in X$ such that $C^\varepsilon(x^\varepsilon) > C^\varepsilon(\bar{x})$. Then x^ε is feasible to the problem (OWE$^\varepsilon$), and, consequently, $f(x^\varepsilon) \geq \alpha(\varepsilon) \geq \alpha$. In view of the compactness of X, by passing to a subsequence if necessary, one can suppose that $x^\varepsilon \to x^* \in X$ as $\varepsilon \searrow 0$. Therefore, $f(x^*) \geq \lim \alpha(\varepsilon) \geq \alpha$. But from $C^\varepsilon(x^\varepsilon) > C^\varepsilon(\bar{x})$ we have $C(x^*) \geq C(\bar{x})$ and since $\bar{x} \in X_E$ it follows that $x^* = \bar{x}$, and hence $f(\bar{x}) \geq \lim \alpha(\varepsilon) \geq \alpha$. Since this holds for arbitrary $\bar{x} \in X_E \setminus X^\varepsilon_{WE}$ we conclude that $\lim \alpha(\varepsilon) = \alpha$.

(iii) We only sketch the proof, referring the interested reader to [14], Proposition 15.1, for the details. Since $C(x)$ is linear we can consider the cones $K := \{x|C_i(x) \geq 0,\ i = 1, \ldots, m\}$, $K^\varepsilon = \{x|\ C^\varepsilon_i(x) \geq 0,\ i = 1, \ldots, m\}$. Let $x^0 \in X_E$ and denote by F the smallest facet of X that contains x^0. Since $x^0 \in X_E$ we have $(x^0 + K) \cap F = \emptyset$, and hence, there exists $\varepsilon_F > 0$ such that $(x^0 + K^\varepsilon) \cap F = \emptyset\ \forall \varepsilon \in (0, \varepsilon_F]$. Then we have $(x + K^\varepsilon) \cap F = \emptyset\ \forall x \in F$ whenever $0 < \varepsilon \leq \varepsilon_F$. If $\mathcal{F}$ denotes the set of facets of X such that $F \subset X_E$, then $\mathcal{F}$ is finite and $\varepsilon_0 = \min\{\varepsilon_F|\ F \in \mathcal{F}\} > 0$. Hence, $(x + K^\varepsilon) \cap F = \emptyset\ \forall x \in F, \forall F \in \mathcal{F}$, i.e. $X_E \subset X^\varepsilon_{WE}$ whenever $0 < \varepsilon \leq \varepsilon_0$.

As a consequence, an approximate optimal solution of (OE) can be obtained by solving (OWE$^\varepsilon$) for any ε such that $0 < \varepsilon \leq \varepsilon_0$.

A second approach consists in solving (OE) as a problem (Q) with some additional manipulations to take account of the additional constraint $\rho(y) = 0$.

Given any point $y \in \Gamma$ the next proposition can be used to compute an efficient point $\bar{x}$ such that $C(\bar{x}) \geq y$.

Proposition 10. *For any $y \in \Gamma$, there exists an $\bar{y} \in \text{ext}^+\Gamma$ such that $\bar{y} \geq y$, and any optimal solution $\bar{x}$ of the problem*

$$\max\{\sum_{i=1}^{m} C_i(x) \mid y \leq C(x),\ x \in X\} \tag{22}$$

is an efficient point in X.

Proof. By Proposition 3, there is an $\bar{y} \in \text{ext}^+\Gamma$ such that $y \in [a, \bar{y}]$, hence $\bar{y} \geq y$. To show the second part of the Proposition, observe that, since $y \in \Gamma$, the feasible set of the problem (22) is nonempty (and compact), so the problem (22) has an optimal solution $\bar{x}$. If there exists $x \in X$ such that $C(x) \geq C(\bar{x})$ then $C(x) \geq y$ (because $C(\bar{x}) \geq y$), i.e. x is feasible to (22), hence $\sum_{i=1}^{m} C_i(x) \leq \sum_{i=1}^{m} C_i(\bar{x})$. This in turn implies that $\sum_{i=1}^{m} C_i(x) = \sum_{i=1}^{m} C_i(\bar{x})$, hence $C(x) = C(\bar{x})$. Therefore, $\bar{x} \in X_E$.

Proposition 11. *For a box $[p, q] \subset [a, b]$ such that $p \in \Gamma$, $q \notin \Gamma$ let $\lambda = \max\{\alpha \mid p + \alpha(q - p) \in \Gamma\}$, $z = p + \lambda(q - p)$. If*

$$z^i := z + (q_i - z_i)e^i \notin \Gamma \quad \forall i = 1, \ldots, m$$

then $[p, q) \cap \text{ext}^+\Gamma \neq \emptyset$.

Proof. The hypothesis implies that for any $i = 1, \ldots, m$, there is no point $y \in [z, q] \cap \Gamma$ with $y_i = q_i$. Indeed, if such an y exists then $z^i := z + (q_i - z_i)e^i$ satisfies $z^i_i = q_i = y_i$, whereas $z^i_j = z_j \leq y_j\ \forall j \neq i$, hence $z^i \leq y$ and since $y \in \Gamma$, it follows that $z^i \in \Gamma$, contradicting the hypothesis. Now, by Proposition 3, since $z \in \Gamma$ there exists a $\bar{y} \in [z, b]$ such that $\bar{y} \in \text{ext}^+\Gamma$, and since, as has just been proved, we must have $\bar{y}_i < q_i\ \forall i$, it follows that $\bar{y} \in [z, q) \subset [p, q)$.

Remark 4.1 If $C : \mathbb{R}^n \to \mathbb{R}^m$ is a concave mapping (i.e. every function $C_i(x), i = 1, \ldots, p$ is concave) then Γ is a convex set. It is then easily seen that in the most favourable case when $f(x) = F(C(x))$, with $F(y)$ a decreasing (strictly decreasing, resp.) function, then the optimization problem

$$\min\{f(x) \mid x \in X_{WE}\}, \qquad (\min\{f(x) \mid x \in X_E\},\ \text{resp.})$$

reduces to the monotonic optimization problem

$$\min\{F(y) \mid y \in \Gamma\},$$

and hence, can be solved efficiently by standard monotonic optimization methods (see e.g. [1]).

5 Solution method for problem (Q)

For small values of m the monotonic optimization problem (Q) can be solved fairly efficiently by the copolyblock approximation algorithm even in its initial version [26] (see e.g. [27], [16]). Below we will use, instead, a BRB (branch-reduce-and-bound) algorithm which is more suitable for problems with larger values of m. This BRB algorithm was initially developed in [26] and significantly improved in [31].

As the name indicates, the BRB algorithm proceeds according to the standard branch and bound scheme with three basic operations: branching, reducing (the partition sets) and bounding.

-Branching consists in a successive rectangular partition of the initial box $M_0 = [a, b]$ following an *exhaustive* subdivision rule, i.e. such that any infinite nested sequence of partition sets generated by the algorithm shrinks to a singleton. A commonly used exhaustive subdivision rule is the *standard bisection.*

-Reducing consists in applying valid cuts to reduce the size of the current partition set $M = [p, q] \subset [a, b]$. The box $[p', q']$ obtained from M as a result of the cuts is referred to as a *valid reduction* of M.

-Bounding consists in estimating a *valid lower bound* $\beta(M)$ for the objective function value $\varphi(y)$ over the feasible portion contained in the valid reduction $[p', q']$ of a given partition set $M = [p, q]$.

5.1 Valid Reduction

At a given stage of the BRB algorithm for (Q), a feasible point $\bar{y} \in \Omega$ is available which is the best so far known. Let $\gamma = \varphi(\bar{y})$ and let $[p, q] \subset [a, b]$ be a box generated during the partitioning procedure which is still of interest. Since an optimal solution of (Q) is attained at a point on the lower boundary of Ω, i.e. in the set $h(y) = 0$, the search for a feasible solution of (Q) in $[p, q]$ such that $\varphi(y) \le \gamma$ can be restricted to the set $B_\gamma \cap [p, q]$, where

$$B_\gamma := \{y |\ \varphi(y) - \gamma \le 0,\ h(y) \le 0 \le h(y)\}. \tag{23}$$

The reduction operation aims at replacing the box $[p, q]$ with a smaller box $[p', q'] \subset [p, q]$ without losing any point $y \in B_\gamma \cap [p, q]$, i.e. such that

$$B_\gamma \cap [p', q'] = B_\gamma \cap [p, q].$$

The box $[p', q']$ satisfying this condition is referred to as a *γ-valid reduction* of $[p, q]$ and denoted by $\text{red}_\gamma[p, q]$.

In the sequel, e^i denotes the i-th unit vector, i.e. a vector with 1 at the i-th position and 0 everywhere else.

Lemma 2. (i) *If* $\min\{\varphi(p)-\gamma,\ h(p)\}>0$ *or* $h(q)<0$, *then* $B_\gamma\cap[p,q]=\emptyset$, *i.e.* $\mathrm{red}_\gamma[p,q]=\emptyset$.

(ii) *If* $\varphi(p)\le\gamma,\ h(p)\le 0$, *and* $h(q)\ge 0$, *then* $\mathrm{red}_\gamma[p,q]=[p',q']$ *with*

$$p'=q-\sum_{i=1}^{n}\alpha_i(q_i-p_i)e^i,\quad q'=p'+\sum_{i=1}^{n}\beta_i(q_i-p'_i)e^i \tag{24}$$

where, for $i=1,\ldots,n$,

$$\alpha_i=\sup\{\alpha|\ 0<\alpha\le 1,\ h(q-\alpha(q_i-p_i)e^i))\ge 0\} \tag{25}$$

$$\beta_i=\sup\{\beta|\ 0<\beta\le 1,\ \varphi(p'+\beta(q_i-p'_i)e^i)\le\gamma,\ h(p'+\beta(q_i-p'_i)e^i)\le 0\}. \tag{26}$$

Proof. (i) If $\theta(p):=\min\{\varphi(p)-\gamma,\ h(p)\}>0$, then, since $\theta(y)$ is increasing, $\theta(y)\ge\theta(p)>0$ for every $y\in[p,q]$. Similarly, if $h(q)<0$, then $h(y)<0$ for every $y\in[p,q]$. In both cases, $B_\gamma\cap[p,q]=\emptyset$.

(ii) Let $y\in[p,q]$ satisfy $\varphi(y)\le\gamma$ and $h(y)=0$. If $y\not\ge p'$ then there is i such that $y_i<p'_i=q_i-\alpha_i(q_i-p_i)$, i.e. $y_i=q_i-\alpha(q_i-p_i)$ with $\alpha>\alpha_i$. In view of (25), this implies that $h(q-\alpha(q_i-p_i)e^i))<0$ and hence, $h(y)<0$), conflicting with $h(y)=0$. Similarly, if $y\not\le q'$ then there is i such that $y_i>q'_i=p'_i+\beta_i(q_i-p'_i)$, i.e. $y_i=p'_i+\beta(q_i-p'_i)$ with $\beta>\beta_i$ and from (26) it follows that either $\varphi(p'+\beta(q_i-p'_i)e^i)>\gamma$, (which implies that $\varphi(y)>\gamma$), or $h(p'+\beta(q_i-p'_i)e^i>0$ (which implies that $h(y)>0$, in either case conflicting with $y\in B_\gamma$). Therefore, $\{y\in[p,q]|\ \varphi(y)\le\gamma,\ h(y)=0\}\subset[p',q']$.

Remark 5.1 It can easily be verified that the box $[p',q']=\mathrm{red}_\gamma[p,q]$ still satisfies $\varphi(p')\le\gamma,\ h(p')\le 0,\ h(q')\ge 0$.

Remark 5.2 When applying the above reduction procedures to problem (OWE) note that in this problem (see (13), (14))

$$\varphi(y)=\min\{f(x)|\ x\in X,\ g(x)\le 0,\ y\le C(x)\},$$
$$h(y)=\min\{t|\ y_i-C_i(x)\le t,\ i=1,\ldots,m,x\in X\}.$$

5.2 Valid Bounds

Let $M:=[p,q]$, be a partition set which is supposed to have been reduced, so that according to Remark 5.1:

$$\varphi(p)\le\gamma,\ h(p)\le 0,\ h(q)\ge 0.$$

Let us now examine how to compute a lower bound $\beta(M)$ for

$$\min\{\varphi(y)|\ y\in[p,q],\ h(y)=0\}.$$

Since $\varphi(y)$ are increasing, an obvious lower bound is $\varphi(p)$. We will shortly see that to ensure convergence of the BRB Algorithm, it suffices that the lower bounds satisfy

$$\beta(M) \geq \varphi((p). \tag{27}$$

We shall refer to such a bound as a *valid lower bound.*

Define $\theta(y) = \min\{\varphi(y) - \gamma,\ h(y)\}, \Delta = \{y \in [p, q] |\ \theta(y) \leq 0\}$.

If $\theta(p) \leq 0 \leq h(p)$, then obviously p is an exact minimizer of $\varphi(y)$ over the feasible points in $[p, q]$ at least as good as the current best, and can be used to update the current best solution. Suppose therefore that $\theta(p) \leq 0,\ h(p) < 0$, i.e. $p \in \Delta \setminus \Omega$.

For each $y \in [p, q]$ such that $h(y) < 0$ let $\pi(y)$ be the first point where the line segment from y to q meets the lower boundary of Ω, i.e.

$$\begin{aligned} \pi(y) = y + \lambda(q - y),\ \text{with} \\ \lambda = \max\{\alpha|\ h(y + \alpha(q - y)) \leq 0\}. \end{aligned} \tag{28}$$

Obviously, $h(\pi(y)) = 0$.

Lemma 3. *If* $z = \pi(p),\ z^i = p + (z_i - p_i)e^i,\ i = 1, \ldots, m$, *and* $I = \{i|\ z^i \in \Delta\}$, *then a valid lower bound over* $M = [p, q]$ *is*

$$\begin{aligned} \beta(M) &= \min\{\varphi(z^i)|\ i \in I\} \\ &= \min\{f(x)|\ x \in D,\ C(x) \geq z^i,\ i \in I\}. \end{aligned}$$

Proof. Let $M_i = [z^i, q]$. From the definition of z it is easily seen that $h(u) < 0\ \forall u \in [p, z)$, i.e. $[p, q] \cap \Delta \cap \Omega \subset [p, q] \setminus [p, z)$. Noting that $\{u|\ p \leq u < z\} = \cap_{i=1}^m \{u|\ p_i \leq u_i < z_i\}$, we can write $[p, q] \setminus [p, z) = [p, q] \setminus \cap_{i=1}^m \{u|\ u_i < z_i\} = \cup_{i=1,\ldots,m} \{u \in [p, q] |\ z_i \leq u_i \leq q_i\} = \cup_{i=1}^m M_i$. Thus, if $I = \{i|\ z^i \in \Delta\}$ then $[p, q] \cap \Delta \cap \Omega \subset \cup_{i \in I} M_i$. Since $\varphi(z^i) \leq \min\{\varphi(y)|\ y \in M_i\}$, the result follows.

Remark 5.3 Each box $M_i := [z^i, q]$ can be reduced by the method presented above. If $[p'^i, q'^i] = \text{red}[z^i, q],\ i = 1, \ldots, m$, then without much extra effort, we can have a more refined lower bound, namely

$$\beta(M) = \min_{i \in I} \varphi(p'^i), \quad I = \{i|\ z^i \in \Delta\}.$$

The above constructed points $z^i, i \in I$, determine a set $Z := \cup_{i \in I}[z^i, q]$ containing $B_\gamma \cap [p, q] := [p, q] \cap \Delta \cap \Omega$. Such a set Z is called a *copolyblock* (reverse polyblock) with vertex set $z^i, i \in I$, see [26]. The above procedure thus amounts to constructing a copolyblock $Z \supset B_\gamma$ - which is possible because $\varphi(y)$ is increasing. To have a tighter lower bound, one can even construct a sequence of copolyblocks $Z_1 \supset Z_2, \ldots$, approximating the set $B_\gamma \cap [p, q]$ more and more closely. For the details of this construction the interested reader is referred to [26], or better, [31]. By using copolyblock approximations one could compute a bound as tight as we wish. Since, however, the computation cost increases rapidly with the accuracy requirement for copolyblock approximation, a trade-off must be made, so practically just one approximating copolyblock as in the above Lemma is used.

5.3 Algorithm and convergence

We are now in a position to state the proposed algorithm for (Q).

BASIC BRB ALGORITHM FOR (Q)

Step 0. Start with $\mathcal{P}_1 = \{M_1\}, M_1 = [a,b], \mathcal{R}_1 = \emptyset$. If a current best feasible solution (CBS) is available let CBV (current best value) denote the value of $f(x)$ at this point. Otherwise, set $CBV = +\infty$. Set $k = 1$.

Step 1. For each box $M \in \mathcal{P}_k$:
- Compute the γ-valid reduction $\text{red}_\gamma M$ of M for $\gamma = CBV$;
- Delete M if $\text{red}_\gamma M = \emptyset$;
- Replace M by $\text{red}_\gamma M$ if $\text{red}_\gamma M \neq \emptyset$;
- If $\text{red}_\gamma M = [p,q]$ then compute a valid lower bound $\beta(M)$ for $\varphi(y)$ over the feasible solutions in M.

Step 2. Let $\mathcal{P}'_k$ be the collection of boxes that results from $\mathcal{P}_k$ after completion of Step 1. From $\mathcal{R}_k$ remove all $M \in \mathcal{R}_k$ such that $\beta(M) \geq CBV$ and let $\mathcal{R}'_k$ be the resulting collection. Let $\mathcal{M}_k = \mathcal{R}'_k \cup \mathcal{P}'_k$.

Step 3. If $\mathcal{M}_k = \emptyset$ then terminate: the problem is infeasible (if $CBV = +\infty$), or CBV is the optimal value and the feasible solution $\bar{y}$ with $\varphi(\bar{y}) = CBV$ is an optimal solution (if $CBV < +\infty$).

Otherwise, let $M_k \in \text{argmin}\{\beta(M) | \; M \in \mathcal{M}_k\}$.

Step 4. Divide M_k into two subboxes by the standard bisection. Let $\mathcal{P}_{k+1}$ be the collection of these two subboxes of M_k.

Step 5. Let $\mathcal{R}_{k+1} = \mathcal{M}_k \setminus \{M_k\}$. Increment k and return to Step 1.

Proposition 12. *Whenever infinite, the Basic BRB Algorithm generates an infinite filter of boxes $\{M_{k_l}\}$ whose intersection yields a global optimal solution.*

Proof. If the algorithm is infinite, it must generate an infinite filter of boxes $\{M_{k_l}\}$ and $\cap_{l=1}^{\infty} M_{k_l} = \{y^*\}$. Since $M_{k_l} = [p^{k_l}, q^{k_l}]$ with $h(p^{k_l}) \leq 0 \leq h(q^{k_l})$, and $y^* = \lim p^{k_l} = \lim q^{k_l}$, it follows that $h(y^*) \leq 0 \leq h(y^*)$, i.e. y^* is a feasible solution. Therefore, if the problem is infeasible, the algorithm must stop at some iteration where no box remains for consideration, giving evidence of infeasibility. Otherwise,

$$\varphi(p^{k_l}) \leq \beta(M_{k_l}) \leq \varphi(q^{k_l}),$$

whence $\lim_{l\to+\infty} \beta(M_{k_l}) = \varphi(y^*)$. On the other hand, since M_{k_l} corresponds to the minimum of $\beta(M)$ among the current set of boxes, we have $\beta(M_{k_l}) \leq \min\{\varphi(y) | \; y \in [a,b] \cap \Omega\}$ and, consequently,

$$\varphi(y^*) \leq \min\{\varphi(y) | \; y \in [a,b] \cap \Omega\}.$$

Since y^* is feasible, it follows that y^* is an optimal solution.

Remark 5.4 As said above, the problem (OE) can be approximated, as closely as desired, by a problem (OWE$^\varepsilon$) with a suitable ε. Therefore, the above Algorithm can be applied for solving (OE), by replacing $C(x)$ with $C^\varepsilon(x)$ defined by (4).

6 Improvements for problems (OWE) and (OE)

In the special cases of problems (OWE) and (OE) the above algorithm can be improved by using more efficient reduction, bounding and branching operations. Also alternative bounding methods can be developed by simultaneously exploiting the monotonic and d.c. structures of the problem.

6.1 Improved Bounds when C is concave

Recall from (14) that the problem (Q) equivalent to (OWE) is

$$\min\{\varphi(y)|\ y \in \Omega\}$$

where $\Omega = \{y \in [a,b] \mid h(y) \geq 0\}$, $h(y) = \min\{t|\ y_i - C_i(x) \leq t, i = 1,\dots,m,\ x \in X\}$ (see (13), and

$$\varphi(y) = \min\{f(x)|\ x \in X,\ g(x) \leq 0,\ C(x) \geq y\}.$$

Lemma 4. *If the functions $C_i(x)$, $i = 1,\dots,m$, are concave, the set $\Gamma = \{y \in [a,b] \mid h(y) \leq 0\}$ is convex.*

Proof. By Proposition 7, $\Gamma = \{y \in [a,b] \mid y \leq C(x)$ for some $x \in X\}$. Let $y, y' \in \Gamma$ and $0 \leq \alpha \leq 1$. Then $y \leq C(x), y' \leq C(x')$, with $x, x' \in X$ and $\alpha y + (1-\alpha)y' \leq \alpha C(x) + (1-\alpha)C(x') \leq C(\alpha x + (1-\alpha)x')$, where $\alpha x + (1-\alpha)x' \in X$ because X is assumed to be convex.

The convexity of Γ permits a simple method for computing a tight lower bound for the minimum of $\varphi(y)$ over the feasible solutions still of interest in $[p,q]$.

For each $i = 1,\dots,m$ let y^i be the last point of Γ on the line segment joining p to $p^i := p + (q_i - p_i)e^i$, i.e. $y^i = p + \lambda_i(q_i - p_i)e^i$ with

$$\lambda_i = \max\{\alpha|\ p + \alpha(q_i - p_i)e^i) \in \Gamma,\ 0 \leq \alpha_i \leq 1\}.$$

Proposition 13. *If the functions $C_i(x)$, $i = 1,\dots,m$, are concave then a lower bound for $\varphi(y)$ over the feasible portion in $M = [p,q]$ is*

$$\begin{aligned}\beta(M) = \min\{f(x)|\ x \in X,\ g(x) \leq 0,\ C(x) \geq p,\\ \textstyle\sum_{i=1}^m (C_i(x) - p_i)/(y_i^i - p_i) \geq 1\}.\end{aligned} \tag{29}$$

Proof. First observe that, by convexity of Γ, the simplex $S = [y^1,\dots,y^m]$ satisfies $[p,q] \cap \partial^+\Gamma \subset S + \mathbb{R}^m_+$, and since $\varphi(y)$ is increasing, we have $\varphi(y) \leq \varphi(\pi(y))\ \forall y \in S$, where $\pi(y) = y + \lambda(q-y)$ with $\lambda = \max\{\alpha|\ h(y + \alpha(q-y)) \leq 0\}$, so that $h(\pi(y)) = 0$. Hence,

$$\begin{aligned}\beta(M) &:= \min\{\varphi(y)|\ y \in S\} \\ &\le \min\{\varphi(y)|\ y \in [p,q] \cap \partial^+\Gamma\} \\ &\le \min\{\varphi(y)|\ y \in [p,q],\ h(y) = 0\},\end{aligned} \tag{30}$$

so (30) gives a valid lower bound. On the other hand, letting

$$E = \{(x,y)|x \in X,\ g(x) \ge 0,\ C(x) \ge y,\ y \in S\},$$
$$E_y = \{x \in X|\ g(x) \le 0,\ C(x) \ge y\} = \{x|\ (x,y) \in E\}$$

for every $y \in S$, we can write

$$\begin{aligned}&\min_{y\in S} \varphi(y) \\ &= \min_{y\in S} \min\{f(x)|\ x \in E_y\} \\ &= \min\{f(x)|\ (x,y) \in E\} \\ &= \min\{f(x)|\ x \in X,\ g(x) \le 0,\ C(x) \ge y,\ y \in S\},\end{aligned}$$

Since $S = \{y \ge p|\ \sum_{i=1}^m (y_i - p_i)/(y_i^i - p_i) = 1\}$, (29) follows.

Remark 6.1 In particular, when X is a polytope, C is a linear mapping, $g(x), f(x)$ are affine functions, as e.g. in problems (OWE) with linear objective and criteria, then (R(y)) as well as (29) are linear programs. Also note that the linear programs (R(y)) for different y differ only by the right-hand side of the linear constraints, and the linear programs (29) for different partition sets M differ only by the last constraint and the right-hand side of the constraints $C(x) \ge p$. Taking account of these facts, reoptimization techniques can be used to save time for solving these linear programs.

6.2 Adaptive Subdivision Rule

Let y^M be an optimal solution of the problem (29) and $z^M = \pi(y^M)$. If it so happens that $y^M = z^M$, then $\beta(M) = \min\{\varphi(y)|\ y \in [p,q] \cap \partial^+\Gamma\}$. This motivates a usually more efficient subdivision method than the standard bisection, which is often referred to as an adaptive bisection (see [25]) and proceeds according to the following rule.

Determine the index $i \in \{1,\ldots,m\}$ such that

$$|y_i^M - z_i^M| = \max_{j=1,\ldots,m} |y_j^M - z_j^M|,$$

and divide M by a hyperplane parallel to the i-th coordinate axis, passing through the middle of the line segment joining y^M to z^M.

With this subdivision rule the convergence of the algorithm is still ensured, because if M_k is a filter of partition sets generated by the algorithm (see the proof of Proposition 12), then, as proved in [25], $y^{M_k} - z^{M_k} \to 0$, hence $y^* = \min y^{M_k}$ will yield an optimal solution. Experience has shown that the adaptive subdivision rule often helps to reduce the gap $\|y^{M_k} - z^{M_k})\|$ faster than the standard bisection rule.

6.3 Special Reduction-Deletion Rule for (OE)

As we saw in Section 3, the problem (OE) can be converted to the form ($\tilde{\mathrm{Q}}$):

$$\min\{\varphi(y)|\ h(y) \geq 0,\ \rho(y) = 0\}, \tag{31}$$

which differs from problem (Q) by the presence of the additional constraint $\rho(y) = 0$.

As was said in Section 4, one way to deal with this additional constraint is to replace the mapping C by a slightly perturbed mapping C^ε, thus transforming (OE) into an (OWE$^\varepsilon$) approximating (OE); then the problem (Q) corresponding to (OWE$^\varepsilon$) can be solved by the Basic BRB Algorithm.

Alternatively, to solve (31), i.e. ($\tilde{\mathrm{Q}}$), the Basic BRB Algorithm can be applied with the following precautions:

(i) A "feasible solution" is an $y \in [a, b]$ satisfying $\rho(y) = 0$, i.e. a point of $\mathrm{ext}^+\Gamma$. The current best solution is the best among all so far known feasible solutions.

(ii) In Step 1, after completion of the reduction operation described in Subsection 5.1, a special supplementary reduction-deletion operation is needed, to identify and to fathom any box $[p, q]$ that contains no point of $\mathrm{ext}^+\Gamma$. This supplementary operation will ensure that every infinite nested sequence of partition sets $\{[p^{k_\nu}, q^{k_\nu}]\}$ collapses to a point corresponding to an efficient point.

Recall that $C_i^\varepsilon = C_i(x) + \varepsilon \sum_{j=1}^m C_j(x),\ i = 1, \ldots, m$, and let $h^\varepsilon(y), \Gamma^\varepsilon$ be defined as in (13) and (11), resp., with $C^\varepsilon(x)$ replacing $C(x)$. Since $C(x) \geq a \in \mathbb{R}^m_+\ \forall x \in X$, we have $C(x) \leq C^\varepsilon(x)\ \forall x \in X$, hence $\Gamma \subset \Gamma^\varepsilon$.

Let $[p, q]$ be a box which has been reduced according to the rules described in Subsection 5.1, so that $p \in \Gamma, q \notin \Gamma$. Since $p \in \Gamma$ we have $\rho(p) \leq 0$ where $\rho(y)$ is defined by (16).

Supplementary reduction-deletion rule

Fathom $[p, q]$ if either of the following events occurs:

a) $\rho(q) \leq 0$ (so that $\rho(y) \leq 0\ \forall y \in \Gamma \cap [p, q] \supset \partial^+\Gamma \cap [p, q]$); if $\rho(q) = 0$ then q gives an efficient point and can be used to update the current best.

b) For some chosen small enough $\varepsilon > 0$

$$\min\{t|\ x \in X,\ q_i - C_i(x) - \varepsilon \sum_{i=1}^{m} C_i(x) \leq t,\ i = 1, \ldots, m\} \leq 0$$

(In that case $q \in \Gamma^\varepsilon$, see (19), so the box $[p, q)$ contains no point of $\partial^+(\Gamma^\varepsilon)$, and hence no point of $\mathrm{ext}^+\Gamma$).

Proposition 14. *If the Basic BRB Algorithm is applied with precautions* (i),(ii), *then, whenever infinite, it generates an infinite filter of boxes whose intersection is a global optimal solution y^* of ($\tilde{\mathrm{Q}}$), corresponding to an optimal solution x^* of (OE).*

Proof. Because of precautions (i) and (ii) the box chosen for further partitioning in each iteration contains either a point of $\mathrm{ext}^+\Gamma$ or a point of $\mathrm{ext}^+\Gamma^\varepsilon$. Therefore, any infinite filter of boxes generated by the Algorithm either collapses to a point $y^* \in \mathrm{ext}^+\Gamma$, (yielding an $x^* \in X_E$) or collapses to a point $y^* \in \mathrm{ext}^+\Gamma^\varepsilon$ (yielding $x^* \in X^\varepsilon_{WE} \subset X_E$).

6.4 Alternative Solution Methods

In the case of the constrained optimization over the weakly efficient set, when $C_i(x)$, $i = 1, \dots, m$, are concave, while $f(x)$, $g(x)$ are convex, each problem (R(y)) is a convex program. Since $\varphi(y)$ is a convex function the problem (Q) thus reduces to a convex minimization problem under the complement of a convex set Γ:

$$\min\{\varphi(y)|\ y \in [a, b] \cap \mathrm{cl}(\mathbb{R}^m_+ \setminus \Gamma)\}. \tag{32}$$

Therefore, the problem can be treated by d.c. optimization methods as developed e.g. in [25].

In fact, the d.c. approach is quite common for solving (OWE) (and (OE)) when X is a polytope, while $g(x) \equiv 0, f(x), C_i(x)$ are linear, convex or concave. In [22] (see also [14]), by noting that, under the stated assumptions, a point $x \in X_{WE}$ can be characterized by a vector $\lambda \in \mathbb{R}^m_+$ such that $\sum_{i=1}^m \lambda_i = 1$ and x is a maximizer of the problem

$$\max\{\sum_{i=1}^m \lambda_i C_i(x')|\ x' \in X\}, \tag{33}$$

the problem (OWE) can be formulated as

$$\min_{x,\lambda}\{f(x)|\ x \in X, \lambda \in \mathbb{R}^m_+,\ \sum_{i=1}^m \lambda_i = 1,$$
$$\langle \lambda, C(x)\rangle \geq \max_{x' \in X}\langle \lambda, C(x')\rangle\}. \tag{34}$$

In view of the compactness of X and the continuity of $C(x)$, without loss of generality we can assume that $C(x) > 0\ \forall x \in X$, so that $\max_{x' \in X}\langle \lambda, C(x')\rangle > 0$. The problem can then be rewritten as

$$\min_{x,\lambda}\{f(x)|\ x \in X, \lambda \in \mathbb{R}^m_+,\ \langle \lambda, C(x)\rangle \geq 1 = \max_{x' \in X}\langle \lambda, C(x')\rangle\},$$

which, in turn, is equivalent to

$$\min\{\Phi(\lambda)|\ \lambda \in \mathbb{R}^m_+,\ \max_{x' \in X}\langle \lambda, C(x')\rangle \leq 1\}, \tag{35}$$

with

$$\Phi(\lambda) = \min\{f(x)|\ x \in X,\ \langle \lambda, C(x)\rangle \geq 1\}.$$

Since, as can easily be verified, the function $\Phi(\lambda)$ is quasiconcave, while $\lambda \mapsto \max_{x\in X}\langle\lambda, C(x)\rangle$ is convex, the problem (35) is a quasiconcave minimization over a convex set. In [22] a cutting plane method has been developed for solving this problem and hence, the problem (OWE). As for (OE), these authors propose to approximate it by an (OWE$^\varepsilon$) for a sufficiently small value of $\varepsilon > 0$ (see Section 4).

In [15] an alternative d.c. formulation of (OWE) and (OE) is used which is closely related to the formulation (35) . Specifically, setting $h(\lambda, x) := \max_{x'\in X}\langle\lambda, C(x'-x)\rangle$, the problem (34) can be reformulated as

$$\min\{f(x)|\ x\in X,\ h(\lambda,x)\le 0,\ \lambda\in\mathbb{R}^m_+,\ \sum_{i=1}^m \lambda_i = 1\},$$

which, by setting

$$\Psi(\lambda) = \min\{f(x)|\ x\in X,\ h(\lambda,x)\le 0\},\ \Lambda = \{\lambda\in\mathbb{R}^m_+|\ \sum_{i=1}^m \lambda = 1\}$$

is equivalent to

$$\min\{\Psi(\lambda)|\ \lambda\in\Lambda\}. \tag{36}$$

Furthermore, it follows from a known result ([20]) that there exists $\delta\in(0,1/m)$ such that $x\in X_E$ if and only if x solves (33) for some $\lambda\in\mathbb{R}^m_+$ satisfying $\lambda_i\ge\delta,\ i=1,\dots,m$. Therefore, setting $\Lambda_\delta = \{\lambda\in\mathbb{R}^m_+|\ \lambda_i\ge\delta,\ \sum_{i=1}^m\lambda_i = 1\}$, we can also write (OE) (with linear C) as

$$\min\{\Psi(\lambda)|\ \lambda\in\Lambda_\delta\}. \tag{37}$$

When $f(x)$ is a concave function as assumed in [15], the value of $\Psi(\lambda)$ is determined by solving a concave minimization under a convex constraint. In the just cited paper this concave minimization is rewritten as a d.c. optimization problem

$$\Psi(\lambda) = \min\{f(x) - th(\lambda,x)|\ x\in X\}$$

with $t>0$ being a suitable penalty parameter. Thus, the problem reduces to (36) or (37), with $\Psi(\lambda)$ computed by solving a d.c. optimization problem.

Also in the case $C_i(x) = (A_ix+\alpha_i)/(B_ix+\beta_i)$ (linear-fractional multicriteria) with $B_ix+\beta_i>0\ \forall x\in X$, that was studied in [17], [18], [19], we have

$$x\in X_E \Leftrightarrow (\exists\lambda>0)\ x\in \operatorname{argmax}\{\langle\lambda, C(x')\rangle|\ x'\in X\}$$

and it is plain to check that the latter condition reduces to

$$x\in\operatorname{argmax}\{\sum_{i=1}^m \lambda_i[B_ix'+\alpha_i)A_i - (A_ix'+\beta_i)B_i]x'|\ x'\in X\}$$

([17]). Based on this property, the problem (OE) with linear-fractional criteria can be reformulated as

$$\min_{x,\lambda} f(x), \text{ s.t. } x \in X, \lambda > 0,$$
$$\max_{x' \in X} \sum_{i=1}^{m} \lambda_i [B_i x + \alpha_i) A_i - (A_i x + \beta_i) B_i](x - x') \leq 0$$

No algorithm is given in [17], [18], while in [19] the just formulated problem is in turn converted to a bilinear program in x, λ and solved by a method that may require computing a large number of vertices of X (assuming, as previously, that X is a polytope and $f(x)$ is linear).

Other methods for solving (OWE) and (OE) can be considered as specializations of global optimization methods developed over the last three decades for problems of convex minimization under complementary convex constraints of the form (32):

$$\min\{\varphi(y)|\ h(y) \geq 0\}$$

(see e.g. [14] or [25]). In particular, the conical algorithm in [23]) for (OE) is just a specialization of the conical branch and bound method for these problems. Observe in passing, however, that, as it stands, the algorithm of Thoai, devised for solving (OE), may converge to a weakly efficient but not efficient point which is, consequently, neither an optimal solution of (OE) nor even an optimal solution of (OWE). In fact, the partition set chosen for subdivision in a current iteration of this algorithm is not guaranteed to contain a feasible solution, so that the intersection of a nested sequence of partition sets is not guaranteed to give an efficient point.

It should be further noticed that, since the function $\Phi(\lambda)$ is decreasing while the set Λ (or Λ_δ) is a normal compact subset of $\mathbb{R}^m_+$, the problem (35) is also a monotonic optimization problem and, consequently, can be solved by the same method as that presented above for problem (Q). Thus, under usual assumptions, there are several alternative d.c. formulations of the problems (OWE), (OE), which could also be treated as monotonic optimization problems. However, the monotonic approach, aside from requiring much less restrictive assumptions and being applicable to a wider range of problems, seems to be more flexible and more easily implementable.

In this connection, it is useful also to mention that in certain cases the best results may be obtained by a hybrid approach combining monotonic with d.c. methods. In particular, local optimization methods may be useful for improving the bounds and accelerating the convergence. Specifically, recall from Proposition 7 that $\Gamma = \{y|\ h(y) \leq 0\}$, where

$$h(y) = \min\{t|\ y_i - C_i(x) \leq t,\ i = 1, \ldots, m,\ x \in X\}$$

is a convex increasing function (assuming $C(x)$ to be concave). Given a feasible solution $\bar{y} \in \partial^+\Gamma$ (the current best) one can take a vector $w \in \partial h(y)$ and compute a minimizer of $\varphi(y)$ over the polytope $\{y \in [p,q]|\ \langle w, y - \bar{y}\rangle \geq 0\} \subset$

$\{y \in [p,q] | \ h(y) \geq h(\bar{y}) = 0\} \subset (p + \mathbb{R}^m_+) \setminus \Gamma$. If y' is such a minimizer then the intersection $\hat{y}$ of $\partial^+ \Gamma$ with the line segment from p to y' satisfies $h(\hat{y}) = 0$ and $\hat{y} \leq y'$, hence $\varphi(\hat{y}) \leq \varphi(y') \leq \varphi(\bar{y})$ (in most cases $\varphi(\hat{y}) < \varphi(\bar{y})$, otherwise $\bar{y}$ is already a local minimizer of $\varphi(y)$ on $\partial^+ \Gamma$). This procedure requires the knowledge of a vector $w \in \partial h(\bar{y})$. But is easy to see that one can always take w to be an optimal solution of the dual to the linear program defining $h(\bar{y})$.

Indeed, since $h(\bar{y}) = \min_{x \in X, t}\{t - \langle w, t + C(x) - \bar{y}\rangle\}$, one can write

$$
\begin{aligned}
& h(y) - h(\bar{y}) \\
& \geq \min_{x \in X, t} \{t - \langle w, t + C(x) - y\rangle\} - \min_{x \in X, t} \{t - \langle w, t + C(x) - \bar{y}\rangle\} \\
& \geq \langle w, y\rangle - \langle w, \bar{y}\rangle = \langle w, y - \bar{y}\rangle \quad \forall y
\end{aligned}
$$

proving that $w \in \partial h(\bar{y})$. Also $\langle w, y - \bar{y}\rangle \geq 0 \Rightarrow h(y) \geq h(\bar{y}) = 0$.

7 Problems with a composite monotonic objective function

A class of problems closely related to the above problem (P) is constituted by problems with a composite monotonic objective function, i.e. problems of the form

$$\min\{F(C(x)) | \ x \in D\} \tag{R}$$

where $F(y)$ is an increasing function, $C : \mathbb{R}^n \to \mathbb{R}^m_+$ a continuous mapping and D a polytope in $\mathbb{R}^n$.

This general class includes many nonconvex optimization problems such as multiplicative programming and related problems, extensively studied over the last fifteen years (see [14], see also [6], [7]). A unified approach to a large subclass of this class that includes linear-fractional programming problems has also been developed in [12] in the framework of monotonic optimization.

Clearly the problem (R) can be rewritten as

$$\min_{x,y}\{F(y) | \ y \geq C(x), \ \ x \in D, y \in \mathbb{R}^m_+\}$$

and by setting $\Omega = \{y \in \mathbb{R}^m_+ | \ y \geq C(x), \ \ x \in D\}$ it can be reduced to a problem in the image space of C, i.e. of much less dimension, namely

$$\min\{F(y) | \ y \in \Omega\}. \tag{38}$$

Since $F(y)$ is an increasing function while Ω is obviously a conormal set in $\mathbb{R}^m_+$ the problem (38) is a standard monotonic optimization problem in y, tractable by well developed methods.

In [12] the problem (38) has been solved by the copolyblock approximation method initially proposed in [26]. Although the computational results with this copolyblock approximation algorithm as obtained in [12] are quite satisfactory on problems with $m \leq 7$, they should be significantly improved, if the BRB algorithm in [31] were used.

8 Illustrative examples and computational results

To show the practicability of the proposed method we present some numerical examples together with the computational results, focussing on the problems (OWE) and (OE) with convex objective function $f(x)$ and/or linear-fractional criteria. The algorithm was coded in C^{++} and run on a PC Pentium IV 2.53GHz, RAM 256Mb DDR, with linear subproblems solved by the LP software CPLEX 8.0. In all cases the algorithm is terminated when η-optimality is achieved, i.e. when the current best lower bound differs from the current best upper bound by no more than η, where $\eta = 0.01$ is the tolerance.

1. *Problem* (OWE) *with linear input functions.*

Example Solve

$$\min\{f(x) \mid x \in X,\ g(x) \leq 0,\ x \in X_{WE}\}$$

with following data:

$f(x) = \langle c, x\rangle,\ c = (4, -8, -3, -1, -7, 0, -6, -2, 9, -3),$

$X = \{x \in R^{10}_{+} \mid A_1 x \leq b_1,\ A_2 x \geq b_2, 0 \leq x_i \leq 10, i = 1, \ldots, 10\}$, where

$$A_1 = \begin{bmatrix} 7 & 1 & -3 & -7 & 0 & 9 & 2 & 1 & -5 & 1 \\ -5 & 8 & -1 & 7 & 5 & 0 & -1 & -3 & 4 & 0 \\ 2 & -1 & 0 & -2 & 3 & -2 & 2 & -5 & -1 & -3 \\ -1 & -4 & 2 & 9 & -4 & 3 & -3 & 4 & 0 & -2 \\ -3 & -1 & 0 & 8 & -3 & -1 & -2 & -2 & 5 & -5 \end{bmatrix}$$

$b_1 = (-66, 150, -81, 79, 53)$

$$A_2 = \begin{bmatrix} -2 & 9 & -1 & -2 & 2 & 1 & 4 & -1 & 5 & 2 \\ -2 & 3 & 2 & 4 & 5 & 4 & 1 & -9 & -2 & -1 \\ -4 & -8 & 1 & 1 & -5 & 3 & -2 & 0 & -2 & 9 \\ 2 & 7 & -1 & -2 & -5 & -9 & 4 & -1 & -2 & 0 \\ 2 & 3 & -1 & -1 & 4 & 3 & -1 & 0 & 0 & -6 \\ -6 & 0 & 0 & 0 & -4 & 3 & -2 & -2 & 4 & -6 \\ 2 & 2 & 3 & -5 & 6 & -4 & 0 & 0 & -1 & -4 \\ -1 & 4 & 4 & 6 & 0 & 3 & -4 & 2 & -4 & -1 \end{bmatrix}$$

$b_2 = (126, 12, -52, -23, 2, -23, -28, 90)$

$g(x) := Gx \leq d$ where

$$G = \begin{bmatrix} -1 & 1 & 2 & 4 & -4 & 4 & -1 & -4 & -6 & 3 \\ 4 & 9 & 0 & -1 & -2 & 1 & -6 & 5 & 0 & 0 \end{bmatrix} \qquad d = (-14, 89)$$

$$C = \begin{bmatrix} 5 & 1 & 7 & 1 & 4 & 9 & 0 & -4 & -3 & 7 \\ -1 & -2 & -5 & -4 & -1 & -6 & -4 & 0 & -3 & 0 \\ -3 & -3 & 0 & 4 & 0 & 1 & -2 & 1 & 4 & 0 \end{bmatrix}$$

Results of computation

Optimal solution:

$x^* = (0, 8.741312, 8.953411, 9.600323, 3.248449, 3.916472,$
$6.214436, 9.402384, 10, 4.033163)$

(found at iteration 78 and confirmed at iteration 316)

Optimal value: -107.321065

Computational time: 1.406 sec.

Maximal number of nodes generated: 32

2. *Problem* (OWE) *with convex objective function.*

Example Solve (OWE) with following data:

$f(x) = \langle c, x\rangle + \langle x, Qx\rangle$ where $c = (6, 0, -3, -2, -6, 3, -4, -4, 3, 1)$

$$Q = \mathrm{diag}[1, 4, 3, 3, 5, 7, 8, 7, 8, 8]$$

$g(x) \equiv 0.$

$X = \{x \in R^{10} |\ A_1x \leq b_1, A_2x \geq b_2, A_3x = b_3, 0 \leq x_i \leq 10, i = 1, \ldots, 10\}$
where

$$A_1 = \begin{bmatrix} 0 & 3 & 2 & 2 & 6 & -1 & 2 & 2 & 4 & -3 \\ -4 & -2 & -8 & 1 & 3 & 3 & 4 & -2 & -7 & -4 \\ -2 & -3 & 0 & 0 & -8 & -7 & -2 & 1 & -3 & 4 \end{bmatrix}$$

$b_1 = (139, -94, -140)$

$$A_2 = \begin{bmatrix} 3 & 4 & 0 & 2 & 3 & 0 & 1 & 3 & -6 & -4 \\ 4 & -6 & -5 & 2 & -4 & 3 & -2 & 4 & -7 & -2 \end{bmatrix}$$

$b_2 = (11, -121)$

$A_3 = [6, 7, -4, -6, -4, -1, 2, 5, -7, 5] \quad b_3 = -92$

$$C = \begin{bmatrix} 2 & 0 & 8 & 4 & -1 & 5 & -8 & -6 & -5 & -7 \\ 6 & 0 & 1 & -4 & 5 & 2 & -1 & 5 & -4 & 3 \\ -9 & -4 & 4 & 4 & 1 & -5 & 0 & -3 & -5 & -3 \\ -5 & 6 & -1 & -6 & -2 & -8 & 2 & -6 & -8 & -5 \\ -3 & 4 & 1 & -4 & -8 & 6 & -7 & 3 & 3 & 0 \end{bmatrix}$$

Results of computation:

Optimal solution:

$x^* = (4.954651,\ 1.816118,\ 10,\ 2.442424,\ 10,\ 4.198649,\ 0,\ 0,\ 5.083933,\ 0)$
(found at iteration 11 and confirmed at iteration 11)

Optimal value: 1148.500433

Computational time: 0.156 sec.

Maximal number of nodes: 8

3. *Problem* (OE) *with convex objective function*

Example Solve

$$\min\{f(x)|\ x \in X,\ g(x) \leq 0,\ x \in X_{WE}\}$$

with $f(x)$, $g(x)$, $X, C(x)$ as in the previous example.

Results of computation:
Optimal solution:
$x^* = (5.343288, 1.482599, 10, 2.803773, 9.260545, 5.081512, 0, 0, 5.070228, 0)$
(found at iteration 39 and confirmed at iteration 39)
Optimal value: 1147.469164
Computational time: 0.39 sec.
Maximal number of nodes: 15

4. *Problem* (OE) *with linear-fractional multicriteria*

Example (taken from [19]) Solve

$$\min\{f(x)|\ x \in X,\ x \in X_E\}$$

with following data:

$f(x) = -x_1 - x_2$

$X = \{x \in R^2_+|\ Ax \leq b\}$ where

$$A = \begin{bmatrix} 1 & -2 \\ -1 & -2 \\ -1 & 1 \\ 1 & 0 \end{bmatrix}, \qquad b = (2, -2, 1, 6)^T$$

$$C_1(x) = \frac{x_1}{x_1 + x_2},\ C_2(x) = \frac{-x_1 + 6}{x_1 - x_2 + 3}$$

Results of computation:
Optimal solution:
$x^* = (1.991962, 2.991962)$ (found at iteration 16 and confirmed at iteration 16)
Optimal value: -4.983923
Computational time: 0.484 sec.
Maximal number of nodes: 6

5. *Computational Experiments*

The algorithm was tested on a number of problem instances with varying n and m. Tables 1 and 2 report the computational results for the problems (OWE) and (OE), resp., with convex quadratic objective function. Table 3 reports the computational results for the problem (OE) with linear-fractional criteria. In all the test problems $g(x)$ is taken to be $\equiv 0$.

Table 1

Prob.	n	m	Iteration	Nodes	Time (in seconds)
1-10	10	3	4483	441	38.311
11-20	20	3	26823	2185	385.647
21-30	30	3	29782	2074	1147.655
31-40	10	4	11679	1220	114.717
41-50	20	4	34783	2885	908.922

Table 2

Prob.	n	m	Iteration	Nodes	Time (in seconds)
1-10	10	3	2209	208	15.742
11-20	20	3	25960	2045	309.975
21-30	30	3	18340	1841	646.466
31-40	10	4	10880	1205	84.864
41-50	20	4	29043	2960	471.725

Table 3

Prob.	n	m	Iteration	Nodes	Time (in seconds)
1-5	10	4	507	59	40.159
6-10	15	3	2644	622	320.375
11-15	15	4	14790	969	1781.613

References

1. L.T. Hoai-An, N.T. Bach-Kim and T.M. Thanh: 'Optimizing a monotonic function over the Pareto set', Preprint, Department of Applied Mathematics and Informatics, Institute of Technology, Hanoi, 2004.
2. H.P. Benson: 'Optimization over the Efficient Set', *Journal of Mathematical Analysis and Applications*, 38(1984), 562-584.
3. H.P. Benson: 'An algorithm for Optimizing over the Weakly-Efficient Set', *European Journal of Operations Research* 25(1986), 192-199.
4. H.P. Benson: 'An all-linear programming relaxation algorithm for optimizing over the efficient set', *Journal of Global Optimization* 1(1991), 83-104.
5. H.P. Benson: 'A bisection-extreme point search algorithm for optimizing over the efficient set in the linear dependence case', *Journal of Global Optimization* 3(1993), 95-111.
6. H.P. Benson: 'An Outcome Space Branch and Bound-Outer Approximation Algorithm for Convex Multiplicative Programming', *Journal of Global Optimization* 15(1999), 315-342.
7. H.P. Benson and G.M. Boger: 'Outcome-Space Cutting-Plane Algorithm for Linear Multiplicative Programming', *Journal of Optimization Theory aznd Applications* 104(2000), 301-322.
8. S. Bolintineanu: 'Minimization of a Quasiconcave function over an Efficient Set', *Mathematical Programming* 61(1993), 89-110.
9. S. Bolintineanu: 'Optimality condition for minimization over the (weakly or properly) efficient set', *Journal of Mathematical Analysis and Applications* 173(1993), 523-541.

10. E.U. Choo and D.R. Atkins: 'Bicriteria linear fractional programming', *Journal of Optimization Theory and Applications* 36(1982), 203-220.
11. J.G. Ecker and J.H. Song: 'Optimizing a linear function over an efficient set' *Journal of Optimization Theory and Applications* 83(1994), 541-563.
12. N.T. Hoai-Phuong and H. Tuy : 'A Unified Monotonic Approach to Generalized Linear Fractional Programming', *Journal of Global Optimization* 23(2002), 1-31.
13. E. Ishizuka: 'Optimality Conditions for Directionally Differentiable Multiobjective Programming Problems', *Journal of Optimization Theory and Applications* 72(1992), 91-112.
14. H. Konno, P.T. Thach and H. Tuy: *Optimization on Low Rank Nonconvex structures*, Kluwer, 1997.
15. H.A. Le Thi, T. Pham Dinh, and L.D. Muu: 'Simplicially-Constrained DC Optimization over Efficient and Weakly Efficient Sets', *Journal of Optimization Theory and Applications* 117 (2003), 503-531.
16. L. T. Luc: 'Reverse Polyblock Approximation for Optimization over the Weakly Efficient Set and Efficient Set', *Acta Mathematica Vietnamica* 26(2001), 65-80.
17. C. Malivert: 'Multicriteria fractional optimization', in *Proc. of the 2nd Catalan days on applied mathematics*, M. Sofonea and J.N. Corvellac eds, (1995), 189-198.
18. C. Malivert and N. Popovici: 'Bicriteria Linear Fractional Optimization', in *Optimization*, V.H. Nguyen, J-J. Strodiot and P. Tossings eds, Lecture Notes in Economics and Mathematical Systems 481, Springer 2000, 305-319.
19. L.D. Muu and H.Q. Tuyen: 'Bilinear programming approach to optimization over the efficient set of a vector affine fractional problem', *Acta Mathematica Vietnamica* 27(2002), 119-140.
20. J. Phillip: 'Algorithms for the Vector Maximization Problem', *Mathematical Programming* 2(1972), 207-229.
21. P.T. Thach and H. Tuy: 'Parametric Approach to a Class of Nonconvex Global Optimization Problems', *Optimization* 19(1988), 3-11.
22. H. Tuy, P.T. Thach, H. Konno and Yokota:'Dual Approach to Optimization on the Set of Pareto-Optimal Solutions', *Journal of Optimization Theory and Applications* 88(1996), 689-707.
23. N.V. Thoai: 'Conical algorithm in global optimization for optimizing over efficient sets', *Journal of Global Optimization* 18(2000), 321-320.
24. H. Tuy: 'Global Minimization of a Difference of Two Convex Functions', *Mathematical Programming Study 30*, North-Holland, 1987, pp. 150-182.
25. H. Tuy: *Convex Analysis and Global Optimization*, Kluwer, 1998.
26. H. Tuy: 'Monotonic Optimization: Problems and Solution Approaches', *SIAM Journal on Optimization* 11(2000), 464-494.
27. H. Tuy and L.T. Luc: 'A New Approach to Optimization Under Monotonic Constraint', *Journal of Global Optimization* 18(2000), 1-15.
28. H. Tuy: 'Convexity and Monotonicity in Global Optimization' , in *Advances in Convex Analysis and Optimization*, N. Hadjisavvas and P.M. Pardalos eds, Kluwer, 2001, 569-594.
29. H. Tuy: 'Monotonicity in the framework of generalized convexity', in *Generalized Convexity, Generalized Monotonicity and Applications*, A. Eberhard, N. Hadjisavvas and D.T. Luc eds, Springer, 2005, 61-85.
30. H. Tuy, P.T. Thach and H. Konno: 'Optimization of Polynomial Fractional Functions' , *Journal of Global Optimization*, 29(2004), 19-44.

31. H. Tuy, F. Al-Khayyal and P.T. Thach: 'Monotonic Optimization: Branch and Cuts Methods', in *Essays and Surveys on Global Optimization*, eds. C. Audet, P. Hansen, G. Savard, Kluwer, to appear.
32. H. Tuy: 'Partly convex and convex-monotonic optimization problems', in *Modelling, Simulation and Optimization of Complex Processes*, eds. H. G Bock, E. Kostina, H. X. Phu, R. Rannacher, Springer, 2005.
33. H. Tuy: 'Polynomial Optimization: A Robust Approach', *Pacific Journal of Optimization*, 2005, to appear
34. H. Tuy, M. Minoux and N.T. Hoai Phuong: 'Discrete Monotonic Optimization With Application to A Discrete Location Problem', *SIAM Journal of Optimization*, to appear.
35. P.L. Yu: '*Multicriteria Decision Making*, Plenum Press,New York, 1995.

On a Local Search for Reverse Convex Problems

Alexander Strekalovsky

Institute of System Dynamics and Control Theory SB of RAS, 134 Lermontov St., Irkutsk-33, 664033 Russia `strekal@icc.ru`

Summary. In this paper we propose two variants of Local Search Method for reverse convex problems. The first is based on well-known theorem of H. Tuy as well as on Linearization Principle. The second variant is due to an idea of J. Rosen. We demonstrate the practical effectiveness of the proposed methods computationally.

Key words: Nonconvex optimization, reverse convex problem, local search, computational testing.

1 Introduction

The present situation in Continuous Nonconvex Optimization may be viewed as dominated by methods transferred from other sciences [1]-[3], as Discrete Optimization (Branch&Bound, cuts methods, outside and inside approximations, vertex enumeration and so on), Physics, Chemistry (simulated annealing methods), Biology (genetic and ant colony algorithms) etc.

On the other hand the classical method [11] of convex optimization have been thrown aside because of its inefficiency [1]-[6]. As is well known, the conspicuous limitation of convex optimization methods applied to nonconvex problems is their inability of escape a local extremum or even a critical point depending on a starting point [1]-[3]. So, the classical apparatus shows itself inoperative for new problems arising from practice.

In such a situation it is desirable to create a global optimization approach aimed at nonconvex problems — in particular to Reverse Convex Problem (RCP) — connected on with Convex Optimization Theory and using the methods of Convex Optimization.

We ventured to propose such an approach [12] and to advance the following principles of Nonconvex Optimization.

I. The linearization of the basic (generic) nonconvexity of a problem of interest and consequently a reducing of the problem to a family of (partially) linearized problems.

II. The application of convex optimization methods for solving the linearized problems and, as a consequence, within special local search methods.

III. Construction of "good" (pertinent) approximations (resolving sets) of level surfaces and epigraph boundaries of convex functions.

Obviously, the first and the second are well known. The depth and effectiveness of the third may be observed in [12]-[23].

Developing these principles we get the solving methodology for nonconvex problems which can be represented as follows.

1. Exact classification of a problem under study.
2. Application of special (for a given class of problems, for instance, RCP) local search methods.
3. Applyication of special conceptual global search methods (strategies).
4. Using the experience of similar nonconvex problems solving to construct pertinent approximations of level surfaces of corresponding convex functions.
5. Application of convex optimization methods for solving linearized problems and within special local search methods.

This approach lifts Classical Convex Optimization to a higher level, where the effectiveness and the speed of the methods become of paramount importance not only for Convex Optimization, but for Nonconvex problems (in particular for RCP, which is discussed below).

Our computational experience suggests that if you follow the above methodology you have more chances to reach a global solution of a nonconvex problem of large size (≥ 1000 variables) than applying the Branch-and-Bound or Cutting Plane methods.

In this paper we decided to focus only on the advantages of Principle I — Linearization applied for Local Search Problem. In order to do this we propose two variants of Local Search Method. The first is based on a well-known theorem of H. Tuy [1] as well as on the Linearization Principle. The second variant is due to an idea of J. Rosen [10], which was only slightly modified by adding the procedure of free descent on the constraint $g = 0$. Finally we demonstrate the practical effectiveness of these methods by a computational testing and propose to unify two methods.

Before this we recall a few facts from Reverse Convex Theory.

2 Some features of RCP

Let us consider the problem

$$h(x) \downarrow \min, \quad x \in S, \quad g(x) \geq 0, \tag{P}$$

where h is a continuous function and the function g is a convex function on $I\!R^n$, $S \in I\!R^n$.

Denote the feasible set of the problem (P) by D

$$D := \{x \in S \mid g(x) \geq 0\} \neq \emptyset. \tag{1}$$

Further, suppose

$$h_* := \inf(h, D) \triangleq \inf_x \{h(x) \mid x \in S,\ g(x) \geq 0\} > -\infty. \tag{2}$$

It can be easily seen that the nonconvexity of (P) is generated by the reverse convex constraints $g \geq 0$ defining the complement of convex open set $\{x \in I\!R^n \mid g(x) < 0\}$. That is why we suppose this constraint to be active at any solution of (P) $(Sol(P))$. Otherwise, by solving the relaxed problem

$$(PW): \qquad h(x) \downarrow \min, \quad x \in S, \tag{3}$$

(which is simpler than (P)) one can simultaneously find a solution to (P).

The regularity conditions (when $g \geq 0$ is substantial) [1]-[6] may be given in different ways. For instance,

$$(G): \qquad \left.\begin{array}{c}\text{There is no any solution } x^* \in D \\ \text{to } (P) \text{ such that } g(x^*) > 0.\end{array}\right\} \tag{4}$$

The latter is equivalent to

$$(G'): \qquad Sol(P) \cap \{x \in I\!R^n \mid g(x) > 0\} = \emptyset, \tag{2.4'}$$

where $Sol(P)$ is the solution set of the problem (P) $Sol(P) = Argmin(P)$.

One can express the regularity condition with the help of the optimal value function for problem (P) and the relaxed problem (PW)–(3)

$$\mathcal{V}(PW) := \inf_x \{h(x) \mid x \in S\} < \mathcal{V}(P) \triangleq \inf_x \{h(x) \mid x \in S,\ g(x) \geq 0\}. \tag{5}$$

One of corollaries of the last condition is the fact that by solving the relaxed problem (PW)–(3), say, with convex h and S it is possible to perform a descent to the constraint $g = 0$ by means, for instance, one of the classical methods of convex optimization. As a consequence, one has

$$Sol(P) \subset \{x \in I\!R^n \mid g(x) = 0\}.$$

The following result is fundamental in RCP theory. Additionally, this theorem establishes a relation between Problem (P) and the problem of convex maximization [1]-[4], [12].

Theorem 1. *(H. Tuy [1, 2]) Let us suppose the assumption (G)–(4) to be fulfilled, and a point z to be a solution to (P). Then*

$$\max_x \{g(x) \mid x \in S,\ h(x) \leq h(z)\} = 0. \tag{6}$$

If the following assumption takes place

$$(H1): \qquad \left.\begin{array}{c} \forall y \in S: \; g(y) = 0, \; \exists \varepsilon > 0, \\ \exists u \in S \cap B(y, \varepsilon): \; g(u) > 0; \end{array}\right\} \tag{7}$$

then the condition (6) becomes sufficient for z to be a global solution to (P).

According to this result, instead of solving the Problem (P) one can consider the convex maximization problem

$$(Q_\beta): \qquad g(x) \uparrow \max, \;\; x \in S, \; h(x) \le \beta. \tag{8}$$

If one got that the value of (Q_β)

$$V(\beta) := \max_x \{g(x) \mid x \in S, \; h(x) \le \beta\}$$

with $\beta = h(z)$ is equal to zero, then $z \in Sol(P)$.

A theorem of H. Tuy generated a stream of interest in Solution Methods Theory for RCP leading to reducing Problem (P) to the dual problem (Q_β). Let us note two properties of such a reduction. First, the basic (generic) nonconvexity of the Problem (P) has not been dissipated. It stays in the goal function of (Q_β)–(8) so that even with convex h and S the problem (8) is nonconvex. Second, it is not clear how to choose the parameter β.

Finally, the question is: is it possible to apply convex optimization methods to solve (8)? It will be shown below that there exists another way to solve the Problem (P) [12], i.e. to employ an effective local search process.

3 Local search methods

Let us suppose the function h and the set S to be convex and, besides, the following regularity condition to be fulfilled (cf. (5))

$$(H_0): \qquad \exists v \in S, \; h(v) < h_* \triangleq \mathcal{V}(P), \;\; g(v) < 0. \tag{9}$$

Under this hypothesis we propose a special local search method consisting of two parts. The first procedure begins at a feasible point $y \in S, \; g(y) \ge 0$, and constructs a point $x(y) \in S$, such that

$$g(x(y)) = 0, \;\; h(x(y)) \le h(y).$$

The second procedure consists in the consecutive solution of the Linearized problems:

$$(LQ(u, \beta)): \qquad \left.\begin{array}{c} \langle \nabla g(u), x \rangle \uparrow \max, \\ x \in S, \;\; h(x) \le \beta; \end{array}\right\} \tag{10}$$

where the parameters u and β will be defined below.

It can be readily seen, that the linearized problems $(LQ(u,\beta))$ are convex. So a convex optimization method can be applied to get an approximate global solution to (10).

Now let us move to a more detailed description of the calculation process.

Procedure 1. [12, 20] Let us consider a point $y \in S, \quad g(y) \geq 0$. If $g(y) = 0$, we set $x(y) = y$. In the case $g(y) > 0$, there exists $\lambda \in]0,1[$ such that $g(x_\lambda) = 0$, where $x_\lambda = \lambda v + (1-\lambda)y$, since $g(y) > 0 > g(v)$ due to (H_0)–(9).

In addition, because of the convexity of $h(\cdot)$ one has

$$h(x_\lambda) \leq \lambda h(v) + (1-\lambda)h(y) < \lambda h_* + (1-\lambda)h(y) \leq h(y). \tag{11}$$

That is why we set $x(y) = x_\lambda$ and obtain

$$h(x(y)) < h(y), \tag{12}$$

which is what we wanted. Usually one calls Procedure 1 “free descent” (that is, free from the constraint $g(x) \geq 0$).

Procedure 2. This starts at a feasible point $\tilde{x} \in S, \quad g(\tilde{x}) = 0$, and constructs a sequence $\{u^r\}$ such that $(r = 0,1,2,\dots)$

$$u^r \in S, \;\; g(u^r) \geq 0, \;\; h(u^r) \leq \beta, \tag{13}$$

where $\beta := h(\tilde{x}), \;\; u^0 := \tilde{x}$.

The sequence $\{u^r\}$ is constructed as follows. If a point u^r, $r \geq 0$ verifying (13) is given, then the next point u^{r+1} is constructed as an approximative solution to the linearized problem $(LQ(u^r,\beta))$, so that the following inequality holds:

$$\langle \nabla g(u^r), u^{r+1} \rangle + \delta_r \geq \sup_x \{ \langle \nabla g(u^r), x \rangle \mid x \in S, \;\; h(x) \leq \beta \}, \tag{14}$$

where the sequence $\{\delta_r\}$ is such that

$$\delta_r > 0, \; r = 0,1,2,\dots, \; \sum_{r=1}^{\infty} \delta_r < +\infty. \tag{15}$$

Theorem 2. *[12] Let us suppose that the optimal value of the dual problem:*

$$(Q_\gamma): \qquad g(x) \uparrow \max, \quad x \in S, \; h(x) \leq \gamma, \tag{16}$$

is finite for some $\gamma \geq \beta$:

$$\mathcal{V}(Q_\gamma) := \sup_x \{ g(x) \mid x \in S, \; h(x) \leq \gamma \} < +\infty. \tag{17}$$

In addition, the function $g(\cdot)$ is convex and continuously differentiable on an open domain Ω containing the set

$$S \cup \{ x \in I\!R^n \mid g(x) = 0 \}. \tag{18}$$

Then:

i) the sequence $\{u^r\}$ generated by Procedure 2 verifies the condition

$$\lim_{r\to\infty}[\sup_x\{\langle\nabla g(u^r), x-u^r\rangle \mid x\in S,\ h(x)\le\beta\}]=0. \tag{19}$$

ii) For every cluster point u_* of the sequence $\{u^r\}$ the following conditions holds:

$$\langle\nabla g(u_*), x-u_*\rangle\le 0\ \forall x\in S:\ h(x)\le\beta, \tag{20}$$

$$g(u_*)\ge 0. \tag{21}$$

iii) If S is closed, then a cluster point u_* turns out to be normally critical (stationary) to the problem (Q_β).

We now show how to construct the point $y(\tilde{x})$ with the help of the sequence $\{u^r\}$. If we consider numbers $\varepsilon>0$ and $r\ge 0$ such that $\delta_r\le\varepsilon/2$ and

$$\langle\nabla g(u^r), u^{r+1}-u^r\rangle\le\varepsilon/2, \tag{22}$$

then we set $y=y(\tilde{x},\varepsilon):=u^r$. It can be shown [12] that the point y verifies the condition

$$\sup_x\{\langle\nabla g(y), x-y\rangle \mid x\in S,\ h(x)\le\beta\}\le\varepsilon, \tag{23}$$

i.e. y turns out to be an ε-solution to the linearized problem $(LQ(y,\beta))$–(10) where $\beta=h(\tilde{x})$.

Let us now unify the procedures 1 and 2 into one method. In what follows, we consider a feasible point $x_0\in S,\ g(x_0)\ge 0$, and number sequences $\{\delta_r\}$ and $\{\varepsilon_s\}$ verifying (15) and the condition:

$$\varepsilon_s>0,\ s=0,1,2,\ldots,\ \ \varepsilon_s\downarrow 0\ (s\to+\infty).$$

Special Local Search Method (SLSM).
Step 0. Set $s:=0,\ x^s:=x_0,\ \beta_s:=h(x^s)$.
Step 1. (Procedure 2) Beginning at the point x^s, construct a point $y^s=y(x^s,\varepsilon_s)$:

$$y^s\in S,\ g(y^s)\ge 0,\ h(y^s)\le\beta_s,$$

which is ε_s-solution to linearized problem $(LQ(y^s,\beta_s))$, i.e.

$$\langle\nabla g(y^s), x-y^s\rangle\le\varepsilon_s\ \forall x\in S:\ h(x)\le\beta_s.$$

Step 2. (Stopping criterion) If $g(y^s)\le 0$, **STOP**.
Step 3. (Procedure 1) With the help of the point y^s construct $u:=x(y^s)$ such that

$$u\in S,\ g(u)=0,\ h(u)<h(y^s)\le\beta_s.$$

Step 4. Set $s:=s+1,\ x^s:=u,\ \beta_s:=h(u)$ and loop to Step 1.

It is easy to see [12] that SLSM described above

(a) either is finite with N iterations, $g(y^N) = 0$;
(b) or generates two sequences $\{x^s\}$ and $\{y^s\}$ with the properties:

$$x^s \in S, \;\; g(x^s) = 0, \;\; y^s \in S, \;\; g(y^s) > 0, \tag{24}$$

$$\beta_{s+1} := h(x^{s+1}) < h(y^s) \leq \beta_s := h(x^s). \tag{25}$$

Besides, the following equalities hold

$$\beta_* := \lim_{s \to \infty} \beta_s = \lim_{s \to \infty} h(y^s). \tag{26}$$

Theorem 3. *Let us consider a convex function $h(\cdot)$ and a convex set S. In addition, suppose the set $F_0 = \{x \in S \mid h(x) \leq h(x_0)\}$ to be bounded and the regularity condition (H_0)–(9) to be fulfilled. Then SLSM:*

(a) either (in the finite case) obtains a point $y^N \in S, \;\; g(y^N) = 0$, that is an ε_N-solution to linearized problem $(LQ(y_N, \beta_N))$ where N is the number of the stopping iteration;
(b) or (in the general case) in addition to the properties (24)–(26) the sequences $\{x^s\}$ and $\{y^s\}$ verify the conditions:

$$0 = g(x^s) = \lim_{s \to \infty} g(y^s), \tag{27}$$

$$x_* = \lim_{s \to \infty} x^s = \lim_{s \to \infty} y^s, \tag{28}$$

with a point $x_* \in I\!R^n, \;\; g(x_*) = 0$.

Furthermore, the point x^* is a solution to the linearized problem $(LQ(x_*, \beta_*))$:

$$\langle \nabla g(x_*), x - x_* \rangle \leq 0 \;\; \forall x \in S, \;\; h(x) \leq \beta_*, \tag{29}$$

and a normal stationary point with respect to the dual problem $(Q(\beta_*))$.

Remark. If one changes the stopping criterion of SLSM $g(y^s) \leq 0$ to the simultaneous fulfilment of the three inequalities below

$$g(y^s) \leq \tau, \;\; \varepsilon_s \leq \tau, \;\; \beta_{s-1} - \beta_s \leq \tau, \tag{30}$$

where τ is a given tolerance, then it is easy to see that SLSM turns out to be finite. Besides, it yields the point y^N with the properties

$$\left.\begin{array}{c} g(y^N) \leq \tau, \;\; h(y^N) \leq \beta_N, \\ \langle \nabla g(y^N), x - y^N \rangle \leq \tau \;\; \forall x \in S, \;\; h(x) \leq \beta_N, \end{array}\right\} \tag{31}$$

which is suitable for a local search.

Note also that SLSM yields an approximate stationary point to the dual problem (Q_β) (for some β), but not for Problem (P), what completely corresponds to the duality Theorem 1 of H. Tuy [1].

Further, in addition to SLSM we consider a variant of a well-known method proposed by Rosen J.B. in 1966 [10]. This method consists in a consecutive solution of linearized problems of type different from $(LQ(u,\beta))$:

$$(PLR_r): \qquad \left.\begin{array}{c} h(x) \downarrow \min, \;\; x \in S, \\ \langle \nabla g(u^r), x - u^r \rangle + g(u^r) \geq 0, \end{array}\right\} \tag{32}$$

where $u^r \in S$ is a given point. The next point u^{r+1} is defined as an exact solution to (32).

In [9, 10], the convergence of the method was investigated. We proposed [12] a modification of Rosen method (MRM) which consists of two procedures.

The first procedure is an approximate solution of problem (PLR_r)–(32) which is obviously convex, if $h(\cdot)$ and S are convex. Then it becomes possible to apply a suitable convex optimization method to find a global (approximate) solution of (32).

The second procedure coincides with Procedure 1 of free descent on the constraint $g = 0$ (see description of SLSM). We were able to prove convergence of the proposed method [12].

In the following section we shall show the extent of the effectiveness of the local search theory proposed so far.

4 Computational testing

In this paragraph we present the results of computational solving by two local search methods presented above of a series of RCP of the following type:

$$\left.\begin{array}{c} \langle c, x \rangle \downarrow min, \\ x \in S \triangleq \{x \in I\!R^n \mid Ax \leq b, \; x \geq 0\}, \\ g(x) \geq 0, \end{array}\right\} \tag{33}$$

with the function g of two forms:

$$g_1(x) \triangleq \| x \|^2 - \langle d, x \rangle - \gamma, \tag{34}$$

$$g_2(x) \triangleq \langle x, Qx \rangle - \langle d, x \rangle - \gamma, \tag{35}$$

Here Q is an $n \times n$ symmetric ($Q = Q^T$) positive definite ($Q > 0$) matrix, with $d \in I\!R^n, \;\; \gamma \in I\!R$.

The result of computational experiments are presented in Table 1. Some comments on the particularities of the computational implementation are in order.

Note that the Linearized problems (LQ)–(10) and (PLR_r)–(32) are in fact Linear Programming (LP) problems and have been solved by one of the standard methods of LP [11].

		SLSM			MRM		
name	h_0	h_f	PL	$T(10)$	h_f	PL	$T(10)$
jm11x5h1	13,78	-12,000000	2	00.00	-12,000000	3	00.00
	6,00	-11,733224	2	00.00	-11,733224	2	00.00
	388,00	-11,725533	4	00.00	-11,725533	3	00.00
jm30x10h1	21,54	-25,998666	4	00.00	-25,998666	2	00.00
	-3,00	-25,998666	4	00.00	-25,998666	2	00.00
	-19,00	-26,000000	1	00.00	-26,000000	2	00.00
jm11x15h1	802,45	-502,000000	17	00.23	-502,000000	3	00.05
	-18,00	-498,348422	17	00.23	-498,348422	3	00.05
	90,79	-502,000000	16	00.19	-502,000000	4	00.05
jm10x20h1	1156,52	-723,996423	8	00.07	-723,996423	3	00.00
	23,00	-723,996423	8	00.07	-723,996423	3	00.00
	-701,00	-724,000000	7	00.06	-724,000000	2	00.00
sr15x10	17,03	-117,309715	29	00.22	-117,309716	5	00.05
	-29,78	-109,218649	9	00.11	-109,218649	4	00.00
	-5,61	-116,691999	23	00.16	-116,692001	5	00.06
sr20x15	22,86	-156,013872	72	01.26	-156,013874	6	00.17
	-25,41	-146,763729	19	00.38	-146,763729	6	00.16
	-69,45	-146,632051	8	00.16	-146,632052	3	00.11
sr25x15	35,31	*-37,911422*	16	00.33	-37,833231	5	00.11
	3,93	-62,031119	176	03.90	-62,031121	8	00.16
	-10,43	-51,415461	52	01.37	-51,415462	5	00.17
sr30x15	40,24	-13,725476	44	01.31	-13,725477	11	00.33
	4,96	-12,712342	31	00.93	-12,712343	5	00.22
	-24,47	*-10,876227*	2	00.00	-10,335201	9	00.55
sr25x18	40,63	-30,511466	184	07.96	-30,511469	8	00.39
	-10,31	-27,049467	59	02.14	-27,049468	5	00.22
	8,02	-7,707654	35	01.16	-7,707655	5	00.22
sr30x18	37,24	-101,765161	16	00.88	-101,765161	5	00.28
	-21,32	-102,354811	17	01.10	-102,354811	5	00.22
	-25,72	-122,043711	102	09.62	-122,043711	6	00.25
sr25x20	84,75	-262,645181	95	07.19	-262,645181	6	00.39
	-15,31	-276,171787	429	29.17	-276,171789	8	00.60
	-126,16	-240,721359	18	01.16	-240,721361	4	00.28
sr10x40	6197,17	-11644,52285	25	05.50	-11644,52285	4	00.77
	-5,23	-11634,37336	24	06.87	-11634,37336	7	00.88
	1736,43	-11634,37336	24	06.87	-11634,37336	8	00.88

Table 1. Computational results.

In Table 1 we use the following notation: *name* is the test problem name, which expresses the size ($m \times n$) of the problem and the type of the function g, so that "$jm \dots h1$" notes the problems with the function $g_1(\cdot)$ while "$sr \dots$" marks the problems with the function $g_2(\cdot)$). Further, h_0 stands for the goal function value at an initial point. For each method (SLSM or MRM) the goal function value at obtained critical points have been denoted by h_f. Besides, PL means the number of solved linearized problems, and $T(10)$ is the solving time for 10 problems (since the solving time for one problem has turned out to be too small).

It can be easily seen from the Table 1 that SLSM and MRM have found the same τ-critical points in almost all test problems. At the same time the number of solved linearized problems is smaller for MRM in some cases (cf. problems $sr25 \times 15$, $sr25 \times 18$, $sr25 \times 20$). On the other hand, in some of the test problems SLSM found critical points which are better than the for MRM (cf. problems $sr25 \times 15$ and $sr30 \times 15$).

In summary, MRM works faster but SLSM sometimes finds a better critical point. Therefore according the results of computational experiments it would be practical, in order to solve similar problems, to apply a combination of SLSM and MRM. For instance, from the beginning MRM can be applied to get a critical point, and afterwards that point is improved on by SLSM.

5 Conclusion

In this paper, after presenting the principles and the methodology of Nonconvex Optimization:

- we discussed some features of RCP;
- further, we proposed two Local Search Methods for RCP and gave a convergence theorem for one of them;
- finally, we presented a computational testing of the Special Local Search Method (SLSM) and Modified Rosen Method (MRM) on a series of special RCPs.

The analysis of computational testing results led the author to propose a combination of SLSM and MRM.

References

1. Horst, R. and Tuy, H. (1993), *Global Optimization. Deterministic Approaches*, Springer-Verlag, Berlin.
2. Horst, R., Pardalos, P.M. and Thoai, V. (1995), *Introduction to Global Optimization*, Kluwer, Dordrecht.
3. Tuy, H. (1995), *D. C. Optimization: Theory, Methods and Algorithms*, in Horst. R., Pardalos P. (eds.), *Handbook of Global Optimization*, Kluwer Academic Publishers, 149–216.
4. Tuy, H. (1987), *Convex programs with an additional reverse convex constraint*, Journal of Optimization Theory and Applications, **52**, 463–485.
5. Tuy, H. (1992), *The Complementary Convex Structure in Global Optimization*, Journal of Global Optimization, **2**, 21–40.

6. Bansal, P.P. and Jacobsen, S.E. (1975), *An algorithm for optimizing network flow capacity under economy of Scale*, Journal of Optimization Theory and Applications, **15**, 565–586.
7. Hiriart-Urruty, J.-B. and Lemaréchal, C. (1993), *Convex Analysis and Minimization Algorithms*, **1–2**, Springer-Verlag, Berlin.
8. Clarke, F.H. (1983), *Optimization and nonsmooth analysis*, John Wiley & Sons, New York.
9. Meyer, R. (1970), *The Validity of a Family of Optimization Methods*, Journal SIAM Control. **8**, 41–54.
10. Rosen, J.B. (1966), *Iterative Solution of Nonlinear Optimal Control Problems*, Journal SIAM Control, **4**, 223–244.
11. Vasil'ev, F.P. (1988), *Numerical methods of extremal problems solving*, Nauka, Moscow (in russian).
12. Strekalovsky, A.S. (2003), *Elements of Nonconvex Optimization*, Nauka, Novosibirsk (in russian).
13. Strekalovsky, A.S. (1994), *Extremal problems on complements of convex sets*, Cybernetics and System Analysis, **1**, Plenum Publishing Corporation, 88-100.
14. Strekalovsky, A.S. (1993), *The search for a global maximum of a convex functional on an admissible set*, Comput. Math. and Math. Physics, **33**, 315–328, Pergamon Press.
15. Strekalovsky, A.S. (1998), *Global optimality conditions for nonconvex optimization*, Journal of Global Optimization, **12**, 415–434.
16. Strekalovsky, A.S. (1997), *On Global Optimality Conditions for D. C. Programming Problems*, Irkutsk University Press, Irkutsk.
17. Strekalovsky, A.S., (2000), *One way to Construct a Global Search Algorithm for d.c. Minimization Problems* In: Pillo, G. Di., Giannessi, F. (eds.), *Nonlinear Optimization and Related Topics*, **36**, Kluwer, Dordrecht, 429–443.
18. Strekalovsky, A.S. and Tsevendorj, I. (1998), *Testing the* $\mathbb{R}$*-strategy for a Reverse Convex Problem*, Journal of Global Optimization, **13**, 61–74.
19. Strekalovsky, A.S. and Kuznetsova, A.A. (1999), *The convergence of global search algorithm in the problem of convex maximization on admissible set*, Russian Mathematics (IzVUZ), **43**, 70–77.
20. Strekalovsky, A.S. and Yakovleva, T.V. (2004), *On a Local and Global Search Involved in Nonconvex Optimization Problems*, Automation and remote control, **65**, 375–387.
21. Strekalovsky, A.S. (2001), *Extremal Problems with D.C.-Constraints* Comp. Math. And Math. Physics, **41**, 1742–1751.
22. Strekalovsky, A.S., Kuznetsova, A.A. (2001) *On solving the Maximum Clique Problem*, Journal of Global Optimization, **21**, 265–288.
23. Strekalovsky, A.S., Kuznetsova, A.A., Tsevendorj, I. (1999) *An approach to the solution of integer optimization problem* Comp. Math. And Math. Physics, **39**, 9–16.

Some Transformation Techniques in Global Optimization

Tapio Westerlund

Process Design Laboratory, Åbo Akademi University, Biskopsgatan 8, FIN-20500 ÅBO, Finland tapio.westerlund@abo.fi

Summary. In this chapter some transformation techniques, useful in deterministic global optimization, are discussed. With the given techniques, a general class of nonconvex MINLP (mixed integer non-linear programming) problems can be solved to global optimality. The transformations can be applied to signomial functions and the feasible region of the original problem can be convexified and overestimated by the transformations. The global optimal solution of the original nonconvex problem can be found by solving a sequence of convexified MINLP sub-problems. In each such iteration a part of the infeasible region is cut off and the algorithm terminates when a solution point is sufficiently close to or within the feasible region of the original problem. The principles behind the algorithm are given in this chapter and numerical examples are used to illustrate how the global optimal solution is obtained with the algorithm.

Key words: Transformation techniques, reformulation, mixed integer nonlinear programming, signomial functions.

1 Introduction

The transformations, discussed in this chapter, are applicable to signomial functions and can be applied to problems where the objective function or some of the constraints are composed of a convex and a signomial function. The transformations are made in two steps. Single variable transformations are first applied term-wise to convexify every signomial term. Secondly, the transformations are selected such that the signomial terms are not only convexified but also underestimated. The latter property is important when developing a global optimization approach and this property is obtained by carefully selecting the transformations such that they can be applied together with piecewise linear approximations of the inverse transformations. This allows us not only to convexify and to underestimate every generalized signomial constraint but also to convexify the entire nonconvex problem and to overestimate the feasible region of it. When generalized signomial constraints are included in the problem, the originally nonconvex feasible region is divided into convex sub-regions, by this approach, each sub-region being defined by the binary variables used in the piecewise linear approximations of the inverse transformations.

2 The MINLP Problem

MINLP problems for which the method guarantees global optimal solutions can, in mathematical terms, be written as follows

$$\begin{aligned} &\min_{\mathbf{z} \in L \cap N \cap S} f(\mathbf{z}) \\ &L = \{\mathbf{z} \,|\, \mathbf{Az} = \mathbf{a}, \mathbf{Bz} \le \mathbf{b}\} \cap X \times Y \\ &N = \{\mathbf{z} \,|\, \mathbf{g}(\mathbf{z}) \le \mathbf{0}\} \\ &S = \{\mathbf{z} \,|\, \mathbf{q}(\mathbf{z}) + \mathbf{s}(\mathbf{z}) \le \mathbf{0}\} \end{aligned} \quad \text{(P)}$$

The class of problems that can be solved to global optimality with the transformation approach is slightly dependent on the MINLP algorithm used to solve the intermediate convexified problems. If the extended cutting plane method discussed in Westerlund and Pörn (2002) is used as a sub-algorithm then the intermediate MINLP problems may be pseudo-convex. In this case we can allow the objective function, f, in the problem (P), to be a differentiable pseudo-convex function and $\mathbf{g}$, $\mathbf{q}$ and $\mathbf{s}$ vectors of differentiable pseudo-convex, convex and signomial functions respectively, all defined on the set L. N and S are sets defined by the pseudo-convex and generalized signomial inequality constraints respectively. The vector of variables, $\mathbf{z}$, is composed of continuous variables $\mathbf{x}$ in X, and integer variables $\mathbf{y}$ in Y, where X is a compact subset of a finite dimensional Euclidean space and Y a finite dimensional integer set, defined by appropriate bounds of the variables. $\mathbf{A}$, $\mathbf{B}$, $\mathbf{a}$ and $\mathbf{b}$ are matrices and vectors of constants respectively.

If the intermediate MINLP sub-problems in the proposed algorithm are solved using a method that guarantees global optimal solution for convex MINLP problems only, (for example the outer approximation method by Duran and Grossmann (1986)) then the objective function, f, and the constraints, $\mathbf{g}$, in (P) need be convex.

A signomial function is composed of a sum with products of power terms, where each product with power terms is multiplied by a real constant according to

$$s(\mathbf{z}) = \sum_{j=1}^{J} c_j \prod_{i=1}^{n} z_i^{p_{i,j}} \tag{1}$$

All constants, c_j, and powers, $p_{i,j}$, may be positive or negative. The special case, when all constants in a signomial function are positive, is called a posynomial. Generally, variables in a signomial function need be positive. Appropriate bounds on the variables included in the signomial functions must, therefore, be defined. A generalized signomial function is defined as a sum of a convex and a signomial function. If each signomial term in a signomial function is convex, the signomial function is obviously convex. This is, of course, not a strict requirement; a signomial function may be convex also when only some of the terms are convex. However, the convexification strategies used in the following are applied on each term in the signomial functions. We thus convexify all signomial functions such that every signomial term will be convex.

Observe, further, that since the considered transformation approach is valid for signomial constraints with both positive and negative constants, c_j, signomial equality constraints can also be handled by the approach. In this case each signomial equality constraint and its negative counterpart may be relaxed into two signomial inequality constraints, being then defined in the set S. If the objective function is defined as a generalized signomial function, then one can obviously rewrite the problem such that an additional variable is minimized and the generalized signomial objective function is rewritten as a generalized signomial constraint.

3 The transformation approach

In the following we will show how signomials terms (and thus also signomial functions) can be transformed into convex signomial form by using single variable power transformations. In order to obtain appropriate properties for the convexified signomial terms we will, further, show how the signomial terms and thus also the signomial functions can be underestimated, by properly selecting the power transformations.

A signomial term in n variables with m positive powers, $p_i > 0,_i = 1, 2, ..., m$ and $n - m$ negative powers, $p_i < 0,_i = m + 1, ..., n$ can be written according to

$$cz_1^{p_1} z_2^{p_2} \cdots z_m^{p_m} z_{m+1}^{p_{m+1}} \cdots z_n^{p_n} \tag{2}$$

Different sets of transformations for both positive and negative signomial terms have been given in Pörn et al. (1999), Pörn (2000), Björk (2002), Westerlund and Björk (2002a and 2002b) and Björk et al. (2003). Power transformations for both positive and negative signomial terms are given below.

3.1 Convexifying negative signomial terms

A negative signomial term ($c < 0$) can be convexified by applying the following power transformations:

$$z_i = Z_i^R \qquad i = 1, 2, ..., m \tag{3}$$

$$z_i = Z_i^{-S} \qquad i = m + 1, ..., n \tag{4}$$

When applying the power transformations (3) and (4) on (2) we obtain,

$$cZ_1^{Rp_1} Z_2^{Rp_2} \cdots Z_m^{Rp_m} Z_{m+1}^{-Sp_{m+1}} \cdots Z_n^{-Sp_n} \tag{5}$$

When applying the transformations to the signomial term it can be convexified since a negative signomial term with positive powers, where the sum of the powers is less or equal to one, is convex (Maranas and Floudas (1995)).

This convexifying requirement is fulfilled if the parameters R and S used above are positive and in addition fulfill the condition

$$R \sum_{i=1}^{m} p_i - S \sum_{i=m+1}^{n} p_i \leq 1 \tag{6}$$

If R and S are positive the left hand side of (6) is composed of a sum of two positive terms (the latter sum is negative). Thus, it is obvious that the given condition can always be satisfied with sufficiently small R and S values.

3.2 Underestimating negative signomial terms

In addition to being convexified by the transformations, the signomial term should also be underestimated with the given transformation technique. Now, if the inverse transformations are approximated by piecewise linear functions and the estimated transformation variables, obtained from the piecewise linear functions, are used in the convexified signomial term, we obtain,

$$c\hat{Z}_1^{Rp_1} \hat{Z}_2^{Rp_2} \cdots \hat{Z}_m^{Rp_m} \hat{Z}_{m+1}^{-Sp_{m+1}} \cdots \hat{Z}_n^{-Sp_n}, \tag{7}$$

where $\hat{Z}_i$ are the estimated transformation variables for each $i = 1, \ldots, n$. The reason for using an estimated transformation variable instead of the transformation variable itself is threefold. First of all, the properties of the convexified signomial term do not change by replacing a transformation variable itself with its corresponding estimate. Secondly, by approximating the inverse transformations, with piecewise linear functions, the original inverse transformations, represented by nonlinear equality constraints, can be represented by linear expressions. This second property is important because it enables us, not only to convexify the signomials but, also to convexify the entire problem, since no parts of the signomials or the transformations will remain nonconvex in the resulting convexified model.

Finally, we are able to underestimate the original signomial term with the convexified signomial term using the estimated transformation variables, when using the piecewise linear approximation of the inverse transformations, if the transformations are properly selected.

The inverse transformations corresponding to the transformations (3) and (4) are simply given by,

$$Z_i = z_i^{\frac{1}{R}} \qquad i = 1, 2, ..., m \tag{8}$$

$$Z_i = z_i^{-\frac{1}{S}} \qquad i = m+1, ..., n \tag{9}$$

The underestimating property means that the following inequality should hold:

$$c\hat{Z}_1^{Rp_1} \cdots \hat{Z}_m^{Rp_m} \hat{Z}_{m+1}^{-Sp_{m+1}} \cdots \hat{Z}_n^{-Sp_n} \leq cZ_1^{Rp_1} \cdots Z_m^{Rp_m} Z_{m+1}^{-Sp_{m+1}} \cdots Z_n^{-Sp_n} \tag{10}$$

Since all powers in (10) are positive, for positive values on R and S, and we are dealing with negative signomial terms ($c < 0$), the inequality (10) is generally valid if,

$$\hat{Z}_i \geq Z_i \qquad i = 1, 2, ..., m, m+1, ..., n \tag{11}$$

Now, since the estimated transformation variables, $\hat{Z}_i$, are obtained from piecewise linear approximations of the inverse transformations (8) and (9), we, find that (11) and, thus, also (10) will be generally valid if (8) and (9) are convex functions.

The condition for (8) to be a convex function is fulfilled if the parameter R is chosen in the interval

$$0 < R \leq 1 \tag{12}$$

For S it is sufficient that $S > 0$ for (9) to be a convex function. Thus no additional requirement on the parameter S is needed to satisfy the underestimating property of the convexified signomial term, while the additional requirement (12) is set on the parameter R.

A negative signomial term will thus be convexified by applying the power transformations, (3) and (4) with the additional condition (6), on the original signomial term. The original signomial term will, further, be underestimated if the transformation variables, in the convexified signomial term, are replaced by their estimates obtained from piecewise linear approximations of the inverse transformations (8) and (9) and the additional condition (12) is set on the transformation (3).

Observe that a signomial term can, in this way, not only be transformed to convex form, but the entire problem can be convexified with the additional property that the feasible region of the convexified problem overestimates the feasible region of the original nonconvex problem.

3.3 Convexifying positive signomial terms

In an almost similar way as with negative signomial terms, positive signomial terms can be convexified and underestimated. In this case single variable power transformations need be applied only on variables with a positive power in the original signomial term. The following power transformations may be used for positive ($c > 0$) signomial terms (5):

$$z_i = Z_i^{-T} \qquad i = 1, 2, ..., m-1 \tag{13}$$

$$z_m = Z_m^W \tag{14}$$

A positive signomial term will be convexified as well as underestimated by using the transformations (13) and (14), in a similar way as negative signomial terms, if the inverse transformations are approximated by piecewise linear functions, and the estimated transformation variables are used in the convexified signomial term. The convexified signomial term that underestimates the original signomial term will then be:

$$c\hat{Z}_1^{-Tp_1}\hat{Z}_2^{-Tp_2}\cdots\hat{Z}_{m-1}^{-Tp_{m-1}}\hat{Z}_m^{Wp_m}z_{m+1}^{p_{m+1}}\cdots z_n^{p_n} \tag{15}$$

$\hat{Z}_i$, $i = 1,\ldots,m$ are the estimated transformation variables. Each estimated transformation variable will be obtained from a piecewise linear approximation of the corresponding inverse transformation. Observe, that the transformations are, in the case with positive signomial terms, only applied on those variables having positive power. The last $n-m$ variables, with negative power are unaffected by the transformations. The variables in the original signomial term may also be arranged, for example, such that the transformation (14) is applied on the variable in the original signomial term with the largest positive power.

In the case with a positive signomial term, there are several alternative ways of convexifying the term (Björk, (2002), Börk et al. (2003)). If all powers in a positive signomial term are negative, then the term is convex. From the properties of power-convex functions, as given in Lindberg (1981), it can also be shown that a positive signomial term, with at most one positive power, is convex if the sum of the powers is greater or equal to one. If the sum is equal to one, the signomial term is a 1-convex function.

Thus, if the powers $T$and $W$are positive, then the latter requirement for convexification will be:

$$-T\sum_{i=1}^{m-1} p_i + Wp_m + \sum_{i=m+1}^{n} p_i \geq 1 \tag{16}$$

Since the first and third terms, in the expression (16), are negative (and finite for finite T) and the second term can be made infinitely large, the expression (16) is valid for any sufficiently large W.

3.4 Underestimating positive signomial terms

In addition to being convexified by the transformations, the positive signomial terms should also be underestimated. The inverse transformations of (13) and (14) are given by:

$$Z_i = z_i^{-\frac{1}{T}} \qquad i = 1, 2, ..., m-1 \tag{17}$$

$$Z_m = z_m^{\frac{1}{W}} \tag{18}$$

If the inverse transformations are approximated by piecewise linear functions and the estimated transformation variables obtained from the piecewise linear functions are used in the convexified signomial term, the inequality

$$c\hat{Z}_1^{-Tp_1}\cdots\hat{Z}_{m-1}^{-Tp_{m-1}}\hat{Z}_m^{Wp_m}z_{m+1}^{p_{m+1}}\cdots z_n^{p_n}\le cZ_1^{-Tp_1}\cdots Z_{m-1}^{-Tp_{m-1}}Z_m^{Wp_m}z_{m+1}^{p_{m+1}}\cdots z_n^{p_n} \quad (19)$$

should hold true for positive (c >0) signomial terms. This will be the case if,

$$\hat{Z}_i \ge Z_i \qquad i = 1, 2, \dots, m-1 \quad (20)$$

$$\hat{Z}_m \le Z_m \quad (21)$$

This is particularly true since the first $m-1$ variables in the convexified signomial term (15) have negative powers and thus correspond to decreasing functions, while the m-th transformation variable in (15) has a positive power and, thus, corresponds to an increasing function. Since the approximate transformation variables are obtained from piecewise linear approximations of the inverse transformations (17) and (18) we find that (20) will be generally valid if the inverse transformations (17) are given by convex functions. In addition (21) will generally be valid if the inverse transformation (18) is given by a concave function.

The, additional, underestimating condition is, thus, fulfilled if $T > 0$ and

$$W \ge 1 \quad (22)$$

A positive signomial term will thus be convexified by applying the power transformations, (13) and (14) with the additional condition (16), on the original signomial term. The original signomial term will, further, be underestimated if the transformation variables, in the convexified signomial term, are replaced by their estimates obtained from piecewise linear approximations of the inverse transformations (17) and (18) and the additional condition (22) is set on the transformation (14).

3.5 Piecewise linear approximations of the inverse transformation

The transformations above, for both negative and positive signomial terms, were selected so that every signomial term can be convexified as well as underestimated if the inverse transformations are approximated by piecewise linear functions. Piecewise linear functions can be modeled in different ways. Some alternatives are given in Floudas and Pardalos (2001).

An appropriate way to model a piecewise linear function in K intervals is given below. An inverse transformation is then represented by K binary variables, b_k, and equal many nonnegative real variables, s_k, by the following linear expressions:

$$\hat{Z} = \sum_{k=1}^{K} (Z_k b_k + (Z_{k+1} - Z_k) s_k) \quad (23)$$

$$z = \sum_{k=1}^{K} (z_k b_k + (z_{k+1} - z_k) s_k) \quad (24)$$

$$\sum_{k=1}^{K} b_k = 1 \tag{25}$$

$$0 \leq s_k \leq b_k \tag{26}$$

$$k = 1, 2, ..., K; b_k \in \{0, 1\}; s_k \in \mathrm{I\!R}^+$$

$\hat{Z}$ is the estimated transformation variable obtained from the piecewise linear function at z. Z_k; k=1, 2,..., K+1 are given values of the transformation variable Z, obtained from the original inverse transformation, at K+1 grid points, $z_1 < z_2 < \ldots < z_{K+1}$.

The binary variables in the expressions used in modeling the piecewise linear functions can be defined as so-called special ordered sets in MILP solvers, since the sum of the binary variables is equal to one. Also observe that no binary variables are initially needed when the linear approximations are made in one step from the lower bound on the variable to its upper bound, since Eq. (25) indicates that the one and only binary variable, corresponding to each piecewise linear function, should be equal to one. Observe, further, that a piecewise linear function in 2^N intervals can be modeled by using only N binary variables, as shown in the appendix. The formulation above is, however, usually computationally equal efficient since the additional requirement (25) can be used as a special ordered set, as mentioned above. The formulation above also has the advantage that each binary variable will correspond to own convex sub-regions when using the formulation in the given global optimization approach.

When applying several transformations in one original variable, z_i to different signomial terms (in the same or different signomial constraints), each transformation needs its own transformation variable Z. However, it should be observed that the same binary variables, b_k, (and real variables, s_k) can be used in all different piecewise linear approximations, being a function of the same original variable z_i which substantially reduces the number of binary variables, in the problem.

Geometrically each binary variable divides the original feasible region into new convex sub-regions. When applying the algorithm discussed below, the original feasible region may, thus, be divided into one new or several new convex sub-regions in a subsequent iteration.

4 Examples of transformations

In the following, a few examples are given in order to illustrate the transformation procedure.

4.1 An example of applying the transformations on a negative signomial term

Consider the negative signomial term:

$$-z_1^{0.2} z_2^{-0.1}$$

In this case m=1 and n=2 and since one of the powers is negative we need to convexify the term. From (6) we obtain the condition,

$$R \sum_{i=1}^{m} p_i - S \sum_{i=m+1}^{n} p_i = 0.2R + 0.1S \leq 1$$

The parameters R and S should be positive, but in order to satisfy the underestimating condition we need, further, to consider the additional requirement,

$$0 < R \leq 1$$

Thus, for example, if R=1 and S=1 all conditions are satisfied. Since R can be set equal to one, we, thus, only need to transform the variable z_2.

Applying the transformation (4) with S=1 on the signomial term, it will be convexified. The convexified signomial term will further underestimate the original signomial term if we replace the transformation variable by its estimate obtained from a piecewise linear approximation of the inverse transformation (9).

The convexified term underestimating the original signomial term is thus given by,

$$-z_1^{0.2} \cdot \hat{Z}_2^{0.1}$$

The estimated transformation variable, $\hat{Z}_2$, is obtained by approximating the inverse transformation

$$Z_2 = z_2^{-1}$$

in the interval $z_{2,\text{min}} \leq z_2 \leq z_{2,\text{max}}$ using a piecewise linear function (Eqs. (23-26)).

4.2 An example of applying the transformations on a positive signomial term

Consider the positive signomial term:

$$z_1^{1.2} z_2^{-0.5} z_3^{-1.5}$$

In this case m=1 and we find that the sum of the powers is equal to -0.8. Thus, we need to convexify the term. From (15) we obtain the condition,

$$W p_m + \sum_{i=m+1}^{n} p_i = 1.2W - 2 \geq 1$$

If the signomial term is convexified such that RHS=1, we obtain W=2.5. The additional underestimating requirement (22) ($W \geq 1$) is, thus, also fulfilled.

Applying the transformations and inserting the estimated transformation variable into the convexified signomial term we obtain,

$$\hat{Z}_1^3 z_2^{-0.5} z_3^{-1.5}$$

We, thus, obtain a transformed signomial term that is both convexified and will underestimate the original signomial term. The approximate transformation variable, $\hat{Z}_1$, is obtained from the approximation of the inverse transformation,

$$Z_1 = z_1^{0.4}$$

with a piecewise linear function, Eqs. (23-26), in the interval $z_{1,\min} \leq z_1 \leq z_{1,\max}$.

4.3 Illustration of the underestimation of a one-dimensional signomial function

Consider the function $f(x) = x^4 - 3x^3 - 1.5x^2 + 10x$. This function should be convexified and underestimated in the interval $-2 \leq x \leq 3$. Since, in this particular case, the variable range includes negative variable values, we first apply a translation such that our function is written in a positive variable. Let our translation be $z = x + 3$. The function can then be written as,

$$f(z) = (z-3)^4 + 25.5z^2 - 62z + 37.5 - 3z^3$$

The interval to be considered is now $1 \leq z \leq 6$.

Observe that only the last term in $f(z)$ is non-convex and since this concave term is a negative signomial term it can be convexified using the transformation (3). The convexification condition (6) is $3R \leq 1$, which directly also fulfills the underestimating condition $0 < R \leq 1$ for this power transformation. Let $R = 1/3$, apply the transformation and replace the transformation variable, in the convexified expression, with the corresponding estimated transformation variable. We then obtain,

$$f(z, \hat{Z}) = (z-3)^4 + 25.5z^2 - 62z + 37.5 - 3\hat{Z}$$

The estimated transformation variable, $\hat{Z}$, is given by a piecewise linear approximation of the inverse transformation $Z = z^3$. A piecewise linear approximation, for example, in two steps with selected grid points at: $(z_1, Z_1) = (1,1)$, $(z_2, Z_2) = (4,64)$ and $(z_3, Z_3) = (6,216)$ is, according to Eqs. (23-26), given by

$$\begin{aligned}
&\hat{Z} = b_1 + 63s_1 + 64b_2 + 152s_2\\
&z = b_1 + 3s_1 + 4b_2 + 2s_2\\
&b_1 + b_2 = 1\\
&0 \leq s_1 \leq b_1\\
&0 \leq s_2 \leq b_2\\
&b_1, b_2 \in \{0,1\}\,;\, s_1, s_2 \in \mathbb{R}^+
\end{aligned}$$

In Figure 1 the original function $f(z)$ as well as the function $f(z, \hat{Z})$ are plotted versus z. From the figure it can be observed that the original function is underestimated by the function $f(z, \hat{Z})$ in the entire interval. From the figure it can further be observed that the function $f(z, \hat{Z})$ is convex in the two sub-regions defined by the binary variables, i.e. when $b_1 = 1$ $(1 \leq z \leq 4)$ and when $b_2 = 1$ $(4 \leq z \leq 6)$.

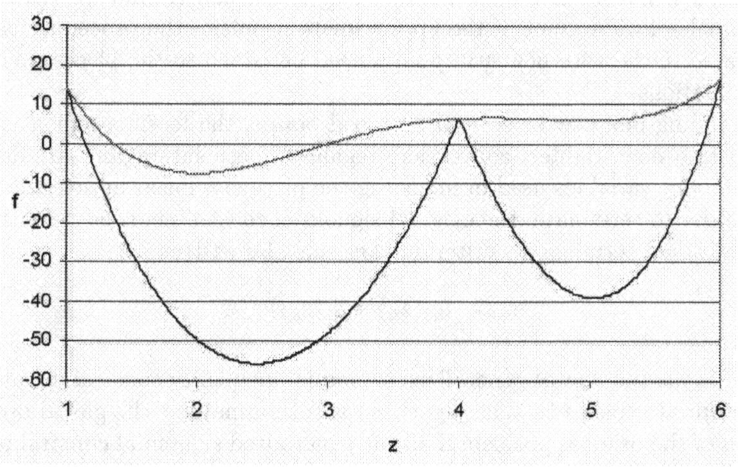

Fig. 1. Illustration of the underestimation of a function in one-dimension.

5 The GGPECP algorithm

The transformations for signomial terms (applicable on signomials and generalized signomials), discussed above, can be used to solve the nonconvex MINLP problem (P) to global optimality. This can be achieved by solving a sequence of convexified MINLP problems, or to include the transformation approach as an integrated part in a MINLP solver. When implementing the transformation approach as an integrated part of a cutting plane approach such as the extended cutting plane (ECP) algorithm (Westerlund and Pörn (2002)) the algorithm can utilize additional information from previous iterations, making the solution approach much more efficient. All techniques, including the use of previous cutting planes and encapsulating the optimal solution by the use of supports from the reduction constraint, can thus be used in the integrated algorithm. These features have been used together with the transformation approach, in the GGPECP algorithm (Westerlund and Westerlund (2003)). For example, cutting planes, created in previous iterations, are used in different ways to improve the solution efficiency of the algorithm.

An initial transformation step, an additional termination criterion and a transformation step, applied after each subsequent iteration, need be added in an algorithm utilizing the transformation techniques.

Generalized signomial functions are included only as constraints in the nonconvex MINLP problem (P) considered. Since the transformation approach allows us to underestimate these constraints, the feasible region of the original problem will be overestimated when solving the convexified MINLP problem. Thus, if the original generalized signomial constraints are satisfied at an optimal solution point of the approximate problem, the procedure can be terminated. Otherwise new grid points need be added to the piecewise linear approximations.

By adding one new or several new grid points, the feasible region will be divided into new, tighter, convex sub-regions. These sub-regions are defined by the binary variables used in modeling the piecewise linear approximations.

In order to terminate, the original signomial constraints need be satisfied. The additional termination criterion can, thus, be written as:

$$\max_{i=1,\ldots,I} (q_i(\mathbf{z}_k) + s_i(\mathbf{z}_k)) \leq \varepsilon \tag{27}$$

where I is the number of generalized signomial inequality constraints and ε is an appropriate tolerance. The algorithm will terminate at the global optimal solution of the original problem if all the generalized signomial constraints are satisfied, with the given accuracy, at the optimal solution of an approximate problem. This is particularly true since the solution of the convexified problem can be verified to be the global optimal solution, with the used sub-algorithm, and this solution is obtained in a region which overestimates the feasible region of the original nonconvex problem. If the termination criterion Eq. (27) is not satisfied, then the termination criterion also indicates in which directions new grid points may be selected. Several alternatives selecting new grid points can be used and are discussed in section 8.

It may be mentioned that the number of binary variables added in a subsequent iteration will not be dependent on the number of signomial constraints, signomial terms, the number of transformations used or the number of original variables. The number of additional binary variables added to the problem in a subsequent iteration will only correspond to the number of appropriate directions (in each of which a new grid point is defined) indicated by the termination criteria. At least one direction, one new grid point, with its corresponding binary variable, must be added in a subsequent iteration. The total number of appropriate directions is, however, of course always less or equal to the number of transformed original variables included in the signomial constraints. In Figure 2 the main iteration loop to handle the signomial constraints is shown.

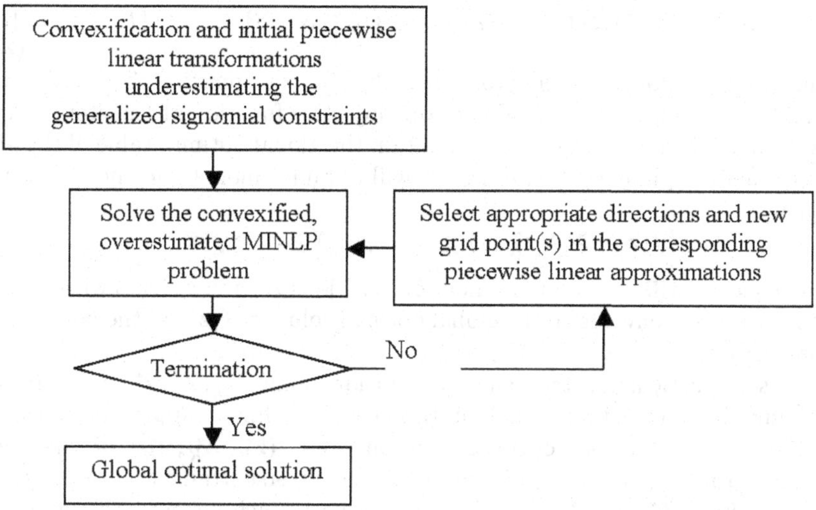

Fig. 2. Iteration loop to handle signomial constraints.

6 Convergence to the globally optimal solution

When applying the transformations approach to the problem (P), the problem is transformed to a convexified problem $(P^{conv})_k$ where, in addition, the feasible region of the original problem is overestimated. By solving a sequence of convexified MINLP problems $(P^{conv})_0$, $(P^{conv})_1$, $(P^{conv})_2$, ... the global optimal solution to the original problem (P) will be obtained.

Let $\hat{S}$ denote the set defined by the convexified generalized signomial constraints, where the transformation variables have been replaced with their estimates, according to $\hat{S} = \{(\mathbf{z}, \hat{\mathbf{Z}}) \mid \mathbf{g}(\mathbf{z}) + \hat{\mathbf{s}}(\mathbf{z}, \hat{\mathbf{Z}}) \leq \mathbf{0}\}$. Furthermore, let $L_k = \{(\mathbf{z}, \hat{\mathbf{Z}}, \tilde{\mathbf{z}}_k) \mid \text{eqs. (23-26)}\} \cap L$ denote the linear set L extended with the linear constraints (23-26) defining the piecewise linear approximations of the inverse transformations, which are used at iteration k. $\hat{\mathbf{Z}}$ is a vector with the estimated, approximate, transformation variables and $\tilde{\mathbf{z}}_k$ is a vector with the additional variables required in the expressions for the piecewise linear approximations of the inverse transformations, at iteration k.

According to the properties of the transformations discussed above the feasible region of a convexified MINLP sub-problem, where the signomial terms are underestimated, overestimates the feasible region of the original nonconvex problem. We thus have,

$$\{L \cap N \cap S\} \subset \{L_k \cap N \cap \hat{S}\} \tag{28}$$

Since a subsequent set $\{L_k \cap N \cap \hat{S}\}$, again, overestimates a previous set $\{L_{k-1} \cap N \cap \hat{S}\}$ we obtain when solving the sequence of convexified problems and inserting new grid points

$$\{L_K \cap N \cap \hat{S}\} \subset \{L_{K-1} \cap N \cap \hat{S}\} \subset \ldots \subset \{L_k \cap N \cap \hat{S}\} \subset \ldots \subset \{L_0 \cap N \cap \hat{S}\}. \tag{29}$$

The solution point of the problem, $(P^{conv})_k$, at iteration k is given by $(\mathbf{z}_k^*, \hat{\mathbf{Z}}_k^*, \tilde{\mathbf{z}}_k^*)$ and the optimal value of the objective function at this iteration is $f(\mathbf{z}_k^*)$. Since the algorithm is able to obtain the global optimal value of the objective function, in every iteration, we will obtain a monotonically increasing sequence

$$f(\mathbf{z}_K^*) \geq f(\mathbf{z}_{K-1}^*) \geq \ldots \geq f(\mathbf{z}_k^*) \geq \ldots \geq f(\mathbf{z}_0^*) \tag{30}$$

This sequence will, in a finite number of iterations K, and at least with an epsilon tolerance, converge to the global optimal solution $f(\mathbf{z}^*)$ of the nonconvex problem (P).

This is particularly true since a solution point $(\mathbf{z}_k^*, \hat{\mathbf{Z}}_k^*, \tilde{\mathbf{z}}_k^*)$ will always be found in a certain interval of the piecewise linear approximations. If all the variables, at an optimal solution point $(\mathbf{z}_K^*, \hat{\mathbf{Z}}_K^*, \tilde{\mathbf{z}}_K^*)$ of the convexified problem, are at grid points of the inverse transformations, then $\mathbf{s}(\mathbf{z}_K^*) = \hat{\mathbf{s}}(\mathbf{z}_K^*, \hat{\mathbf{Z}}_K^*, \tilde{\mathbf{z}}_K^*)$. In this case the value of each generalized signomial constraint will be exactly equal to the value of its corresponding convexified constraint. Since the convexified constraints are satisfied with an epsilon tolerance, at this solution point, this is obviuously also the case for the generalized signomial constraints. We have thus found the global optimal solution to the original problem, within an epsilon tolerance defined by Eq. (27), and the global optimal value on the objective function is given by $f(\mathbf{z}_K^*)$.

If, on the other hand, some of the variables are not at the grid points of the piecewise linear approximations and Eq. (27) is not satisfied, new tighter approximations can always be obtained for the signomials. New grid points, for the actual variables, can, in this case, be selected at the actual solution point or, for example, in the middle of the actual interval. In this case the solution procedure continues. Since adding a new grid point to a piecewise linear approximation always makes the approximation tighter, we will eventually obtain a solution when Eq. (27) is satisfied. In this case we have again obtained the global optimal solution with at least an ε tolerance of the generalized signomial constraints. It may be observed that the sequences are finite since the new grid points can always be selected so that the maximum distance between two grid points of any of the variables in a piecewise linear function, form a Cauchy sequence.

Thus, in the first case we directly obtained the global optimal solution to the problem (P) within the tolerance defined by Eq. (27). In the latter case, the signomial constraints need be accurate enough in order to fulfill the criterion Eq. (27) at termination. In both cases we obtain the global optimal solution to the problem (P) with a finite number of iterations and at least an ε tolerance on the generalized signomial constraints Q.E.D.

7 A numerical example

In the following a numerical example including one linear constraint and one generalized signomial inequality constraint is considered. The problem can be written as follows:

$$\begin{aligned}
&\min_{x,y} y - 2x \\
&s.t. \\
&y + 5x \leq 36 \\
&(2y^2 - 2y^{0.5} + 11y + 8x - 35) + x^{0.5} - 1.5x^{1.1}y^{1.5} \leq 0 \\
&1 \leq x \leq 7; \quad 1 \leq y \leq 6 \\
&x \in \mathbb{R}; \quad y \in Z^+
\end{aligned}$$

The generalized signomial constraint consists of a sum of a convex function, given within the parenthesis, and a signomial function, the latter being composed of one positive and one negative signomial term. In Figure 3 the (integer relaxed) feasible region of the problem is illustrated.

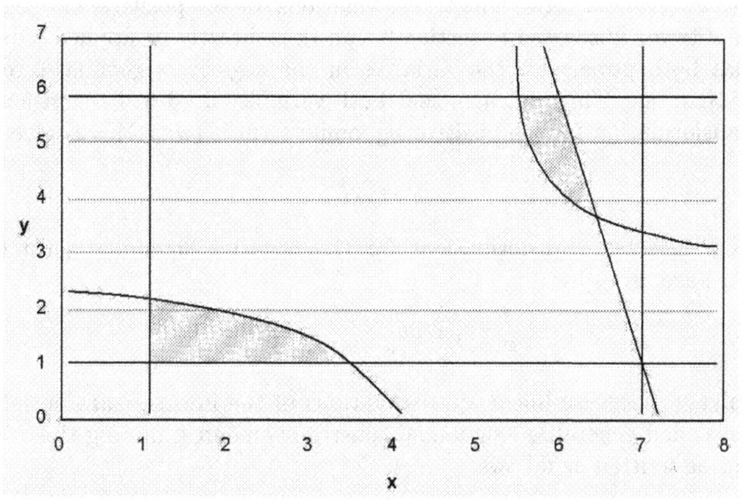

Fig. 3. Integer relaxed feasible region of the nonconvex problem.

From the figure, it can be observed that the feasible region of the problem is divided into two disjoint regions (considering the regions as integer relaxed regions) by the generalized signomial constraint. One of the disjoint regions is in the lower left corner. In this region the feasible values of the integer variable are 1 and 2 while the continuous variable x should be greater or equal to one and less or equal to a value constrained by the signomial constraint. In the upper left corner we find the other disjoint region with feasible values 4, 5 and 6 on the integer variable while the continuous variable is constrained, from below, by the signomial constraint and its upper value is constrained by the linear constraint.

If the problem is solved with a local MINLP solver, then the global optimal solution may not be obtained. If the problem is, for example, solved with the Outer Approximation method given in Duran and Grossmann (1989) then the solution (x=3.513, y=1) is obtained. At this solution point the value of the objective function is -6.025, but this solution is, however, not the global optimal one. The OA method is based on solving an alternating sequence of MILP masters and NLP problems. The method can be started in different ways. If a first MILP masters problems is solved, we obtain the solution x=7 and y=1. As an alternative we could start the algorithm with the candidate solution y=1. However, in both these alternatives, the value of the integer variable is fixed, at this point, and the solution of the following NLP problem result in x=3.513, which is the point where the OA algorithm terminates. With other local MINLP solvers, similar non-optimal results may be obtained.

We will, however, in the following, examine how the given transformations approach and the corresponding global optimization approach solves the problem. As mentioned above the nonlinear constraint contains both a positive and a negative signomial term. The transformations for the positive and negative signomial terms, discussed in section 3, can thus directly be applied. Observe also that both powers for the variables in the negative signomial term are greater than one. This indicates that both variables need to be transformed. The transformation for the positive signomial term is (with W=2) given by,

$$x = X_1^2$$

whist our selected transformations for the negative signomial term (with R=1/2.6) are given by,

$$x = X_2^{1/2.6}; \quad y = Y^{1/2.6}$$

By applying piecewise linear approximations of the inverse transformations, the convexified generalized signomial constraint underestimating the original one, can be written as follows:

$$(2y^2 - 2y^{0.5} + 11y + 8x - 35) + \hat{X}_1 - 1.5\hat{X}_2^{0.4231}\hat{Y}^{0.5769} \leq 0$$

Using the given bounds on the variables x and y, initial piecewise linear approximations of the inverse transformations can be written as:

$$\hat{X}_1 = b_{x,1} + 1.6458 s_{x,1}$$
$$\hat{X}_2 = b_{x,1} + 156.49 s_{x,1}$$
$$x = b_{x,1} + 6 s_{x,1}$$
$$b_{x,1} = 1$$
$$0 \leq s_{x,1} \leq b_{x,1}$$
$$\hat{Y} = b_{y,1} + 104.49 s_{y,1}$$
$$y = b_{y,1} + 5 s_{y,1}$$
$$b_{y,1} = 1$$
$$0 \leq s_{y,1} \leq b_{y,1}$$
$$b_{x,1}, b_{y,1} \in \{0,1\}\,; \qquad s_{x,1}, s_{y,1} \in \mathbb{R}^+$$

Initially no binaries are needed, although each piecewise linear function in one interval uses one binary variable in the given formulation. This is because the value of the binary variable is fixed to one by the sum of the binaries in the expression for the piecewise linear function. In addition, initially no additional continuous variables, s, are needed either. But in order to illustrate the given form of the piecewise linear expressions, the s variables are also included in the expressions above. From the formulation it can, furthermore, be found that the different piecewise linear approximations in the same original variable, can use the same additional variables. In Figure 4 the convexified and initially overestimated feasible region of the problem is illustrated.

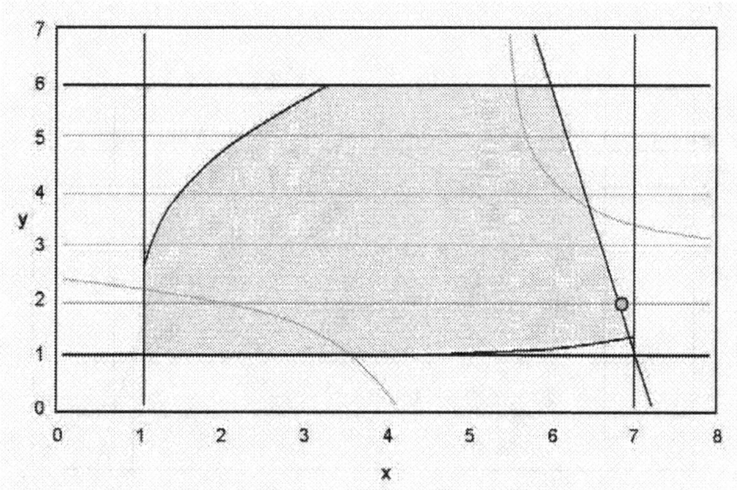

Fig. 4. Convexified and overestimated feasible region at the first iteration.

In Figure 4 the boundaries defined by the original generalized signomial constraint are also illustrated. The boundaries of the convexified overestimated feasible region are illustrated by the darker continuous curves. From Figure 4 it can be observed that the integer relaxed feasible region of the transformed problem is both convex and that it overestimates the feasible region of the nonconvex problem.

The optimal solution to the convexified problem is further illustrated in the figure. The optimal solution to this sub-problem is obtained at x=6.8 and y=2. At this solution point the value of the original generalized signomial constraint is 14.233. One additional or several additional grid points may thus be added to the piecewise linear approximations to be used in the following iteration. At this point we can make different choices, which are discussed in the next section. However, if new grid points are selected at the solution, x=6.8 and y=2, we obtain the following updated piecewise linear approximations,

$$\begin{aligned}
&\hat{X}_1 = b_{x,1} + 1.6077s_{x,1} + 2.6077b_{x,2} + 0.038070s_{x,2} \\
&\hat{X}_2 = b_{x,1} + 145.06s_{x,1} + 146.06b_{x,2} + 11.433s_{x,2} \\
&x = b_{x,1} + 5.8s_{x,1} + 6.8b_{x,2} + 0.2s_{x,2} \\
&b_{x,1} + b_{x,2} = 1 \\
&0 \leq s_{x,1} \leq b_{x,1}; \quad 0 \leq s_{x,2} \leq b_{x,2} \\
&\hat{Y} = b_{y,1} + 5.0629s_{y,1} + 6.0629b_{y,2} + 99.423s_{y,2} \\
&y = b_{y,1} + 1s_{y,1} + 2b_{y,2} + 4s_{y,2} \\
&b_{y,1} + b_{y,2} = 1 \\
&0 \leq s_{y,1} \leq b_{y,1}; \quad 0 \leq s_{y,2} \leq b_{y,2} \\
&b_{x,i}, b_{y,i} \in \{0,1\}; \quad s_{x,i}, s_{y,i} \in \mathbb{R}^+
\end{aligned}$$

In Figure 5 the convexified and overestimated feasible region at iteration 2 is illustrated.

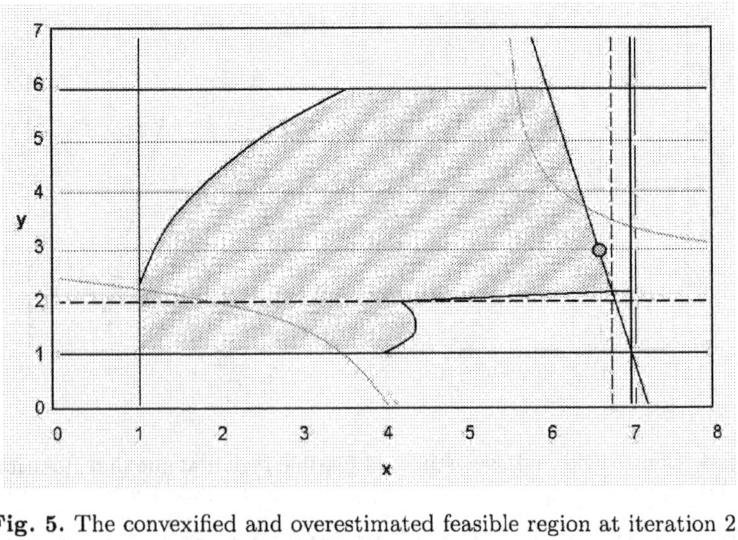

Fig. 5. The convexified and overestimated feasible region at iteration 2.

Since, grid points were added for both variables the feasible region is, in this particular case, divided into four sub-regions defined by the new grid-points and the binary variables. In the figure the borders of the sub-regions are illustrated by the dashed lines and the boundaries of the convexified over-estimated feasible region are illustrated by darker continuous curves. From the figure it can be observed that there are no feasible solutions in the convex sub-regions to the right, whilst feasible solutions can be found in both of the convex sub-regions to the left. The optimal solution to the convexified overestimated problem at iteration two is found at x=6.6 and y=3 which is in the upper left convex sub-region. The optimal solution is also illustrated in the figure. The value of the original generalized signomial constraint can be calculated at this point and is 5.7792. This point is infeasible to the original problem. Thus, one new or several new grid points need be added to the piecewise linear approximations. Adding one grid point at the solution x=6.6 and one grid point at y=3 to the piecewise linear approximations of the inverse transformations result in the following expressions,

$$
\begin{aligned}
&\hat{X}_1 = b_{x,1} + 1.569 s_{x,1} + 2.569 b_{x,2} + 0.038634 s_{x,2} + 2.608 b_{x,3} + 0.03807 s_{x,3} \\
&\hat{X}_2 = b_{x,1} + 134.15 s_{x,1} + 135.15 b_{x,2} + 10.908 s_{x,2} + 146.06 b_{x,3} + 11.433 s_{x,3} \\
&x = b_{x,1} + 5.6 s_{x,1} + 6.6 b_{x,2} + 0.2 s_{x,2} + 6.8 b_{x,3} + 0.2 s_{x,3} \\
&b_{x,1} + b_{x,2} + b_{x,3} = 1 \\
&0 \le s_{x,1} \le b_{x,1}; \quad 0 \le s_{x,2} \le b_{x,2}; \quad 0 \le s_{x,3} \le b_{x,3} \\
&\hat{Y} = b_{y,1} + 5.0629 s_{y,1} + 6.0629 b_{y,2} + 11.336 s_{y,2} + 17.399 b_{y,3} + 88.087 s_{y,3} \\
&y = b_{y,1} + 1 s_{y,1} + 2 b_{y,2} + 1 s_{y,2} + 3 b_{y,3} + 3 s_{y,3} \\
&b_{y,1} + b_{y,2} + b_{y,3} = 1 \\
&0 \le s_{y,1} \le b_{y,1}; \quad 0 \le s_{y,2} \le b_{y,2}; \quad 0 \le s_{y,3} \le b_{y,3} \\
&b_{x,i}, b_{y,i} \in \{0,1\}; \quad s_{x,i}, s_{y,i} \in \mathbb{R}^+
\end{aligned}
$$

In Figure 6 the convexified and overestimated feasible region at iteration 3 is illustrated.

In the figure the borders of the sub-regions are again illustrated by dashed lines while the boundaries of the convexified overestimated feasible region are illustrated by darker continuous curves. From the figure it can be observed that now there are nine convex sub-regions, which are defined by the grid-points and the binary variables. From the figure it can be observed that feasible solutions for the convexified problem can only be found in the three sub-regions to the left. In all these sub-regions $b_{x,1} = 1$. The optimal solution to the convexified overestimated problem at iteration 3 is found at x=6.4 and y=4. This point is feasible to the original problem, the value of the original generalized signomial constraint being -1.7356 at this solution point. Since the solution is feasible in both the overestimated and the original problem and the solution is optimal in the overestimated convexified problem, the solution is consequently the global optimal solution to the original nonconvex problem. The global optimal solution is illustrated in the figure. At this point the value of the objective function is -8.8. In Table 1 the solution values obtained at iterations 1-3 are further given.

From Table 1 and Figure 6 it can be verified that the global optimal solution is obtained after the third main iteration. With the GGPECP solver

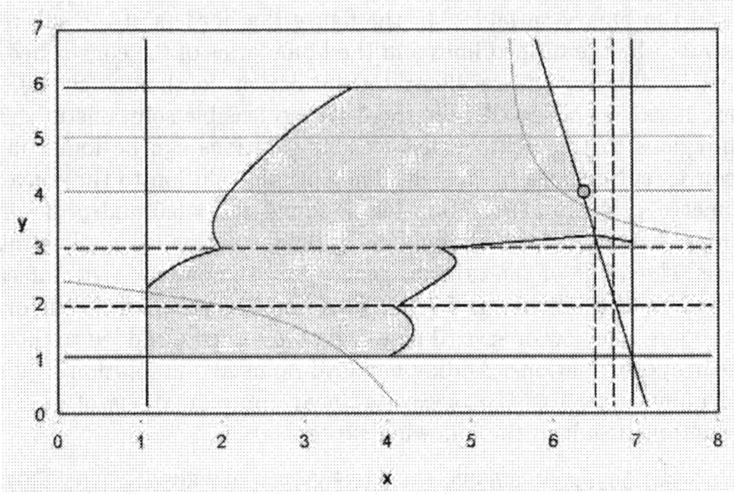

Fig. 6. The convexified and overestimated feasible region at iteration 3.

Iteration	x	y	Objective function	$q(x,y)+s(x,y)$	Solution Status
1	6.8	2	-11.6	14.233	Subprob-opt
2	6.6	3	-10.2	5.7792	Subprob-opt
3	6.4	4	-8.8	-1.7356	Glob-opt

Table 1. Solution data obtained for the example problem. $q(x,y)+s(x,y) = (2y^2 - 2y^{0.5} + 11y + 8x - 35) + x^{0.5} - 1.5x^{1.1}y^{1.5}$.

described in Westerlund and Westerlund (2003), the solution to the above problem could be obtained in 0.2CPUsec on a 3GHz PC.

8 Some aspects on the numerical solution approach

As indicated above there are different alternative ways to implement the transformation approach in a numerical algorithm. These alternatives are, for example, connected to the way new grid points are selected and how the information from the solution of the previous convexified MINLP problems can be utilized when solving a subsequent MINLP problem in the algorithm. In the following some aspect in connection to these issues are discussed. Other important issues, such as choices connected to different transformations, are left out of this discussion.

8.1 The selection of new grid points

After the solution of a convexified MINLP sub-problem the numerical values of the generalized signomial constraints should be evaluated, as indicated in the algorithm in Figure 2. The termination criterion is given by Eq. (27). If the termination criterion is satisfied then the solution is the global optimal solution to the original non-convex problem. On the other hand, if the termination criterion is not satisfied then appropriate directions and new grid points for the piecewise linearizations should be selected for the following iteration. At this point there are several alternative choices that can be made.

Selection of appropriate directions

When evaluating if the original generalized signomial constraints are satisfied, Eq. (27) is used. Since Eq. (27) is defined such that the generalized signomial constraint that violates the most should be satisfied with a certain tolerance, then consequently all other generalized signomial constraints are also satisfied at the considered solution point. If considering the most violating signomial constraint, one may evaluate which of the original variables are involved in transformations in the signomial terms in this constraint. In the example problem we find that both variables are involved. However, generally one could, when Eq. (27) is not valid, select the new appropriate directions and thus the grid points, only in those variables involved in transformation in the most violating signomial constraint. The reason for not selecting all variables involved in transformations for the problem, and new grid points for them, would be to minimize the additional number of binary variables needed in the piecewise linear approximations in the following iteration.

Another possibility, for selecting the appropriate directions, would be to consider the signomial terms in the most violating signomial constraint more carefully. In the example problem the solution of the first convexified MINLP problem is $x = 6.8$ and $y = 2$. The value of the generalized signomial constraint is 14.233 at this solution point (and the value of the convexified underestimated generalized signomial constraint is -25.45 at the same point). Since this constraint has two signomial terms we may calculate which of the signomial terms is underestimated the most. Then we could select the appropriate directions only from those variables that are involved in transformations in the signomial term which is underestimated the most. In our example problem, the first signomial term is given by$x^{0.5}$and this term is underestimated by the convexified term $\hat{X}_1$. The second signomial term is given by $-1.5x^{1.1}y^{1.5}$and this term is underestimated by the convexified term $-1.5\hat{X}_2^{0.4231}\hat{Y}^{0.5769}$.

The solution of the first convexified MINLP sub-problem resulted in the values $\hat{X}_1 = 2.5909$, $\hat{X}_2 = 152.27$ and $\hat{Y} = 21.897$ on the estimated transformation variables. From the solution point we, thus, find that

$$x^{0.5} - \hat{X}_1 = 0.0168$$

and that

$$(-1.5x^{1.1}y^{1.5}) - (-1.5\hat{X}_2^{0.4231}\hat{Y}^{0.5769}) = 39.67$$

Thus, we find that the second signomial term, in the most violating generalized signomial constraint, is underestimated the most. Since the second signomial term includes transformations for both x and y, then we could select new grid points only for these variables. In this particular example, however, x and y are the only variables in the original problem. But, if the first term would have been underestimated the most, then a possible choice would have been to select only $x$as the new direction in which a grid point is added to the piecewise linear approximations in the following iteration.

A more focused alternative to select the appropriate direction(s) would be to make the choice by considering the impact, of the underestimation of the generalized signomials, from each variable, included in the transformations, separately. Also in this case one could focus on the impact of the underestimation only on the generalized signomial constraint, violating the most. In this case the new appropriate direction(s) could be selected from the variable (or some of the variables) giving rise to the largest underestimation. The impact of the underestimation, from a certain variable, can be calculated from the difference between the convexified signomial function using the (true) transformation variables and the corresponding convexified signomial function using the estimated transformation variables that corresponds to the studied variable. In our numerical example, the impact on the underestimation from the $x$variable would, after the first iteration, be

$$(X_1 - 1.5X_2^{0.4231}Y^{0.5769}) - (\hat{X}_1 - 1.5\hat{X}_2^{0.4231}Y^{0.5769}) = 0.6386$$

while the impact on the underestimation from the $y$variable would be

$$(X_1 - 1.5X_2^{0.4231}Y^{0.5769}) - (X_1 - 1.5X_2^{0.4231}\hat{Y}^{0.5769}) = 38.36$$

Since the impact from the y variable is far greater than the impact from the x variable, we could, if selecting only one new appropriate direction, select it as the y-direction.

In the expressions above it may be noted that, by definition

$$(x^{0.5} - 1.5x^{1.1}y^{1.5}) = (X_1 - 1.5X_2^{0.4231}Y^{0.5769})$$

although the LHS is nonconvex and the RHS is convex. This is because the convexified expression is defined, exactly, by the transformations $x = X_1^2$, $x = X_2^{1/2.6}$ and $y = Y^{1/2.6}$. In the above expression, and the considered solution point, the numerical value of both the LHS and the RHS is equal to -32.34.

The underestimation is, again, a result of that a transformation variable is replaced by its corresponding estimated transformation variable obtained from the piecewise linear approximation of the corresponding inverse transformation.

Different other alternatives to select the appropriate directions may also be used.

Selection of appropriate grid points

While the selection of appropriate directions is connected to the selection of original variables for which new grid points are added, the selection of appropriate grid points is connected to the way the grid points, for a certain variable, are selected. Also in the selection of the grid points there are several alternative choices that can be made. The simplest choice is to select a new grid point at the solution point of the variable itself. This alternative choice was used in every iteration in the example problem illustrated previously.

However, since the solution point always corresponds to a certain interval in the actual piecewise linear approximation, an alternative choice would be to select the new grid point as the midpoint of the corresponding interval.

A further choice would be to select a grid point in the opposite way. The value of the estimated transformation variable is also obtained at the solution of a sub-problem. If the estimated value of the transformation variable is inserted in the original transformation, then (if the solution point is not at an old grid point) another estimate (other than the solution value) of the corresponding original variable is obtained. This new estimate of the original variable could also be used as an alternative grid point. Here one should, however, observe that a certain original variable may be involved in several transformations. Thus several new estimates of the same original variable may, in this way, be obtained at the same iteration. However, the mean value of all the estimates of the original variable could, for example, be selected as an alternative new grid point. Different other alternative choices of new grid points may also be used.

8.2 Utilizing information from the solution of previous sub-problems

As mentioned in the beginning, different solution approaches may be used to solve the intermediate convexified MINLP problems. If the Extended Cutting Plane (ECP) method given in Westerlund and Pörn (2003) is used, then the solution of a previous MINLP problem does not only give the solution to the actual sub-problem, but it will also give a linear problem formulation overestimating the feasible region of the convexified MINLP sub-problem. The linear formulation contains the original linear constraints as well as cutting planes generated from the non-linear, including the convexified, constraints. Since the cutting planes underestimates the convexified constraints and the convexified constraints underestimates the original generalized signomial constraints, all cutting planes generated when solving previous MINLP sub-problems are valid cutting planes in the subsequent iterations. Thus all previously generated cutting planes can be utilized when solving a subsequent convexified MINLP problem.

The numerical example – utilizing information from previous cutting planes

If the convexified MINLP sub-problems are solved with the ECP method, then a sequence of MILP problems is solved. Consider our numerical example and the solution of the first MINLP sub-problem, by the ECP method. The first MILP problem, in the ECP method, is now given by the linear constraints in the considered MINLP problem. When solving the MILP problem, in the first ECP-iteration, it will result in the solution (x=7, y=1). At this solution point the value of the convexified signomial constraint overestimating the original generalized signomial constraint is 21.890. This solution point is, thus, not the optimal solution to the first convexified MINLP sub-problem. A cutting plane is generated, in the ECP method, at this solution point. The generated cutting plane is given by,

$$14.002y + 8x + \hat{X}_1 - 0.034267\hat{X}_2 - 7.3574\hat{Y} \leq 38.004$$

When adding this cutting plane to the second MILP problem, to be solved in the ECP method, the optimal solution of the first convexified MINLP sub-problem will, however, directly be obtained. The first convexified MINLP problem is thus solved to optimality in only two MILP iterations by the ECP method. As indicated in Figure 4, the solution to the first convexified MINLP problem is obtained at (x=6.8, $\mathbf{y}$=2). In Figure 7 the feasible region of the first convexified MINLP sub-problem overestimated by the above cutting plane, generated in the ECP method, is illustrated. Observe that the cutting plane, indicated by a dark continuous line in figure 7, overestimates the convexified feasible region and again, that the convexified feasible region overestimates the feasible region of the original non-convex problem, as illustrated both in Figure 4 and Figure 7. From Figure 7, one finds that the solution of the first convexified MINLP problem can be solved to optimality in only two iterations with the ECP method.

Since the solution of the convexified MINLP problem does not satisfy Eq. (27) (the value of the generalized signomial constraint is 14.233 at this point) an additional cut, can be generated, from the convexified signomial constraint, at the solution point of the first convexified MINLP sub-problem. This cut can then be utilized when solving the following MINLP sub-problem. The cut generated at the solution point (x=6.8, $\mathbf{y}$=2) is given by

$$18.295y + 8x + \hat{X}_1 - 0.10123\hat{X}_2 - 3.3252\hat{Y} \leq 44.419$$

Adding this cut, and all previous cutting planes (in this case only one) generated when solving the first convexified MINLP problem, to the following convexified MINLP sub-problem result in that the problem can be solved more efficiently.

Adding these two cuts and the updated piecewise linear approximations of the inverse transformations, at iteration two, the second convexified problem

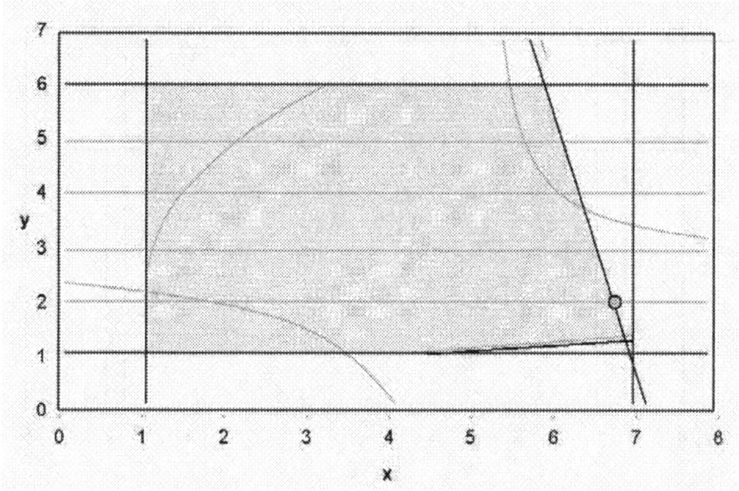

Fig. 7. The overestimated feasible region (with the cutting plane from the ECP method) when solving the first convexified MINLP sub-problem.

can be solved by the ECP-method in only one MILP iteration. The solution of the second convexified MINLP sub-problem is obtained at (x=6.6, y=3). In Figure 8 the feasible region overestimated by the two cutting planes is illustrated. From the figure it is observed that the solution of the second convexified MINLP sub-problem can be obtained in only one MILP-iteration by the ECP method. Observe, further, that the projection of the cutting planes, in the $x - y$ space, changes when new grid points are added to the piecewise linear approximations. Thus the cutting planes are differently projected into the $x - y$ space at iteration two, than at iteration one. In Figure 8 the first cutting plane is projected as a dark continuous straight line, while the second cutting plane is projected as a dark continuous broken line.

At the solution point (x=6.6, y=3), the value of the generalized signomial constraint is 5.7792. This solution point does not satisfy the termination criteria Eq. (27). In a similar way, as in the previous iteration, a cut can be generated, from the convexified signomial constraint, at this point in order to make the solution of the following convexified MINLP sub-problem more efficient. The cut generated at the solution point (x=6.6, **y**=3) is given by

$$22.425y + 8x + \hat{X}_1 - 0.19448\hat{X}_2 - 2.0600\hat{Y} \leq 54.739$$

When adding this cut, the two previous cuts and the updated piecewise linear approximations to the following MINLP sub-problem it can be solved to optimality in only one ECP-iteration. The total number of MILP iterations to solve all three convexified MINLP sub-problems with the integrated ECP approach were, in this particular problem, thus only 2+1+1=4. Thus, the

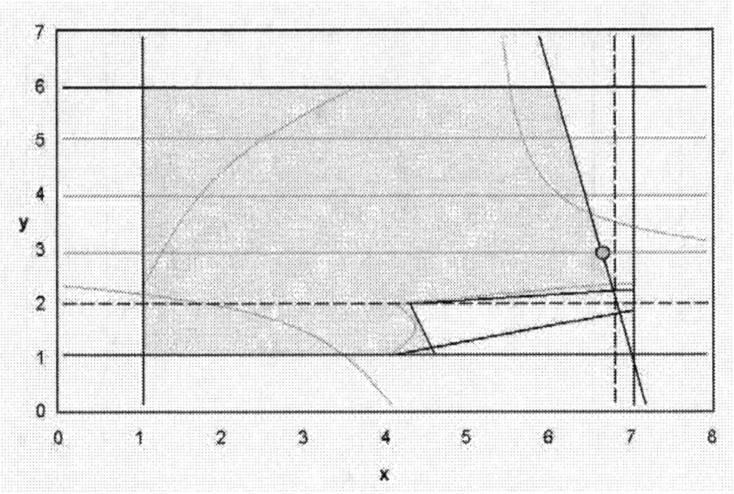

Fig. 8. The final overestimated feasible region (with the cutting planes from the ECP method) when solving the second convexified MINLP sub-problem.

global optimal solution of the nonconvex MINLP problem could be found by solving only 4 MILP problems.

In Figure 9 the feasible region overestimated by the three cutting planes, given above, is illustrated. From the figure it is observed that the solution of the third convexified MINLP sub-problem can be obtained in one MILP-iteration by the ECP method. Observe, again, that the projection of the cutting planes, in the $x - y$ space, changes when new grid points are added to the piecewise linear approximations. Thus the cutting planes above are again differently projected into the $x - y$ space at iteration three, than they were at iterations one and two. In Figure 9 the first cutting plane is projected as a dark continuous straight line, while the second and the third cutting planes are projected as dark continuous broken lines.

At the solution point (x=6.4, y=3), the termination criterion Eq. (27) is satisfied. This solution is, thus, the global optimal solution to the considered non-convex MINLP problem.

9 Conclusions

Transformation techniques, useful when solving nonconvex MINLP problems including signomial functions to global optimality, were discussed in the actual chapter. The transformation techniques can handle both positive and negative signomial functions and can simply be used to extend the applicability of many MINLP solvers to new broader classes of nonconvex MINLP problems. Some features of the transformation approaches were discussed and

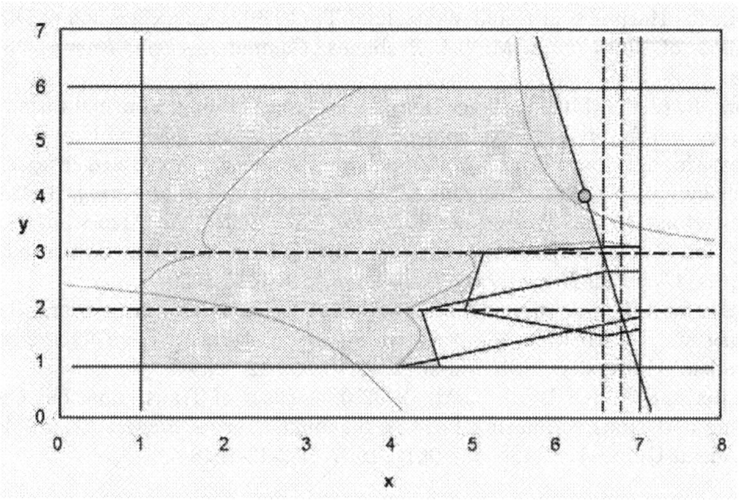

Fig. 9. The final overestimated feasible region (with the cutting planes from the ECP method) when solving the third convexified MINLP sub-problem.

some numerical examples were finally given. The optimization example illustrated, for example, how a disjunctive feasible region of the original problem was divided into convex sub-regions in subsequent iterations, from where the global optimal solution of the original problem finally was obtained.

Acknowledgments

Financial support from the Finnish Technology Agency, TEKES, is gratefully acknowledged.

References

1. Björk, K.-M. (2002). A Global Optimization Method with some Heat Exchanger Network Applications. *PhD-thesis.* Åbo Akademi University.
2. Björk K.-M., Lindberg P. O. and Westerlund T. (2003). Some Convexifications in Global Optimization of Problems Containing Signomial Terms. *Computers chem.. Engng.,* **27**, 669-679.
3. Duran M. A. and Grossmann I. E. (1986). An Outer Approximation Algorithm for a Class of Mixed Integer Nonlinear Programs. *Math. Prog.*, **36**, 307-339.
4. Floudas C. A. and Pardalos P. M. (eds). (2001). *Encyclopedia of Optimization.* Kluwer Academic Publishers. Dordrecht.
5. Lindberg P.O. (1981). Power Convex Functions. Generalized Concavity in Optimization and Economics. *Academic Press.* ISBN 0-12-621120-5, pp. 153-168.
6. Maranas C. D. and Floudas C. A. (1995). Finding all Solutions of Nonlinearly Constrained Systems of Equations. *Journal of Global Optimization*, **7**, 143-182.

7. Pörn R., Harjunkoski I. and Westerlund T. (1999). Convexification of Different Classes of Non-Convex MINLP Problems. *Computers chem. Engng.*, **23**, 439-448.
8. Pörn, R. (2000). Mixed Integer Non-Linear Programming: Convexification Techniques and Algorithm Development. *PhD-thesis.* Åbo Akademi University.
9. Westerlund T. and Pörn R. (2002). Solving Pseudo-Convex Mixed-Integer Problems by Cutting Plane Techniques. *Optimization and Engineering,* **3**, 253-280.
10. Westerlund T. and Westerlund J. (2003). GGPECP – An Algorithm for Solving Non-Convex MINLP Problems by Cutting Plane and Transformation Techniques. *Chemical Engineering Transactions*, **3**, 1045-1050.
11. Westerlund T. and Björk K.-M. (2002a). A Class of Transformations Convexifying and Underestimating Negative Signomial Terms. *Report 02-182-A,* Åbo Akademi University, ISSN 156-9612, ISBN 952-12-0968-2.
12. Westerlund T. and Björk K.-M. (2002b). A Class of Transformations Convexifying and Underestimating Positive Signomial Terms. *Report 02-183-A,* Åbo Akademi University. ISSN 156-9612, ISBN 952-12-1018-4.

Appendix

A piecewise linear function in K intervals can be represented in K binary variables, b_k, and equal many additional continuous variables, s_k, by the linear expressions (23-26) in section 3.5. In the expressions, (23-26), the sum of the K binary variables should be exactly equal to one, reducing the total number of combinations of the K variables to K. However, it is possible to reduce the number of binary variables, in the model, substantially.

Below it is shown that it is possible to replace the K binary variables, b_k, with a corresponding number of continuous variables, b_k, having the same property that exactly one of the continuous variables, b_k, takes the value one whereas all the others take the value zero. In this case the piecewise linear function can still be modeled by the constraints (23, 24 and 26), the constraint (25) can, however, be omitted, but some additional constraints are required to model the K zero one continuous variables, b_k. In this formulation we need N binary variables, β_i, where $2^{N-1} < K \leq 2^N$ to model these K zero-one continuous variables, b_k. Thus, for example, a piecewise linear function in 512 intervals can be modeled with 9 binary variables β_i. Expressions to model the K, continuous variables, b_k, with the mentioned properties, are given below.

First observe that an integer index k, where $k = 1, 2, \ldots, K$, can be expressed by N, zero-one parameters, $\bar{\beta}_i$, as follows,

$$k = 1 + \sum_{i=1}^{N} 2^{i-1} \bar{\beta}_i \tag{31}$$

In the opposite way, since each index k corresponds to different parameters, $\bar{\beta}_i$, the ith zero-one parameter corresponding to a given integer value k can be expressed as $\bar{\beta}(i,k) = \bar{\beta}_i$, where $\bar{\beta}_i$ is defined by (31) for a certain k. For example, if k=9, then $\bar{\beta}(1,9) = 1$, $\bar{\beta}(2,9) = 0$, $\bar{\beta}(3,9) = 0$, $\bar{\beta}(4,9) = 1$ and $\bar{\beta}(i,9) = 0; i \geq 5$.

By using such binary parameters, $\bar{\beta}(i,k)$ (uniquely given by the above relation), together with N binary variables, β_i, we will, in the following, define some linear inequality constraints by which exactly one of K continuous variables b_k; k=1,2,. . . , K will obtain the value one, whereas all other continuous variables must obtain the value zero.

First define N linear inequality constraints for each continuous variable k such that the continuous variable, b_k, is constrained to be less or equal to one only for its unique set of binary parameters, $\bar{\beta}(i,k)$. Otherwise the continuous variable, b_k, is constrained to be less or equal to zero by the same set of constraints. The constraints can be written as follows,

$$\left.\begin{array}{l} b_k + (1 - 2\bar{\beta}(i,k))\beta_i \leq 1 - \bar{\beta}(i,k) \\ i = 1, 2, ..., N \end{array}\right\} \quad k = 1, 2, ..., K \tag{32}$$

Then define one additional linear inequality constraint for each continuous variable, b_k, such that the continuous variable, b_k, is constrained to be greater or equal to one, only for its unique set of binary parameters, $\bar{\beta}(i,k)$, otherwise the continuous variable, b_k, is constrained to be greater or equal to zero, or the constraint is redundant. The constraints are given by,

$$\left.\begin{array}{l} b_k + \sum\limits_{i=1}^{N} (1 - 2\bar{\beta}(i,k))\beta_i \geq 1 - \sum\limits_{i=1}^{N} \bar{\beta}(i,k) \\ b_k \geq 0 \end{array}\right\} \quad k = 1, 2, ..., K \tag{33}$$

Since the continuous variable, b_k, is defined to be nonnegative, each variable b_k, can now obtain only zero-one values. In the case when $K < 2^N$ the variables b_k are not uniquely defined. In this case the combinatorial space for the binary variables, β_i, should be reduced, since in this case, fewer unique sets of binary parameters, $\bar{\beta}(i,k)$, are used, than the possible combinations of the binary variables, β_i. In this case, the requirement that exactly one of the K continuous variables should be equal to one is, thus, not generally satisfied by (32-33). However, the following linear inequality constraint is therefore used in order to eliminate all combinations of the binary variables, β_i, that do not correspond to any of the used unique sets of binary parameters, $\bar{\beta}(i,k)$ (i.e. combinations corresponding to k values $K < k \leq 2^N$).

$$K - \sum_{i=1}^{N} 2^{i-1}\beta_i \leq 1 \tag{34}$$

Now each combination of binary variables, β_i corresponds to a unique set of binary parameters, $\bar{\beta}(i,k)$. The sum of the continuous variables, b_k, is thus also constrained to be exactly equal to one.

Given N, binary variables, β_i, and the $(N+1)K+1$ linear constraints (32)-(34) we have now modeled K continuous variables, b_k, k=1,2,. . . , K that can only obtain zero-one values, and in addition the sum of the variables, b_k must be equal to one. The constraints, (32)-(34), can now be used together with the $K+2$ constraints (23, 24 and 26) when modeling a piecewise linear function in K steps. K continuous variables s_k, and equal many continuous variables b_k as well as N binary variables, β_i, where $2^{N-1} < K \leq 2^N$, are required in the model.

In the special case of only one interval, $K = 1$, the number, N, of binary variables, β_i, is zero ($N = 0$), since $2^{-1} < \quad K \quad \leq 2^0$. In this special case no constraints of the form (32)-(34) are, thus, required. Since the sum of the continuous variables, b_k, should be equal to one, the variable, b_1 in (23, 24 and 26) is, thus, defined to be equal to one (b_1=1) and the expressions (23), (24), (26) and (32)-(34) reduces to

$$\hat{Z} = Z_1 + (Z_2 - Z_1)s_1$$

$$x = x_1 + (x_2 - x_1)s_1$$

$$0 \leq s_1 \leq 1$$

Remarks

Finally, it should be mentioned that although the number of binary variables can substantially be reduced, the combinatorial space is the same in both models. There are K possible combinations of binary variables, both in the model (23-26) with b_k as binary variables and in the model (23, 24, 26 and (32)-(34) with the fewer β_i binary variables. This is because in the first model the constraint (25) reduces the combinations of b_k and in the second model the feasible solutions of the binary variables, β_i are reduced by the constraints (32)-(34) to the same number of combinations. MILP solvers using optional branching strategies can utilize the information in (25) by so-called special ordered sets. The reduction of binary variables, from K to N does, thus, not in general, need to improve the performance of the formulation, if the MILP solver is able to use branching strategies for special ordered sets of zero-one variables.

Solving Nonlinear Mixed Integer Stochastic Problems: a Global Perspective

Maria Elena Bruni

Dipartimento di Elettronica, Informatica e Sistemistica, Università degli Studi della Calabria, 87030 Rende (CS), Italy mebruni@deis.unical.it

Summary. In this paper, we present a novel approach for solving nonlinear mixed integer stochastic programming problems. In particular, we consider two stage stochastic problem with nonlinearities both in the objective function and constraints, pure integer first stage and mixed-integer second stage variables. We formulate the problem by a scenario based representation, adding linear nonanticipativity constraints coming from splitting the first stage decision variables. In the separation phase we fully exploit the partial decomposable structure of SMINLPs. This allows to deal with a separable nondifferentiable problem, which can be solved by Lagrangian dual based procedure. In particular, we propose a specialization of the Randomized Incremental Subgradient Method- proposed by Bertsekas(2001)- which takes dynamically into account the information relative to the scenarios. The coordination phase is aimed at enforcing coordination among the solutions of the scenario subproblems. More specifically, we use a branch and bound in order to enforce the feasibility of the relaxed nonanticipativity constraints. In order to make more efficient the overall method, we embed the Lagrangian iteration in a branch and bound scheme, by avoiding the exact solution of the dual problem and we propose an early branching rule and a worm start procedure to use within the Branch and Bound tree. Although SMINLPs have many application contexts, this class of problem has not been adequately treated in the literature. We propose a stochastic formulation of the Trim Loss Problem, which is new in the literature. A formal mathematical formulation is provided in the framework of two-stage stochastic programming which explicitly takes into account the uncertainty in the demand. Preliminary computational results illustrate the ability of the proposed method to determine the global optimum significantly decreasing the solution time. Furthermore, the proposed approach is able to solve instances of the problem intractable with conventional approaches.

Key words: Stochastic Programming, MINLP, Trim Loss, Lagrangian decomposition, Branch-and-Bound.

1 Introduction

Stochastic programming deals with a class of optimization models and algorithms in which some of the data may be subject to significant uncertainty. Such models are appropriate when data evolve over time and decisions need to be made prior to observing the entire data stream. Such inherent uncertainty is amplified by technological innovation and market forces. Under these circumstances it pays to develop models in which plans are evaluated against a variety of future scenarios that represent alternative outcomes of data. Such models yield plans that are better able to hedge against losses and catastrophic failures. Because of these properties, stochastic programming models have been developed for a variety of applications, including electric power generation, financial planning, telecommunications network planning and supply chain management, to mention but a few. The widespread applicability of stochastic programming models has attracted considerable attention from the OR/MS community, resulting in several recent books [7, 53, 39] and survey articles [6, 57]. Nevertheless, stochastic programming models remain amongst the more challenging optimization problems. While stochastic programming grew out of the need to incorporate uncertainty in linear and other optimization models [15, 12], it has close connections with other paradigms for decision making under uncertainty. For instance, decision analysis, dynamic programming and stochastic control all address similar problems, and each is effective in certain domains. Stochastic programming (SP) provides a general framework to model path dependence of the stochastic process within an optimization model. For this rich class of models a variety of algorithms can be developed. On the downside of the ledger, SP formulations can lead to very large scale problems, and methods based on approximation and decomposition become paramount. In this article, we will present a decomposition based method for mixed integer nonlinear SP models. The paper is organized as follows. In Section 2 we briefly explain the important role played by stochastic mixed integer nonlinear programming in the development of optimization based models. In Section 3 we will review the state of the art for this general class of problems. In section 4 we present the formulation of stochastic nonlinear mixed integer programs with discrete distribution. In Section 5 we discuss specific issues related to the solution of stochastic programs. In Section 6 we formulate a stochastic version of the well known deterministic Trim Loss problem and we present some computational results. Finally, in Section 7 we give our conclusions.

2 Motivations

The decision making processes that take place during the selection of an optimal design in engineering problems can be made more rational and efficient thanks to the use of mathematical models within a global optimization framework. The most significant contribution of mathematical approaches comes from their ability to incorporate many alternative structures within a single problem. This is achieved through the introduction of uncertainties, nonlinearities ad integrality restrictions. In particular, an explicit consideration of these issues leads to more realistic and reliable formulations.

There have been considerable advances in the theory of deterministic global optimization, resulting in the development of very efficient algorithmic procedures and mathematical programming techniques for identifying the global minimum of nonconvex optimization problems. Furthermore, there have been great advances in the capability of solving very large deterministic problems.

On the contrary stochastic mixed integer nonlinear programming (SMINLP for short) has received very little attention from the optimization community; only few works, which we will discuss in detail further on, deal with this class of problems. This despite the great importance of MINLP problems in engineering processes. Starting from the progress in the deterministic case, it should be clear that there are several challenges and opportunities in the area of optimization under uncertainty. Mixed-Integer nonlinear stochastic optimization problems are encountered in a variety of applications in various branches of engineering and applied science, applied mathematics, and operations research.

Most of the MINLP applications under uncertainty are the contributions from the area of chemical engineering community. These represent very important and active research areas that include process synthesis (heat exchanger networks, distillation sequencing, batch plant design [34] etc..), planning under uncertainty [35, 32], design of distillation sequences [51], optimization of core reload patterns for nuclear reactors. Others applications are unit commitment problem [4], nonlinear multiple objective asset-liability models airline crew scheduling problem [64]. The process systems engineering community has long been involved in the development of tools for the solution of design and operational problems under uncertainty. These efforts have been motivated by applications and, in many cases, yielded general-purpose algorithms. In the next section we review some of these developments. It is worth noting that almost all the works deal with continuous probability distribution functions to describe uncertainty.

3 SMINLP: state of the art

In the design of chemical plants, there are usually a number of technical and commercial parameters which are subject to significant uncertainty. These uncertainties can correspond to variations either in internal process parameters such as transfer coefficients, reaction constants, efficiencies and physical properties, or external parameters such as quality of the feedstreams, product demands, environmental conditions and economic cost data. It is easy, thus, to understand the significant developments that have been made in the design and scheduling of batch plants and, more generally, in the optimal process design under uncertainty. A general representation of the problem of design under uncertainty has the following form:

$$\begin{aligned} \max_{y,d,z} \ & P(y,d,z,x,\theta) \\ s.t. \ & h(y,d,z,x,\theta) = 0 \\ & g(y,d,z,x,\theta) \leq 0 \\ & d \in D, z \in Z, x \in X, y \in \{0,1\}^m, \theta \in \mathbb{R}^n \end{aligned} \tag{1}$$

where the binary variables y are primarily associated with the existence or nonexistence of a unit, or in general with a decision concerning the design of the plant. Structurally, they are similar to the the design variables d; z and x are vector of control and state variables (operating conditions) and θ represents the vector of uncertain parameters which is assumed to follow a continuous distribution function $J(\theta)$. The set of equalities h denote process equations (equilibria relation, heat and mass balances), the set of inequalities g correspond to design specifications and logical constraints, and P represents a scalar objective function, typically an economic performance index which must be maximized or minimized. Several approaches have been reported in the literature addressing the problem of design under uncertainty in the form of (1). Essentially three main directions can be broadly distinguished:

(i) stochastic framework;
(ii) parametric programming;
(iii) deterministic equivalent.

As far as the stochastic framework is concerned, we mention the work in [33]. Design feasibility and economic optimality are simultaneously obtained without requiring an a priori discretization of the uncertainty. A very attractive feature of this approach is that since most optimization tasks can be performed independently the algorithm has a highly parallel structure that can be further exploited. There are two major difficulties in directly addressing problem (1): the evaluation of the expected profit requires integration over an (inner) optimization problem; this integration ought to be considered within the feasible region of the plant, which is unknown at the first (design) stage. Furthermore, the integrands for the integration are only implicitly defined through the solution of the inner optimization problem with fixed design variables. To overcome the above difficulties a solution strategy is proposed based on the following ideas:

- The multiple integral for the expected profit evaluation over the feasible region of the plant is approximated through a Gaussian quadrature formula with unknown quadrature points;
- The unknown quadrature points are determined as part of the optimization procedure through the solution of a sequence of feasibility subproblems;

- A modified (generalized) Benders decomposition is employed where the first-stage design variables are selected as complicating variables.

A combined multiperiod/stochastic modelling framework for process design problems under uncertainty has been proposed in [29, 52, 3]. In these works the vector of uncertain parameters is partitioned in two subsets according to the mathematical model used to describe their behavior. The deterministic uncertain parameters (θ_d) are modelled through a series of periods or scenarios in a multiperiod form forcing the structure/design to be feasible at every period/scenario selected. The stochastic uncertainties (θ_s), on the other hand, are described by probabilistic distribution functions ($\theta_s | \theta_s \in J(\theta_s)$) and incorporated in a two-stage stochastic form. In this context, the soft uncertainties can be considered part of the vector θ_s whereas the hard uncertainties will be typically associated with θ_d. Following these considerations, the process synthesis under uncertainty problem can be reformulated in a multiperiod/stochastic optimization framework by considering as an objective function the maximization of a function given by the cost of the design selected and the expected optimal profit as follows:

$$\begin{aligned} \max_{y,d,z} \; & E_{\theta_s} \bar{P}(y,d,z,x,\theta_s,\theta_d^i) \\ s.t. \quad & h^i(y,d,z^i,x^i,\theta_s,\theta_d^i) = 0 \; i = 1,\ldots,N \\ & g^i(y,d,z^i,x^i,\theta_s,\theta_d^i) \leq 0 \; i = 1,\ldots,N \\ & d \in D, z^i \in Z_i, x^i \in X_i, y \in \{0,1\}^m, \theta \in \mathbb{R}^n \end{aligned} \tag{2}$$

where

$$\bar{P}(y,d,z,x,\theta_s,\theta_d^i) = \sum_i w^i P(y,d,z^i,x^i,\theta_s,\theta_d^i).$$

The iterative procedure for solving problem (2) is based on the following steps:

1. Define the deterministic and stochastic uncertain parameters according to the desired design objectives and/or information available.
2. Select initial structural and design values.
3. Determine the feasible region in the span of the soft uncertain parameters of the selected structure and design.
4. Place the integration points within the feasible region and calculate the optimal profit of the plant at each of these points.
5. Obtain a new structure and design for the process by solving a master problem given by the dual representation of the original problem.
6. If the convergence criterion is not meet, return to step 3 with the new values for the design variables.

In this framework three alternative integration schemes are proposed in [3] for the evaluation of the expectancy. In the first integration scheme, the integration points are chosen from the initial uncertain space without any knowledge of the feasible region. There is no control over the number of integration

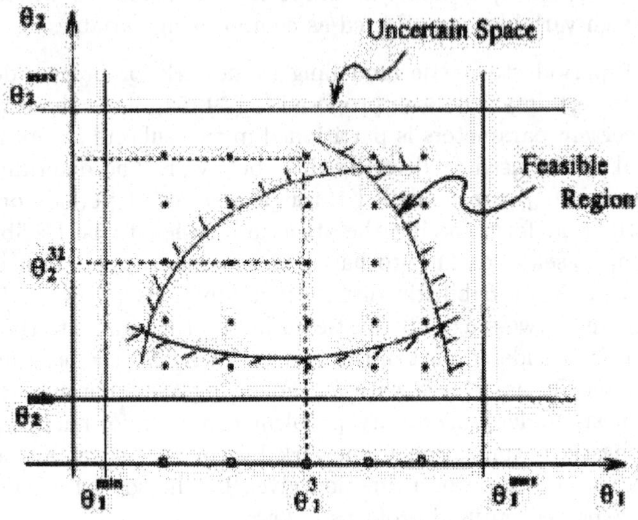

Fig. 1. Quadrature points placed in the entire space

points per parameter that will result in feasible (or infeasible) problems. The quadrature points are depicted in Figure 1.

In the second integration scheme the feasible range for each uncertain parameter is first defined through a series of feasibility optimization subproblems. Within this range the quadrature points are appropriately placed as shown in Figure 2. By placing the integration points within the feasible region, this procedure should in principle give a more accurate approximation of the expected profit with a smaller number of points.

A stochastic Monte Carlo integration for the expectancy can also be considered by randomly generating values of the uncertain parameters from uniform distribution functions as depicted in Figure 3. A practical advantage of the Monte Carlo based integration schemes is that, typically, the number of samples required for the approximation will not increase dramatically when the number of uncertain parameters increases, compared to the number of quadrature points of the numerical integration methods, thereby allowing the solution of problems with large number of uncertainties.

Wei and Realff in [63] proposed two new algorithms (the optimality gap and the confidence level method) to effectively solve convex stochastic MINLPs with continuous probability distribution. The proposed approach is a sampling-based method which approximate the problem (1) by sample average approximation. Hence, once the sample has been generated, the stochastic problem becomes a deterministic one, which can be solved by existing deterministic algorithms. In fact, the method proposed in [63] is wrapped around a traditional method to solve deterministic MINLPs: the Outer Approxima-

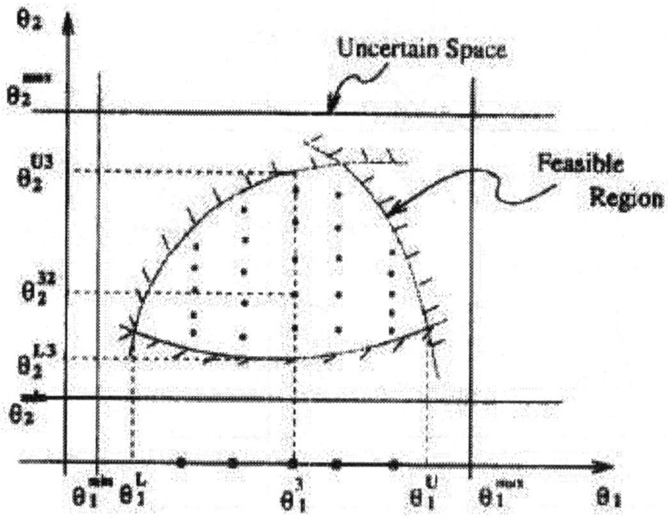

Fig. 2. Quadrature points placed in the feasible space

tion (OA) Method [21]. Both the algorithms (the optimality gap and the confidence level method) involve solving at each iteration a stochastic NLP subproblem (with fixed integer variables) to provide an upper bound, and a stochastic MILP master problem to provide a lower bound and new values for the integer variables. These problems are solved by applying the sample average approximation method. A stochastic version of the Branch and Bound method is proposed in [55] for solving stochastic global optimization problem. This work proposes an internal sampling algorithm opposed to the external sampling algorithms mentioned above. Stochastic upper and lower bounds are generated through the partitioning process ad subsets are not fathomed at each iteration until a sufficiently large number of iterations are carries out. Almost sure convergence of the method is proved.

The aim of the parametric programming approach is to obtain the optimal solution as a function of the parameters. In other words, the theory of parametric programming specifically aims to provide basic tools for the analysis of the effect of parameter changes to the optimal solution of problems, defining a parametric profile of this optimal solution as a function of the uncertainty. The key advantage of using parametric programming to address these applications is that the optimal solution is obtained as a function of the varying parameters without exhaustively enumerating the entire space of the varying parameters. This feature has tremendous potential for on-line control and optimization problems. While, for linear programs under uncertainty, parametric programming theory and tools are readily available, their extension to mixed-integer and nonlinear programs, suitable for the solution of process

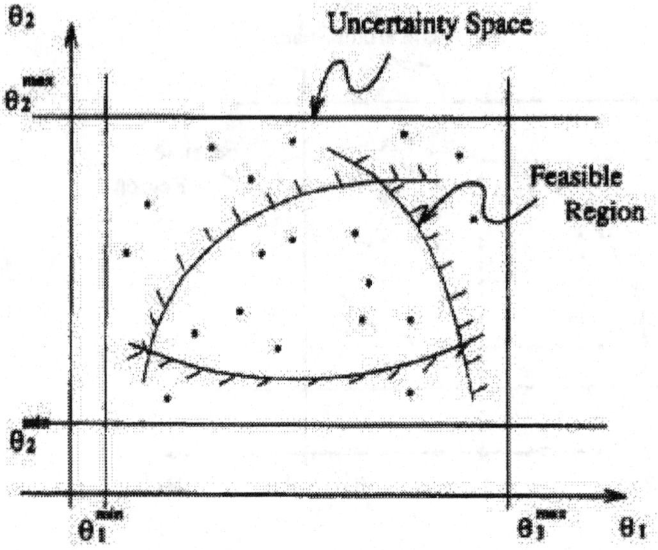

Fig. 3. Random placements of samples

synthesis and design problems, is very limited. In his Ph.D. work on process systems applications of parametric programming [50], Pertsinidis presented two algorithms for the solution of right-hand side MINLPs. The first one is based on Jenkins algorithm [36] for the solution of linear problems, while the second one applies similar principles to the OA/ER algorithm.

In [2],in the context of process synthesis problems under uncertainty, an implementable algorithm for the solution of scalar right hand side parametrizations of the parametric mixed integer optimization problem (1), based on the OA/ER algorithm is proposed, by extending and completing the original ideas of Pertsinidis. The procedure, based on the outer approximation/equation relaxation (OA/ER) algorithm of Kocis and Grossmann [37], involves the iterative solution of NLP subproblems and a parametric MILP master problem. The basic idea of the proposed strategy is to successively alternate between a set of nonlinear primal programs and a single parametric MILP master problem so as to establish appropriate upper and lower bounds of the optimal solution. The primal subproblem is first constructed by fixing the design variables (supposed binary and involved only in linear constraints). The direct parametric solution of the parametric nonlinear problem obtained can effectively be avoided and an approximate solution can instead be obtained by computing valid parametric upper and lower bounds.

The basic idea of the deterministic-based approaches is to transform the original optimization problem into a deterministic approximation, by specifying in advance a number of uncertain parameter realizations in a sequence of time periods or stages. The optimal design problem can then be approximated by the following multi-period (multistage) program:

$$\begin{array}{ll} \max_{y,d,z^i,x^i} & \sum_{i=1}^{N} P(y,d,z^i,x^i,\theta^i) \\ s.t. & h^i(y,d,z^i,x^i,\theta^i) = 0 \; i = 1,\ldots,N \\ & g^i(y,d,z^i,x^i,\theta^i) \leq 0 \; i = 1,\ldots,N \\ & d \in D, z^i \in Z_i, x^i \in X_i, y \in \{0,1\}^m, \theta \in \mathbb{R}^n \end{array} \tag{3}$$

Problem (3) exhibits a block diagonal structure which can be exploited for computational efficiency. In Paules and Floudas [51] a nested solution procedure that combines the Generalized Benders Decomposition [25] and the Outer Approximation/Equality Relaxation is proposed. The decomposition of the overall large scale stochastic problem with MINLP recourse is achieved by first partitioning the variables into structural and periodic variables. Each independent subproblem for the structural variables fixed is a MINLP problem. The solution of these subproblems will partition the variable set into a discrete set and a continuous set and will require the iterative solution of a purely nonlinear subproblem and a mixed-integer linear master problem. These problems are referred in [51] as inner primal problem and inner master problem to distinguish them form the outer master problem controlling the stochastic decomposition and the full scenario subproblems. The inner primal problems include only constraints of the scenario subproblems involving only continuous variables. Also constraints of the scenario subproblems involving only discrete variables are included in the inner master problem. The master problem may be either a MINLP or a NLP. If the problem is a MINLP a Generalized Benders Decomposition method has to be applied.

A branching algorithm has been proposed in [64]. It has been designed to solve a specific application and exploits the particular structure of the problem. In particular, the branching algorithm branches simultaneously on multiple variables without invalidating the optimality condition. The paper focuses on the key idea that scheduling decisions only create additional delays when crews are assigned to switch planes. When crew assignments follow plane assignments, no additional delays are created in the system due to crew schedules. In order to create the branching constraints, a specific type of delay, called the switch delay, is described. A switch delay is a delay due to a plane change. A hierarchy for the flight pairs based on the delay costs, is constructed. Along one branch, the algorithm forcibly includes the identified flight pair in a pairing selected in the optimal solution at that node. On the other branch, it excludes said flight pair from any pairing selected in the solution for that node. Since the algorithm has been designed to solve a specific problem, its applicability to other problems is doubtful. Furthermore the nonlinear recourse function reduces to a linear one when the long-range planning first stage variables are fixed.

This last contribution focuses on nonconvex stochastic problems with discrete probability distribution function, and thus seems to be the only existing work dealing with SMINLPs. To the best of our knowledge, no other contribution dealing with SMINLPs has been published.

4 Problem formulation

We consider the following two-stage stochastic (mixed) integer nonlinear problem (SMINLP):

$$\begin{aligned} \min_x \quad & f^1(x) + Q(x) \\ & g^1(x) = 0, \\ & h^1(x) \leq 0, \\ & g^1 : \mathbb{R}^{n\mathbb{Z}_1} \to \mathbb{R}^{m_e}, \quad h^1 : \mathbb{R}^{n_1} \to \mathbb{R}^{m_i}, \\ & x \in \mathbb{Z}_+^{n_1}, \end{aligned} \tag{4}$$

where

$$\begin{aligned} Q(x) \quad &= E_\xi Q(x, \xi(\omega)) \\ Q(x, \xi(\omega)) &= \min_y \; f^2(y(\omega), \omega) \\ & g^2(x, y(\omega), \omega) = 0, \\ & h^2(x, y(\omega), \omega) \leq 0, \\ & g^2 : \mathbb{R}^{n_1+n_2} \times \Omega \to \mathbb{R}^{t_e}, \quad h^2 : \mathbb{R}^{n_1+n_2} \times \Omega \to \mathbb{R}^{t_i}, \\ & y \in Y, \end{aligned} \tag{5}$$

where Ω is a certain probability space equipped with a σ-algebra F and with a probability measure, ξ is a random variable whose probability measure is available, and $f^1, f^2, g^1, g^2, h^1, h^2$ are general nonlinear functions. x denotes the first stage variables, whereas $y(\omega)$ represent the second stage variables. The set Y is the union of two subsets Y_R and Y_Z with $Y_R \in \mathbb{R}_+^{n_2}$ and $Y_Z \in \mathbb{Z}_+^{n_2}$. That is, in the above formulation, some of the second stage variables (those indexed by the set Y_Z) are constrained to take discrete values. Formulation 5 encompasses nonlinear integer problems (if $Y_R = \emptyset$) as well as mixed integer nonlinear problems. We can recognize the general form of problem (1) with in addition the specification of a two-stage structure. In this framework we can think of the first stage variables as design variables and the second-stage variables as operating variables. The model is a slight generalization of problem (1) allowing first stage decisions to be integer. Second stage decisions can be purely continuous or purely integer as well as mixed integer.

For the sake of clarity we briefly recall the meaning of a stochastic program with recourse (or two-stage stochastic program). The key feature of the Two-Stage stochastic models is the presence of some *recourse actions*. The set of decisions is divided into two groups. A number of decisions have to be taken before the problem parameters are known: these decisions are first-stage decisions and they are taken in the first stage. Other decisions can be taken after the uncertainty is disclosed. These recourse decisions are functions of the actual realization of the uncertain parameters and of the first stage decisions. The sequence of the events characterizes the models as *recourse models*. We maintain here the additive form of the recourse function.

There is a severe shortage of nice properties such as convexity and continuity in two-stage nonlinear integer stochastic problems. This is mainly due to the integer restriction. If the only integer variables are the first stage ones, the properties of the recourse function are the same as in the continuous case. In the continuous nonlinear case if f, h are convex and g is affine for all ξ, the problem is convex. When integrality restrictions are present in the second stage, even for the linear case the recourse function is in general nonconvex. The nature of the SMINLP suggests to solve it as a global optimization problem. It is worth noting that this class of problems is challenging for the inherent difficulty and also for the dimension which strongly depends on the number of scenarios.

The expectation in (5) involves multidimensional integration. To make the problem tractable, uncertainty is usually expressed in terms of an approximate discrete distribution. However, the need for accuracy in modeling inevitably leads to the explosion of dimension in the size of the corresponding mathematical program. This imposes additional limits on the way of modeling stochastic programming problems and further complicates the management of such models. In consequence, there still does not exist a standard way of modeling stochastic programming problems and solution methods are still in infancy. In the next section we address the difficulties of modeling and solving stochastic programs and discuss in detail a decomposition approach developed in [10] to deal with this problem.

The assumption of a discrete probability space allows the objective to be written as a finite sum and the constraints to be replicated for each element in Ω. Assume that ξ has a discrete probability distribution over $\Omega = 1, .., S$, with $\mathrm{P}[\xi = \xi_i] = \pi_i$. We may consider the set of possible outcomes, which is finite. Thus, the problem can be restated as follows:

$$
\begin{aligned}
&\min f^1(x) + \sum_{s=1}^{S} \pi_s f^2(x, y_s, \xi_s) \\
&g^1(x) = 0 \\
&h^1(x) \leq 0 \\
&h_s^2(x, y_s, \xi_s) = 0 \;\; \forall \;\; s = 1, \ldots, S \\
&g_s^2(x, y_s, \xi_s) \leq 0 \;\; \forall \; s = 1, \ldots, S \\
&x \in \mathbb{Z}_+^{n_1}, y_s \in Y_s \;\; \forall \;\; s = 1, \ldots, S \\
&g^1 : \mathbb{R}^{n_1} \rightarrow \mathbb{R}^{m_e} \quad h^1 : \mathbb{R}^{n_1} \rightarrow \mathbb{R}^{m_i} \\
&g_s^2 : \mathbb{R}^{n_1+n_2} \rightarrow \mathbb{R}^{t_e} \;\; h_s^2 : \mathbb{R}^{n_1+n_2} \rightarrow \mathbb{R}^{t_i} \qquad s = 1, \ldots S
\end{aligned}
\tag{6}
$$

where π_s denotes the probability that scenario s occurs. The deterministic equivalent formulation is a large scale nonlinear integer problem with $n_1 + n_2 S$ variables and $m_e + m_i + t_e S + t_i S$ nonlinear constraints. Depending on the number S of scenarios this problem becomes intractable. In fact, due to the integer requirements the recourse function is in general nonconvex and discontinuous. This suggests that this class of problems represents a connection between global optimization and optimization under uncertainty.

It is easy to recognize that problem (6) can be decomposed into two interesting subproblems with common variables x. The first stage variables x act here as complicating variables because they link the first stage with the second stage and, more importantly for our purposes, they make the problem nonseparable with respect to the scenarios. This suggests that we apply efficient relaxation schemes. In the following we propose a coordination-decomposition approach to tackle this general class of problems.

5 The two-phase solution approach

Our method is based on two phases. The first phase consists in exploiting the structure of the problem by decomposing it into smaller subproblems. The decomposition approach breaks the very large problem into smaller manageable optimization problems. This has several advantages. First, the peak memory requirement (needed to generate and then to read the deterministic equivalent problem) can be avoided. Additionally, the problem can be passed to the solver in pieces that are suitable for the decomposition approach. The second phase is based on a Branch and Bound scheme. This global phase works by partitioning the solution space into smaller sets, one for each scenario. The global solution of the problem restricted to each subset is carried out in the decomposition phase. This two-phase structure reflects the usual form of a Global Optimization algorithm in which the solution space is partitioned into smaller sets and then the problem is solved in each subset. The global optimality of the solution found through a Branch and Bound algorithm can be guaranteed only if the bounding step generates valid upper and lower bounds on the mixed integer nonconvex problem. In order to obtain a valid lower bound, a relaxed problem must be constructed from problem (6). This relaxed problem has to be solved to global optimality. Thanks to the decomposition scheme we apply, the solution of this problem can be split into easier and smaller subproblems with a common structure in term of constraints and objective function. Thus, in theory very efficient algorithmic procedures for solving particular classes of nonconvex problems can be used (see for example [34, 19]).

In the following, we give a detailed description of the our decomposition-based Branch and Bound method (DCB&B).

5.1 The Decomposition Phase

The inherent difficulty arising in stochastic programming is mainly due to the high dimension of the equivalent deterministic formulation which can become unmanageably large. If the problem was fully separable the mentioned difficulties could be avoided thanks to the possibility of parallelization. Unfortunately the presence of global variables x leads to a nonseparable problem. The model (6) has a block diagonal structure, in which distinct blocks of variables and constraints are linked by means of global variables x that can be viewed as complicating variables. For large-scale problems with a block separable structure and a reasonably small number of coupling constraints Dual Decomposition methods are often successful. For a comprehensive review on decomposition methods for stochastic programming we refer the reader to [55]. A good review of decomposition methods for deterministic mathematical programming can be found in [43].

Because separability is not perfect, to decouple the submodels we can derive an equivalent deterministic problem by means of a variable splitting representation. This splitting scheme [49] introduces copies $x_1, \ldots, x_S$ of the first-stage variable x and adds simple linking constraints. This set of constraints have a specific meaning in the context of stochastic optimization. In fact they represent the nonanticipativity principle, which states that the first-stage decisions should not depend on the scenario which will occur in the second stage. In other words the *here and now* decision should not depend on the future information.

By applying this scheme, the problem can be reformulated as a scenario block separable MINLP with linear coupling constraints:

$$\min \sum_{s=1}^{S} p_s((f^1(x_s) + f^2(x_s, y_s, \xi_s)) \tag{7}$$
$$h_s^1(x_s) = 0 \quad \forall s = 1, \ldots, S \tag{8}$$
$$g_s^1(x_s) \leq 0 \quad \forall s = 1, \ldots, S \tag{9}$$
$$h_s^2(x_s, y_s, \xi_s) = 0 \quad \forall s = 1, \ldots, S \tag{10}$$
$$g_s^2(x_s, y_s, \xi_s) \leq 0 \quad \forall s = 1, \ldots, S \tag{11}$$
$$x_s = x_{s+1} \qquad \forall s = 1, \ldots, S-1 \tag{12}$$
$$x_s \in \mathbb{Z}_+^{n_1}, y_s \in Y_s \quad s = 1, \ldots S \tag{13}$$

The variable splitting method was originally applied in conjunction with Lagrangian relaxation [44] to optimization problems with "hard" and "soft" set of constraints and it is equivalent to what is termed Lagrangian Decomposition in [26]. Carøe and Schultz [11] and Hemmecke and Schultz [27] used a similar decomposition approach for two stage linear integer problems. We also mention Takriti and Birge [58]. For an impression on Lagrangian approaches for multistage stochastic integer programming developments we refer to Römisch and Schultz [54]. Different Lagrangian decomposition schemes and the resulting decomposition approaches have been proposed in [17] for multistage stochastic programming (the focus of the paper is on comparing the duality gap for different decomposition technique). We observe here that even if we are in presence of a duality gap coming from the nonconvexities in the constraints and the integrality restrictions on the second stage, this gap tends to diminish when the number of scenario increases [5]. The nonanticipativity constraints can be expressed in form of a linear constraint $Ax = 0$ with a suitable matrix A. The coefficients of the constraints (12) define a giant matrix $A = [A_1, \ldots, A_S]$ again scenario separable. The relaxation of these linear linking constraints splits the problem into $|S|$ independent MINLPs.

$$D(\lambda) = \min \sum_{s=1}^{S} \pi_s((f^1(x_s) + f^2(x_s, y_s, \xi_s)) + \sum_{s=1}^{S} \lambda(A_s x_s) \tag{14}$$
$$h_s^1(x_s) = 0 \quad \forall\ s = 1, \ldots, S \tag{15}$$
$$g_s^1(x_s) \leq 0 \quad \forall\ s = 1, \ldots, S \tag{16}$$
$$h_s^2(x_s, y_s, \xi_s) = 0 \quad \forall\ s = 1, \ldots, S \tag{17}$$
$$g_s^2(x_s, y_s, \xi_s) \leq 0 \quad \forall\ s = 1, \ldots, S \tag{18}$$
$$x_s \in \mathbb{Z}_+^{n_1}, y_s \in Y_s \quad s = 1, \ldots S \tag{19}$$

The above minimization is scenario separable and we have:

$$D(\lambda) = \sum_{s=1}^{S} D_s(\lambda) \tag{20}$$

where

$$D_s(\lambda) = \min\{\pi_s((f^1(x_s) + f^2(x_s, y_s, \xi_s)) + \lambda(A_s x_s) : (x^s, y^s) \in X^S\}$$

and X^s denotes the scenario s constraints. The Lagrangian dual is

$$\max_{\lambda} D(\lambda) \tag{21}$$

¿From a formal view point the problem of finding the best lower bound (dual problem) for (20) leads to a non smooth non differentiable concave problem that can be tackled with subgradient methods and their variants. Model (20) is now decomposable into $|S|$ subproblems and, for any choice of λ also yields a lower bound to the optimal solution of the original problem. Nevertheless in presence of nonlinearities in the Lagrangian function for fixed λ may not be convex. As the recent survey Neumaier [47] of complete solution techniques in global optimization documents, there are now about a dozen solvers for constrained global optimization that claim to solve global optimization and/or constraint satisfaction problems to global optimality by performing a complete search. Within the COCONUT project [56], many of the existing software packages for global optimization and constraint satisfaction problems [46, 48, 45, 40, 22] are evaluated and compared. The purpose of a global code is to check whether there is indeed no better point; this may well be the most time-consuming part of a complete search. Thus, it easy to understand the crucial role that the problem dimension plays in the search for the global optimum. It has been shown that the success rate for models of small dimension is greater than the same rate for larger models. In our opinion the good performance of the method we propose reflects this feature of the global solvers. In the following we give a description of the solution phase for the problem (21).

5.2 The solution of the Lagrangian dual problem

Lagrangian relaxation of coupling constraints leads to a nondifferentiable optimization problem. This dual problem consists of the sum of a large number of component functions each over one of the scenarios constraint sets. It is worth observing that in this dual decomposition scheme the Lagrangian subproblems keep all the original constraints. This implies that each subproblem has the same constraints structure of the original SMINLP problem. Thanks to the separability, both, subgradients and dual functions, can be calculated by solving a number of a relatively small subproblems $D_s(\lambda)$, one for each scenario. Furthermore our SMINLP belongs to the class of problems for which the Incremental Subgradient Method has been studied. This method has been proposed in [9] for minimizing a convex function, sum of a large number of component functions. The incremental subgradient method is similar to the standard subgradient method [5]. The main difference is that at each iteration, the multiplier vector is changed incrementally, through a sequence of steps.

Each step is a subgradient iteration for a single component function and there is one step per component function. It has been experimentally observed that incremental subgradient approaches have a good practical rate of convergence.

We consider the general framework of this method and propose some modifications which exploit the specificity of our problem.

The Incremental Subgradient Method performs the subgradient iteration incrementally, by sequentially taking steps along the subgradients of the component functions, with intermediate adjustment of the variables after processing each component function. Following our notation, and recalling that we are in presence of a concave nondifferantiable problem instead of a convex one, the basic step of the method can be formulated. The subgradient of (14) at λ is $g(\lambda) = \sum_{s=1}^{S} A_s x_s(\lambda)$, where $x_s(\lambda)$ are optimal solutions of the scenario subproblems. The subgradient is a vector of dimension $n_1(S-1)$ and is the sum of $g_i(\lambda)$, where $g_i(\lambda)$ is a subgradient of D_i at λ.

We let the superscript k be the iteration count of the standard subgradient method. Each step is a subgradient iteration for a single component function (single scenario in our setting), and there is one step per component function. Thus, an iteration can be viewed as a cycle of S subiterations.

At a generic iteration k of the subgradient method $\lambda^k = \phi_S^k$, where ϕ_s^k is obtained after the S steps

$$\phi_s^k = [\phi_{s-1}^k - \alpha_k g_{s(\lambda)}^k], \quad s = 1, \ldots, S \tag{22}$$

and

$$\phi_0^k = \lambda^k \,, \tag{23}$$

The updates described in (22) are referred to as the S subiteration of the kth cycle. In all subiterations of a cycle we use the same stepsize α_k.

The rows of our giant matrix A have only two nonzero components equal to 1 and -1. Giving the particular structure of the nonanticipativity constraints, at each subiteration of the method sketched above, the subgradient vector $g_{i(\lambda)}$ is worked up by taking into account the term relative to one scenario $A_s x_s(\lambda)$. For the first scenario the only nonzero in the subgradient vector are those relative to the first n_1 rows of the matrix A_1 which has the form:

$$\begin{pmatrix} 1 & 0 & \ldots & 0 \\ 0 & \ddots & \ldots & \vdots \\ \vdots & \vdots & \ddots & \vdots \\ 0 & \ldots & \ldots & 1 \\ 0 & \ldots & \ldots & 0 \\ \vdots & \vdots & \vdots & \vdots \end{pmatrix}$$

and thus has a diagonal block of 1 and all other elements zero. This implies that only the portion of multiplier λ associated with the first scenario will be changed during the first update. For the second scenario the matrix has two diagonal blocks due to the fact that variables associated with the second scenario are present in two constraints of type (12). A_2 has the form:

$$\begin{pmatrix} -1 & 0 & \dots & 0 \\ 0 & \ddots & \dots & 0 \\ \vdots & \vdots & \ddots & \vdots \\ 0 & \dots & \dots & -1 \\ 1 & 0 & \dots & 0 \\ 0 & \ddots & \dots & 0 \\ \vdots & \vdots & \ddots & \vdots \\ 0 & 0 & \dots & 1 \\ 0 & 0 & \dots & 0 \\ \vdots & \vdots & \vdots & \vdots \end{pmatrix}.$$

In this case only the components of the multiplier λ associated with the first and the second scenario will be changed. Similar considerations can be drawn for the remaining scenarios. For all the scenarios except that for the first one, only two components of the multiplier vector are updated with the rule (22) because only two blocks have nonzero elements. The idea is, once the subproblem for a scenario s is solved and the multiplier vector has been updated according with (22),we keep this value fixed for the remaining $S-1$ subiteration. This case allows us to deal with a restricted dimension of the dual vector λ in the subproblem, speeding the solution process.

It can be verified that the order used for processing the component functions $D_s(\lambda)$ can significantly affect the rate of convergence of the method. A randomized version of the incremental subgradient method has been proposed in [9], where the component function to be processed is chosen randomly. At each step a component function is chosen randomly according to a uniform distribution. In our case each component function $D_s(\lambda)$ has an associated probability, namely the probability of the scenario s. So we can reformulate the dual problem as follows: $\max_\lambda E_\omega\{\widehat{D_\omega(\lambda)}\}$ where $\widehat{D_\omega(\lambda)} = \min_{x,y} D_\omega(\lambda)$. Recalling the assumptions made afterwards we can restated the problem as: $\max_\lambda \sum_{s=1}^{S}\{\widehat{\pi_s D_s(\lambda)}\}$. In other words we select the scenario to be processed according to the probability distribution of the different scenarios.

5.3 The Coordination Phase

Relaxing nonanticipativity constraints leads to a block separable problem structure. Once the relaxed problem is solved, the resulting solution might not coincide in their x components. In order to enforce the relaxed nonanticipativity constraints a Branch & Bound procedure is used. This coordination approach, introduced by Carøe in [11] for linear mixed integer stochastic problems, uses Lagrangian relaxation of nonanticipativity constraints as bounding procedure. To come up with a candidate for feasible first-stage solutions, the average

$$\bar{x} = \sum_{s=1}^{S} \pi_s x_s \tag{24}$$

combined with some rounding heuristic in order to fulfill the integrality restriction is used. In the following, we shall denote by P the list of candidate problems p together with an associated lower bound z_{LD}. The outline of the algorithm is as follows:

Step1 (Inizialization). Set $\bar{z} = +\infty$ and let P be a list of problems which initially only contains the original problem.

Step2 (Termination). If P=$\emptyset$ then the solution (x, y) that yielded $\bar{z}$ is optimal.

Step3 (Node Selection). Select and delete a problem p from P, solve the corresponding Lagrangian dual whose optimal value yields the bound $z_{LD}(p)$. If p is infeasible go to Step 2.

Step4 (Bounding). If $z_{LD}(p) \geq \bar{z}$ go to Step 2. Otherwise, if the scenario solutions x_s are identical, update the best known solution and its function value $\bar{z}$. Delete from P all problems with $z_{LD}(p) \geq \bar{z}$. Go to Step 2 . Else if the scenario solutions differ, compute the average $\bar{x}$ and round it by some heuristic to obtain $\bar{x}^R$. If $\bar{x}^R$ is feasible and $z_{LD}(p) \leq \bar{z}$, then update the best known solution and its function value $\bar{z}$. Delete from P all problems with $z_{LD}(p) \geq \bar{z}$. Go to Step 5.

Step5 (Branching). Select a component x_i of x and add two new problems to P obtained from P by adding the constraints $\bar{x}_i \leq \lfloor \bar{x}_i \rfloor$ and $\bar{x}_i \geq \lfloor \bar{x}_i + 1 \rfloor$ respectively. Go to Step 2.

Here nonanticipativity requirements are relaxed and feasibility is obtained when the scenario solutions are identical. At each node of the Branch and Bound tree the proximal bundle method defined in [38] is used in [11] for solving the Lagrangian dual.

The efficiency of the basic approach introduced above depends on different issues. First of all we observe that instead of solving the Lagrangian dual to optimality, we may stop the iterations as soon as the Lagrangian value rises above the best known upper bound $\bar{z}$. Such a combined approach embeds the solution of the Lagrangian dual within the Branch and Bound tree as proposed in [8] and [41]. The basic idea underlying the approach in [41] is to branch early, possibly after a single iteration of the Sequential Quadratic Programming solver. The drawback of the early branching rule is that since the nonlinear problems are not solved to optimality, there is no guarantee that the value function at each step of the SQP solver provides a lower bound. As a consequence, in [41] the author proposed a new fathoming rule in order to derive lower bounds needed in the Branch and Bound process. In a similar way as in the integrated SQP- Branch & Bound, branch early, after a few iterations of the subgradient-based procedure used for solving the dual. Even if the dual is not solved to optimality, the evaluation of the dual function value is a lower bound that can be used within the Branch and Bound framework. Clearly, the tree-search and the iterative solution of the dual are interlaced.

A heuristic for deciding when to branch early is introduced. By exploiting the solutions of the incremental subgradient subproblems, we may define an early branching rule. It is performed only if the nonanticipativity gap $\tau \geq \varepsilon$, where $\tau = \max|x_{(i+1)} - x_{(i)}|$ and ϵ depends on the problem at hand. If such a case, the solution process is stopped before the S subiterations are completely performed.

The Lagrange multiplier vector is initialized at each node of the Branch and Bound tree with the multiplier vector of the parent node. This warm start procedure is motivated by the fact that subproblems generated at a given node of the Branch and Bound tree differ only in a bound constraint from the father.

Finiteness of the algorithm

Consider a node that is unfathomed. Suppose we solve the Lagrangian dual obtaining a solution infeasible with respect to the nonanticipativity constraints. Thus, the branching step can further refine it, branching on a integer first stage variable and creating two subproblems in both of which that variable is fixed. Since our Branch and Bound is an enumeration procedure, the algorithm terminates with an optimal solution.

5.4 Implementation

Most of the difficulties to model uncertainty through stochastic programming originate from the lack of an agreed standard of its representation. Indeed, stochastic programming problems usually involve dynamic aspects of decision making which combined with uncertainty inevitably leads to a complicated model. In consequence there still does not exist a standard way of modeling stochastic programming problems in algebraic modeling languages (AML). AML enables a modeler to express the problem in an index-based mathematical form with abstract entities: sets, indices, parameters, variables and constraints. The key notion in the AML is the ability to group conceptually similar entities into a set. The presence of two different sets associated with stages and uncertainty dimensions in stochastic programs creates a difficulty to an algebraic modeling language. The lack of standardization of modeling stochastic programs in AMLs has at least two reasons. Firstly, there is not yet a widely accepted syntax for a description of stochastic programs. Secondly, there is not yet a compact and flexible format in which AMLs could send the stochastic program to the specialized solver. Although several attempts have been made to standardize this process of modeling linear stochastic programs directly in AMLs [42], for the nonlinear case the process is still in an early phase (SMPS can handle nonlinearities only in the objective function). We should emphasize that the possibility of modeling stochastic programming problems directly in AMLs is not the only issue in this field. Indeed, the size of these problems tend to explode since it grows exponentially with the number of scenarios. Another important point is that these huge problems are structured and they can only be solved by specialized optimization techniques if their structure is exploited. Several powerful codes have been developed for linear stochastic programs such as DECIS, MSLiP, but they still need to be linked with modeling languages. At the moment of writing this chapter, the only option available in AMLs is to generate the full deterministic equivalent. The only alternative left is thus to use the general purpose solvers that by default would use a direct solution method to tackle the problem. This approach is quite efficient as long as the problem is small to medium size and can be generated within memory limits. Even if the user is satisfied with the accuracy of the generated problem, and the general purpose solver can solve this problem efficiently, there is a danger that the generation of the problem significantly contributes to the overall solution process. An alternative would consist in implementing simple decomposition technique directly within AMLs. The interested reader can consult the library of examples of algorithms implemented through AMPL [20]. We underline that there is a definite need to improve the links between the AMLs and the solvers. Stochastic programming solution techniques accessible from modeling systems certainly need further development to reach industry standard. We expect that this progress will be made in the next few years and the integrated modeling system for stochastic programming will enable the modelers to popularize the stochastic programming technology through relevant applications.

Given this lack of a standard for stochastic optimization problems, our implementation does not have the restrictions arising for a specific input format. We have used object-oriented programming techniques to facilitate the management of such models. In particular in order to embed optimization functionality in applications, we have used the Lindo Api [31] callable library and the C++ to write the implementation of the model. The Lindo Application Programming Interfaces is a full-featured callable solver library for software developers. It allows users to incorporate optimization solvers into their own programs. It is able to solve a wide range of optimization problems. In particular the Global solver available in Lindo Api employs branch and cut methods to break a problem into many subregions. Branching is used together with bounding to get a valid lower bound on the optimal objective value in each subregion. A promising subregion may be subdivided further in order to get a more accurate bound. The global solver combines a series of range bounding (e.g., interval analysis and convex analysis) and range reduction techniques (e.g., linear programming and constraint propagation) within a branch-and-bound framework to find proven global solutions. The

Lindo Api Global solver returns a provably global optimal solution and is one of the the fastest and most robust one among the currently available global solvers. The characterization of general nonlinear models is not easy as linear or quadratic models. In fact the constraints and the objective function cannot be represented with matrices and vectors alone. Lindo Api offers two basic interface styles for this purpose: the "black-box" style and "instruction list" style. The drawback of using this global optimizer is that it works only with the"instruction list" input format. Under this style, the programmers have to write a set of instruction lists, one instruction for each row of the model. An instruction list is a vector of integers that encodes the original mathematical model. It represents the model in a variant of Reverse Polish notation. For Lindo Api a postfix expression is simply a list of integers. Each operator has a unique integer associated with it. This flat form problem definition can be very time consuming even for medium size problems. The separation of the problem into many subproblems facilitates this task. At the present, the implementation of the DCB&B method is not hooked with a modelling language. A small parser has been developed which reads a problem in Lingo [30] format and transforms it into in the data structure required by the implemented method. The information about the problem (variables, constraints, objective function) is stored in the class *Dataprob*. This class in addition to the standard constructor and destructor, has a function which allows the parser to build the problem as an object of the *Dataprob* class. We recall that the main task of our DCB&B method is to split the original problem into small problems. That is, we need to create auxiliary problems from the original problem, each of them corresponding to a single scenario. This reformulation is carried out by means of a function which has access to the *Dataprob* class. Lindo Api offers callable functions to add/delete constraints and variables, to modify the constraint type, the variable type (continuous, integer or binary), the upper and lower bounds and the right-hand side of a given model. Furthermore, we may give a name to the variables and constraints. These functionalities are, thus, simply imported in our class. All the characteristic of a given problem can be easily accessed. Because we are interested in using the problem within a Branch and Bound scheme, we equipped the class *Dataprob* with a function able to clone the problem. In the DCB&B method described in section 5 the solution of the dual function is embedded within a Branch and Bound algorithm. The flowchart shown in figure 4 highlights the main features of the algorithm. For the sake of clarity the Solve block will be further exploded in figure 5.

The reformulator function reads an object of the *Dataprob* class and creates $|S|$ subproblem, one for each scenario. Each of these subproblems is still an object of the *Dataprob* class, but with small dimension. This decomposition in small problems but with the same structure of the original problem, leads to considerable reduction in the solution time. Even thought we have to repeat the solution for the subproblems many times, the algorithm shows good performance. This is manly due to the highly nonconvex nature of the prob-

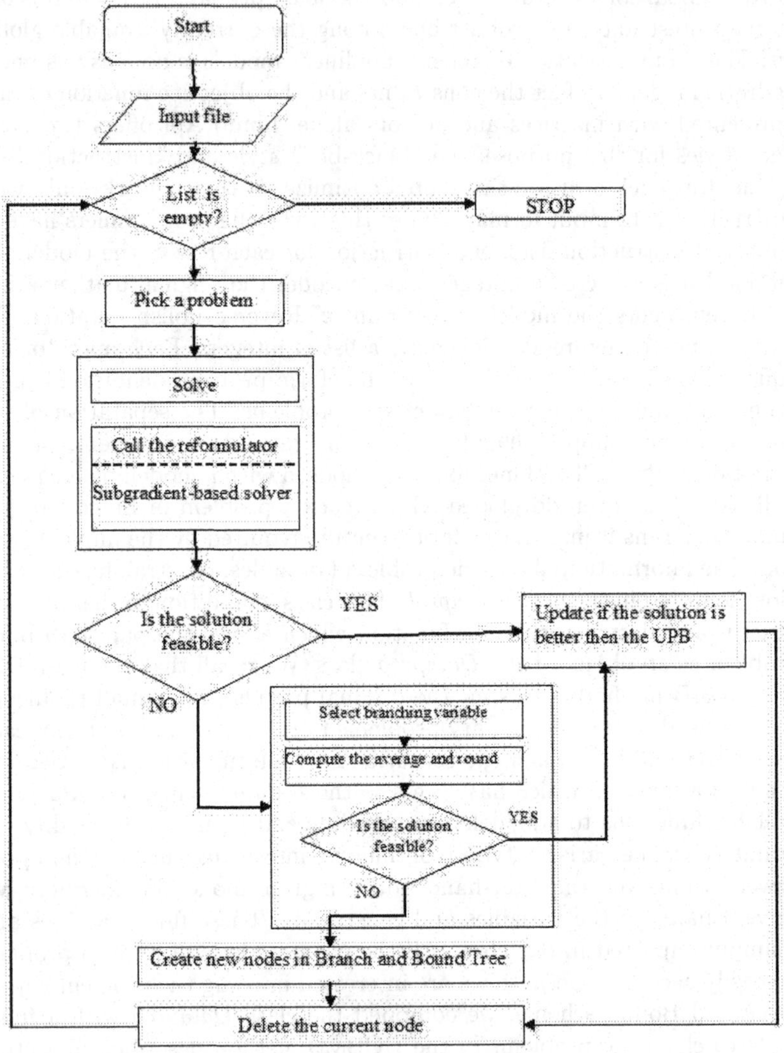

Fig. 4. Flowchart for the DCB&B method.

lem. In fact, the problem of finding a globally optimal solution of a nonconvex problem is a NP-hard task and the time to find a global optimum may increase exponentially with problem size. Thus decomposition seems to be the best way to tackle this kind of problem. The reformulator function utilizes Lindo functions in order to detect a series of independent block structures. It

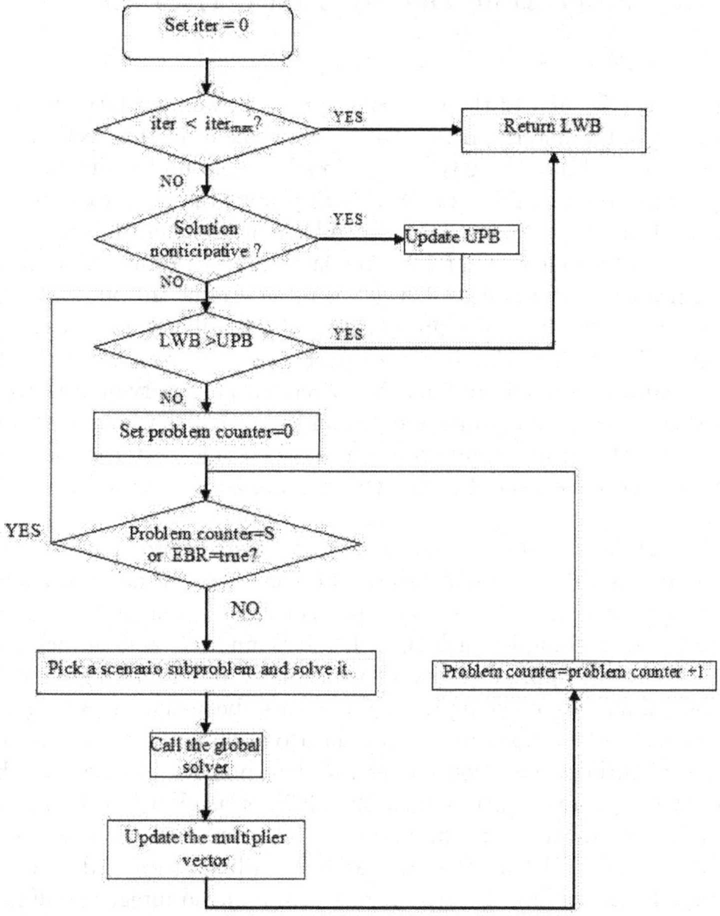

Fig. 5. Flowchart for the Solve block.

is worth noting that the Lindo Api solver offers the possibility to set a user adjustable parameter (`LS_IPARAM_DECOMPOSITION_TYPE`) to check whether a model can be broken into smaller independent problems. If total decomposition is possible, the solver will solve the independent problems sequentially to reach a solution for the original model. This decomposition strategy applied to the problem (7)-(13) without the DCB&B decomposition, doesn't produce substantial improvements in the overall solution time.

6 Illustrative application: the Stochastic Trim Loss Problem

In order to test the efficiency of the proposed approach, we have formulated a stochastic version of the deterministic Trim Loss problem. In this section, we develop a mathematical model for the Trim Loss Problem under uncertainty. The cutting stock problem (CSP) or Trim Loss problem is one of the oldest and most studied problems in the field of combinatorial optimization. Much of the interest in this problem stems from the large number of manufacturing problems to which the CSP applies. The goal is to determine the optimal plan to cut an inventory of standard size materials (rolls of paper, wire) to satisfy a set of customers' demands. Cutting stock problems may involve cuts being made in one or more dimensions and may be solved with a variety of objectives (e.g., minimizing trim loss, maximizing profit and so on). The typical solution approach involves attempting to determine a set of cutting patterns that will produce the required collection of items with a minimal waste production or trim loss.

A general description and classification of cutting stock problems is given in [18]. The one-dimensional cutting stock problem is a NP-hard problem. Because of its complexity, solutions to the Trim-Loss problem have often been generated using techniques such Branch and Bound [16] and Branch and Price [59, 60]. Heuristic approaches such evolutionary programming, genetic algorithms and simulated annealing [13, 24] have also been shown to be quite effective at generating solutions to the standard one dimensional Trim-Loss problem. Various authors have considered the deterministic problem in the context of integer nonconvex optimization [28], [62], or as global optimization test problem [23]. A linear programming approach combined with heuristic rules has been used in [61] in order to handle non-linearities and discrete decisions. The solution of Trim-Loss problems with a mixed integer nonlinear programming algorithm is considered in [1].

We address one of the most interesting generalizations of the CSP, the stochastic Trim Loss problem. In particular we formulate a model that explicitly incorporates uncertainty in customers order via a set of scenarios. The resulting formulation is a stochastic nonlinear integer program. The preliminary computational experience presented in section 6.2 shows that the proposed method allows us to achieve a considerable reduction of the overall solution time, and that the algorithm is quite insensitive to scenario growth.

6.1 Problem formulation

In this section we formulate a mathematical model for the one dimensional Trim Loss Problem under uncertainty. We first describe a deterministic formulation, and later extend this formulation to a stochastic setting by introducing a set of scenarios.

Consider the problem of cutting different product paper rolls from raw paper roll. There are different types of product rolls to be cut. Each type of product roll corresponds to a certain width. The length of the product paper rolls is assumed to be equal to the length of the raw paper rolls. The sum of the widths of the product paper rolls at each type of cut must be between a given width range. The product order specifies the total number of a specific product roll that have to be cut. In general it is not possible to cut out an entire order without throwing away some of the raw paper. The optimum cutting scheme aims at minimizing the waste paper or trim loss. In order to identify the best cutting scheme, a maximum number of different cutting patterns is postulated, where a pattern is defined by the position of the knives. Each cutting pattern may have to be repeated several times. The Trim Loss problem is characterized by the following notation.

Parameters:

I : the total number of different types of product rolls to be cut indexed by i;
J : the total number of different types of cutting patterns indexed by j;
b_i: the width of product i for each i in I ;
$B_{\max}$: the maximum width allowed for a cutting pattern j for each j in J;
Δ the width tolerance for cutting pattern j for each j in J;
N_{max} a physical restriction of the number of knives that can be used in the cutting process;
c_j the cost of the raw material for the cutting pattern j for each j in J;
C_j the cost of the change of cutting pattern j for each j in J;
M_j an upper bound on repeats of pattern j for each j in J.

Variables:

m_j the multiple of cutting pattern j used for each j in J;
y_j a binary variable that indicates wether a pattern j is used or not for each j in J;
n_{ij} number of a product i in cutting pattern j for each i in I and j for each j in J.

The following mathematical model formally describes the problem of determining the optimal cutting scheme.

$$\min_{m_j,n_{ij},y_j} \sum_{j=1}^{J} c_j m_j + C_j y_j \tag{25}$$

$$\text{subject to} \sum_{j=1}^{J} m_j n_{ij} \geq d_i \qquad i = 1, \ldots, I \tag{26}$$

$$(B_{\max} - \Delta) y_j \leq \sum_{i=1}^{I} b_i n_{ij} \leq B_{\max} y_j \quad j = 1, \ldots J \tag{27}$$

$$y_j \leq \sum_{i=1}^{I} n_{ij} \leq N_{max} y_j \qquad j = 1, \ldots, J \tag{28}$$

$$y_j \leq m_j \leq M_j y_j \; j = 1, \ldots, J \tag{29}$$

$$\sum_{j=1}^{J} m_j \geq max \left(\left\lceil \frac{\sum_{i=1}^{I} d_i}{N_{max}} \right\rceil, \left\lceil \frac{\sum_{i=1}^{I} d_i b_i}{B_{\max}} \right\rceil \right) \tag{30}$$

$$y_{j+1} \leq y_j \qquad j = 1, \ldots, J-1 \tag{31}$$

$$m_{j+1} \leq m_j \qquad j = 1, \ldots, J-1 \tag{32}$$

$$y_j \in \{0, 1\}, \qquad j = 1, \ldots, J \tag{33}$$

$$m_j \in \mathbb{Z}, \qquad j = 1, \ldots, J \tag{34}$$

$$n_{ij} \in \mathbb{Z}, \qquad i = 1, \ldots, I, \; j = 1, \ldots, J \tag{35}$$

The change of a cutting pattern involves a cost since the cutting machine has to be stopped before repositioning the knives. The objective function minimizes both the number of cutting patterns used and the number of pattern changes.

Constraints (26) imposes the satisfaction of the customer demands, constraints (27) prevent the patterns to exceed the given width limits. Constraints (28) limit the maximum number of products that can be cut from one pattern (this is due to practical constraints in cutting and winding: exceeding the limit would give rise to difficulties in separating the paper rolls after the winding). In constraints (29) the binary variable y_j is related to the cutting pattern. Constraints (30) impose a lower bound on total number of patterns made. Constraints (31) and (32) introduce an order on y and m variables to reduce degeneracy. It is worth noting that this formulation is the standard one used in [23] and [14]. Because of the bilinear inequality (26) the problem is both nonlinear and nonconvex. In this setting, if the problem parameters, such as demands, are known with complete certainty, the trimloss problem is a deterministic mixed integer nonlinear program.

In practice, the problem parameters associated with the Trim Loss problem are rarely known with complete certainty. To incorporate uncertainty in the decision making process, we adopt a two-stage stochastic programming approach which leads to a stochastic mixed integer nonlinear problem (SMINLP, for short). We assume that the cutting pattern on/off decisions have to be made here and now, with incomplete knowledge of future scenarios of customer order. Once the demands become known, a certain scenario of the problem parameters realizes, and then the optimal decision regarding the number of products in the pattern and the number of repeats of the pattern can be made. In other words, after taking the first stage decisions we can correct our cutting plan by means of recourse decisions on the number of repeats of the pattern and the number of products. The overall objective is to determine the optimal cutting scheme such that the sum of cutting change cost and the expected raw material cost are minimized. To incorporate uncertainty in the demands, we assume that these parameters can be realized as one of the S scenarios. If we denote with p_s the probability of scenario and with d_i^s the customers order for the product i under scenario s, we can extend the deterministic Trim Loss problem to the following two-stage SMINLP.

$$\begin{aligned} &\min_{y_j} \sum_{j=1}^{J} C_j y_j + \sum_{s=1}^{S} p^s Q^s(y) \\ &y_j \in \{0,1\},\ j = 1, \ldots, J \end{aligned} \tag{36}$$

where for all s,

$$\begin{aligned} &Q^s(y) = \min \sum_{j=1}^{J} c_j m_j \\ &\sum_{j=1}^{J} m_j n_{ij} \geq d_i^s \quad i = 1, \ldots, I \\ &(B_{\max} - \Delta) y_j \leq \sum_{i=1}^{I} b_i n_{ij} \leq B_{\max} y_j \ j = 1, \ldots J \\ &y_j \leq \sum_{i=1}^{I} n_{ij} \leq N_{max} y_j \ j = 1, \ldots, J \\ &y_j \leq m_j \leq M_j y_j \ j = 1, \ldots, J \\ &\sum_{j=1}^{J} m_j \geq max \left(\left\lceil \frac{\sum_{i=1}^{I} d_i^s}{N_{max}} \right\rceil, \left\lceil \frac{\sum_{i=1}^{I} d_i^s b_i}{B_{\max}} \right\rceil \right) \\ &y_{j+1} \leq y_j \ j = 1, \ldots, J-1 \\ &m_{j+1} \leq m_j \ j = 1, \ldots, J-1 \\ &y_j \in \{0,1\},\ j = 1, \ldots, J \\ &m_j \in \mathbb{Z},\ j = 1, \ldots, J \\ &n_{ij} \in \mathbb{Z},\ i = 1, \ldots, I \quad j = 1, \ldots, J \end{aligned} \tag{37}$$

Problem (36) represents the first-stage Trim Loss problem where the objective is to minimize the sum of fixed costs and expected variable costs. The function

$Q^s(y)$ represents the optimal variable cost under scenario s for a given pattern configuration. The assumption of complete recourse [7] ensures that the function $Q^s(y)$ is well-defined for any y. The two-stage structure of problem (36)-(37) is justified, since the decision concerning the existence of a pattern needs to be taken in advance in order to set the cutting knives, while the number of repeats can be decided when additional information is available. To our knowledge the stochastic Trim Loss problem has not been previously addressed in the literature.

6.2 Computational Experience

By introducing the variables n_{ij} and m_j for each scenario in problem (36)-(37), we come up with the deterministic equivalent problem of the stochastic Trim Loss problem.

$$
\begin{aligned}
\min_{m_{j,s},n_{ijs},y_j} & \sum_{s=1}^{S}\sum_{j=1}^{J}(C_j y_j + p_s c_j m_{js}) \\
& \sum_{j=1}^{J} m_{js} n_{ijs} \geq d_i^s \quad \forall\, i,\ \forall s \\
& (B_{\max} - \Delta) y_j \leq \sum_{i=1}^{I} b_i n_{ijs} \leq B_{\max} y_j \ \forall j \forall s \\
& y_j \leq \sum_{i=1}^{I} n_{ijs} \leq N_{max} y_j \ \forall j,\ \forall s \\
& y_j \leq m_{js} \leq M_j y_j \ \forall j,\ \forall s \\
& \sum_{j=1}^{J} m_{js} \geq max\left(\left\lceil \frac{\sum_{i=1}^{I} d_i^s}{N_{max}} \right\rceil, \left\lceil \frac{\sum_{i=1}^{I} d_i^s b_i}{B_{\max}} \right\rceil \right) \ \forall s \\
& y_{(j+1)} \leq y_j \ \ j = 1, \ldots, J-1 \\
& m_{(j+1)s} \leq m_{js} \ \ j = 1, \ldots, J-1 \forall s \\
& y_j \in \{0,1\} \ \forall j \\
& m_{js} \in \mathbb{Z},\ \forall j, \quad \forall s \\
& n_{ijs} \in \mathbb{Z},\ \forall i, \quad \forall j, \quad \forall s
\end{aligned}
\tag{38}
$$

The above problem is a large-scale nonconvex integer nonlinear program. We note that all the recourse constraints except the demand constraints link the first stage with the second stage by means of the binary variables y. Depending on the number S of scenarios this problem becomes intractable.

Furthermore the highly nonconvex nature of the Trim Loss problem makes it intractable even for small instances. In order to compare our decomposition method with the commercial software, we have considered randomly generated instances of the stochastic Trim Loss problem. The test problems are full deterministic equivalent problems in the form of (38). Each deterministic equivalent problem contains $(P+PS+PIS)$ variables all integer. The number of constraints is $[I(S)+6(PS)+S]$.

Table 1 summarizes the deterministic parameters in the optimization model. In particular, we have considered two deterministic instances (Trimloss2 and Trimloss4) as starting basis. Then, by varying the number of scenarios of the customer demands, we have generated 9 stochastic instances. The details are reported in Tables 2 and 3. Thus, the smallest instance solved has 32 variables and 85 constraints, whereas the largest has 2004 variables and 3500 constraints. This last test is a very large nonlinear integer problem that can not be solved by a straightforward approach. It is worth noting that for the deterministic case [1], the biggest instances of the trimloss problem has 48 variables and 27 constraints, 6 of which nonlinear.

In order to evaluate the performance of the implemented algorithm, we have measured the solution times and the number of major iteration of the solver (for the DCB&B algorithm we report the sum of iteration over the tree). To evaluate our method and the commercial solver on the same basis, an upper bound of 3600 seconds for the running time has been set.

Table 1. Problem parameters

	I	P	$C(j)$	$c(j)$	N_{max}	$B_{\max}$	B	Δ	b_i
Trimloss2	2	2	1	$\frac{1}{10}j$	5	1900	3	200	$\{330, 360\}$
Trimloss4	4	4	1	$\frac{1}{10}j$	5	1900	15	200	$\{360, 385, 415, 330\}$

Table 2. Problem dimension. Trimloss2

number scenarios	Problems dimension	
	n. variables	n. constraints(nonlinear)
5	32	85(10)
10	62	170(20)
15	90	255(30)
30	182	510(60)
50	302	850(100)
100	602	1700(200)
200	1202	3500(400)

As far as the numerical results are concerned, we report in Figure 1 the CPU time in seconds for solving the Trimloss2, by varying the number of scenarios. It is worth while to remark that the Trimloss2 instance with 5

Table 3. Problem dimension. Trimloss 4

number scenarios	Problems dimension	
	n. variables	n. constraints(nonlinear)
50	1004	1650(200)
100	2004	3500(400)

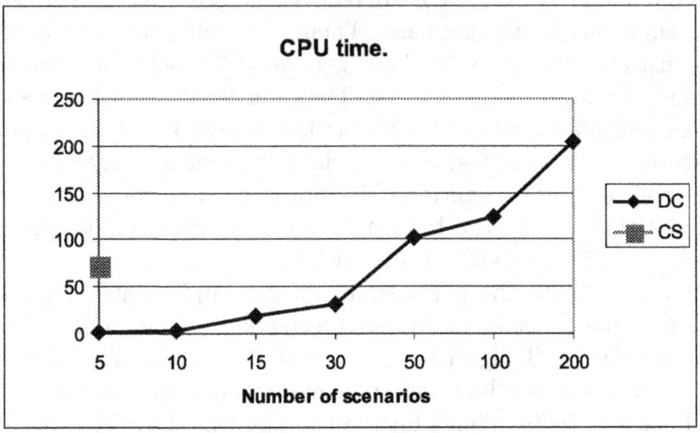

Fig. 6. CPU time in seconds for the trimloss2 as function of the number of scenarios. DC: proposed approach, CS: commercial solver.

scenarios was the only one (among the all 9 stochastic instances) solved by the Lindo Api solver within the allotted time.

Figure 2 reports the same for the Trimloss4 problem. In this case, it is interesting to observe that doubling the number of scenarios has not substantial impact on the solution time.

In Figures 3 and 4, we show the total number of iterations, respectively for the Trimloss2 and Trimloss4, performed by the DCB&B algorithm.

7 Concluding Remarks

In this paper we have proposed a solution method for the class of nonlinear mixed integer stochastic problems. In order to test the efficiency of the proposed method, we have formulated a stochastic version of the well known Trim Loss problem. In particular we dealt with the inherent uncertainty of the product demand by formulating the problem within the framework of stochastic two-stage recourse model. This makes the resulting model more suitable and versatile in terms of better handling the real cases.

The resulting stochastic mixed integer nonlinear problem has been effectively and efficiently solved by the application of a novel algorithmic approach,

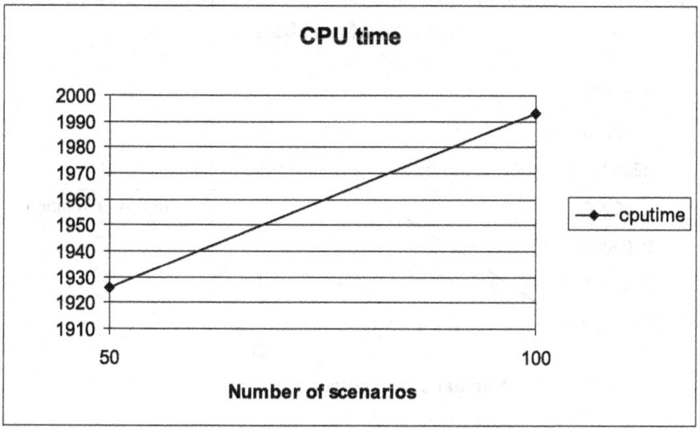

Fig. 7. CPU time in seconds for the trimloss4 as function of the number of scenarios.

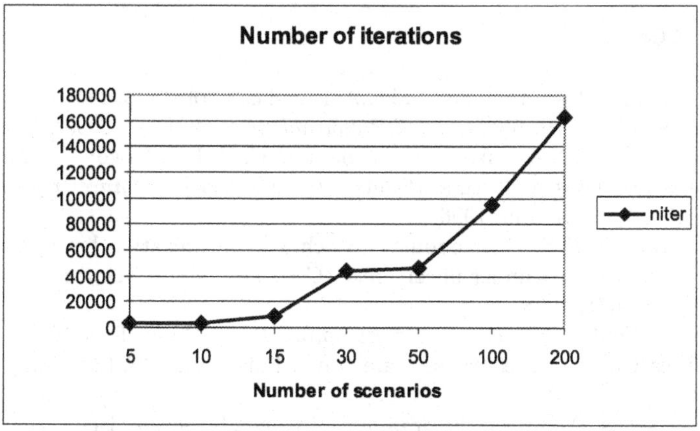

Fig. 8. Number of iterations for the trimloss2 as function of the number of scenarios.

which is able to fully exploit the peculiar structure of the problem. The decomposition procedure made it possible to successfully tackle the increased computational requirements in order to identify the global minimum of a stochastic nonlinear mixed integer problem in computationally realistic times. In fact, the preliminary numerical results demonstrated the efficiency of the proposed approach.

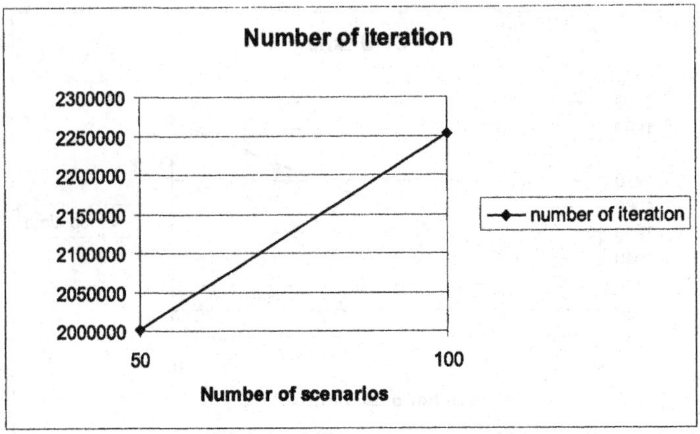

Fig. 9. Number of iterations for the trimloss4 as function of the number of scenarios.

References

1. C.S Adjiman, I.P Androulakis, and C.A. Floudas. Global Optimization of Mixed Integer Nonlinear Problems. *AIChE Journal*, 46:1769–1797, 2000.
2. J. Acevedo and E.N. Pistikopoulos. A parametric MINLP algorithm for process synthesis problems under uncertainty. *Industrial and Engineering Chemistry Research*, 35(1):147–158, 1996.
3. J. Acevedo and E.N. Pistikopoulos. Stochastic optimization based algorithms for process synthesis under uncertainty. *Computers and Chemical Engineering*, 22(4/5):647–671, 1998.
4. F. Bastin. *Nonlinear stochastic Programming.* Ph.D. thesis, Faculté des Sciences – Facultés Universitaires Notre–Dame de la Paix, NAMUR (Belgium), August 2001.
5. D.P. Bertsekas. *Nonlinear Programming.* Athena Scientific, Belmont, MA, second edition, 1999.
6. J.R. Birge. Stochastic Programming Computation and Applications. *INFORMS Journal on Computing*, 9(2):111–133, 1997.
7. J.R. Birge and F.V. Louveaux. *Introduction to Stochastic Programming.* Springer Series on Operations Research, Ney York, Berlin, Heidelberg, 1997.
8. B. Borchers and J.E. Mitchell. An improved Branch and Bound Algorithm for Mixed Integer Nonlinear Programs. *Computers and Operations Research*, 21(4):359–367, 1994.
9. D.P. Bertsekas and A. Nedic. Incremental Subgradient Methods for Nondifferentiable Optimization. *SIAM Journal on Optimization*, 12:109–138, 2001.
10. Maria Elena Bruni. *Mixed Integer nonlinear Stochastic Programming.* Ph. D. dissertation, University of Calabria, Cosenza (Italy), 2005.
11. C.C. Carøe and R. Schultz . Dual Decomposition in Stochastic Integer Programming. *Operations Research Letters*, 24:37–45, 1999.
12. A. Charnes and W.W. Cooper. Chance-constrained programming. *Management Science*, 5:73–79, 1959.

13. C. Chen, S. Hart, and V. Tham. A Simulated Annealing heuristic for the one-dimensional cutting stock problem. *European Journal of Operational Research*, 93(3):522–535, 1996.
14. GAMS Development Corporation. *GAMS IDE Model Library.* 1217 Potomac Street, NW Washington, DC 20007, USA., 2005. web : www.gams.com.
15. G.B. Dantzig. Linear programming under uncertainty. *Management Science*, 1:197–206, 1955.
16. J.M. Valerio de Carvalho. Exact solution of one-dimensional cutting stock problems using column generation and branch and bound. *International Transactions in Operational Research*, 5(1):35–44, 1998.
17. D. Dentcheva and S. Romisch. Duality in nonconvex stochastic programming. Stochastic Programming E-Print series 2002-13, 2002.
18. H. Dyckhoff. A typology of cutting and packing problems. *European Journal of Operational Research*, 1(44):145–159, 1990.
19. Thomas G.W. Epperly, M.G. Ierapertritou, and E.N. Pistikopoulos. On the global and efficient solution of stochastic batch plant design problems. *Computers and Chemical Engineering*, 21(12):1411–1431, 1997.
20. R. Fourer and D. Gay. Implementing algorithms through AMPL scripts. http://www.ampl.com/cm/cs/what/ampl/NEW/LOOP2/index.html, 1999.
21. R. Fletcher and S. Leyffer. Solving Mixed Integer Nonlinear Programs by Outer Approximation. *Mathematical Programming*, 66:327–349, 1994.
22. Deutsche Forschungsgemeinschaft. http://www-iam.mathematik.hu-berlin.de/ eopt/, June 2003.
23. C. A. Floudas, P. M. Pardalos, C .S. Adjiman andW. R. Esposito, Z .H. Gumus, S .T. Harding, J. L . Klepeis, C .A. Meyer, and C. A. Schweiger. *Handbook of Test Problems in Local and Global Optimization.* Kluwer, Dordrecht, 1999.
24. H. Foerster and G. Wascher. Simulated Annealing for order spread minimization in sequencing cutting patterns. *European Journal of Operational Research*, 110(2):272–282, 1998.
25. A.M. Geoffrion. Generalized Benders Decomposition. *Journal of Optimization Theory and Applications*, 10(4):237–260, 1972.
26. M. Guignard and S. Kim. Lagrangean Decomposition: A Model Yielding Stronger Lagrangean Bounds. *Mathematical Programming*, 39:215–228, 1987.
27. Raymond Hemmecke and Rudiger Schultz. Decomposition methods for two-stage stochastic integer programs. In *Online optimization of large scale systems*, pages 601–622. Springer, Berlin, 2001.
28. I. Harjunkoski, T. Westerlund, R. Porn, and H. Skrifvars. Different formulations for Solving Trim Loss Problems in a Paper-Converting Mill with ILP. *Computers and Chemical Engineering*, 20:s121–s126, May 1996.
29. M.G. Ierapetrirou, J. Acevedo, and E.N. Pistikopoulos. An optimization approach for process engineering problems under uncertainty. *Computers and Chemical Engineering*, 6-7, June-July 1996.
30. Lindo System Inc. *Optimization Modeling with Lingo.* North Dayton Street Chicago, Illinois, fourth edition, 2000.
31. Lindo System Inc. *Lindo API. The premier Optimization Engine.* North Dayton Street Chicago, Illinois 60622, July 2003.
32. M.G. Ierapetritou and E.N. Pistikopoulos. Simultaneus incorporation of flexibility and economic risk in operational planning under uncertainty. *Industrial and Engineering Chemistry Research*, 35(2):772–787, 1994.

33. M.G. Ierapetritou and E.N. Pistikopoulos. Novel approach for optimal process design under uncertainty. *Computers and Chemical Engineering*, 19(10):1089–1110, 1995.
34. M.G. Ierapetritou and E.N. Pistikopoulos. Batch Plant Design and Operations Under Uncertainty. *Industrial and Engineering Chemistry Research*, 35:772–787, 1996.
35. M.G. Ierapetritou, E.N. Pistikopoulos, and C.A. Floudas. Operational Planning Under Unceratinty. *Computers and Chemical Engineering*, 20(12):1499–1516, 1996.
36. L. Jenkins. Parametric Mixed-Integer Programming: An Application to Solid Waste Management. *Management Science*, 28:1270–1284, 1982.
37. G. R. Kocis and I. E. Grossmann. Relaxation Strategy for the Structural Optimization of Process Flow Sheets. *Industrial and Engineering Chemistry Research*, 26:1869–1880, 1987.
38. K.C. Kiwiel. Proximity Control in Bundle Methods for Convex Nondifferentiable Optimization. *Mathematical Programming*, 46:105–122, 1990.
39. P. Kall and S.W. Wallace. *Stochastic Programming*. John Wiley and Sons, Chichester, NY, Brisbane,Toronto, Singapore, 1994.
40. Sandia National Laboratories. http://www.cs.sandia.gov/opt/survey/main.html, March 1997.
41. S. Leyffer. Integrating SQP and Branch and Bound for Mixed Integer Nonlinear Programming. *Computational Optimization and Applications*, 18:295–309, 2001.
42. A. Leuba and D. Morton. Generating stochastic linear programs in S-MPS format with GAMS. In *Proceedings INFORMS Conference, Atlanta.*, 1996.
43. P. Mahey. Decomposition methods for mathematical programming. In P. Pardalos and Resende M., editors, *Handbook of Applied Optimization*. Oxford University Press, 198 Madison Avenue, Ney York, 10016, 2002.
44. L. Marshall and Fisher. The Lagrangian relaxation method for solving integer programming problems. *Management Science*, 27(1):1–18, 1981.
45. Hans D. Mittelmann. http://plato.asu.edu/topics/benchm.html, February 2005.
46. I. Nowakz, H. Alperin, and S. Vigerske. LAGO -an Object Oriented Library for solving MINLP. URL:http://www.mathematik.hu-berlin.de/ eopt/papers/LaGO.pdf, 2003.
47. A. Neumaier. Complete search in continuous global optimization and constraint satisfaction. In A. Iserles, editor, *Acta Numerica*, pages 271–369. Cambridge University Press, 2004.
48. A. Neumaier. http://solon.cma.univie.ac.at/ neum/glopt.html., 2005.
49. M. Näsberg, K.O. Jönstern, and P.A. Smeds. Variable Splitting - a new Lagrangian relaxation approach to some mathematical programming problems. Report, Linköping University., 1985.
50. A. Pertsinidis. *On the parametric optimization of mathematical programs with binary variables and its application in the chemical engineering process synthesis*. Ph.D. dissertation, Carnegie-Mellon University, Pittsburgh, PA, 1992.
51. E.G. Paules and C.A. Floudas. Stochastic programming in process synthesis: a Two-Stage model with MINLP recourse for multiperiod heat-integrated distillation sequences. *Computers and Chemical Engineering*, 16(3):189–210, 1992.
52. E.N. Pistikopoulos. Uncertainty in process design and operations. *Computers and Chemical Engineering*, 19:s553–s563, 1995.
53. A. Prékopa. *Stochastic Programming*. Kluwer, Dordrecht, 1995.

54. W. Römisch and R. Schultz. Multi-stage stochastic integer programs: An introduction. In M. Grtchel, S.O. Krumke, and J. Rambau, editors, *Online Optimization of Large Scale Systems*, pages 581–600. Springer, Berlin, 2001.
55. A. Ruszczyński. Decomposition Methods in Stochastic Programming. *Mathematical Programming*, 79:333–353, 1997.
56. H. Schichl. The coconut environment. web site. http://www.mat.univie.ac.at/coconut-environment/, 2004.
57. S. Sen. Stochastic programming: Computational issues and challenges. In S. Gass and C. Harris, editors, *Encyclopedia of OR/MS*. Kluwer, Dordrecht, 2001.
58. S. Takriti and J.R. Birge. Lagrangean Solution Techniques and Bounds for Loosely Coupled Mixed-Integer Stochastic Programs. *Operations Research*, 48(1):91–98, 2000.
59. F. Vance. Branch-and-price algorithms for the one-dimensional cutting stock problem. *Computational Optimization and Applications*, 9(3):212–228, 1998.
60. F. Vanderbeck. Computational study of a column generation algorithm for bin packing and cutting stock problems. *Mathematical Programming*, 86:565–594, 200.
61. G. Washer. An LP-based approach to cutting stock problems with multiple objectives. *European Journal of Operational Research*, 44:175–184, 1990.
62. I. Harjunkoski T. Westerlund, J. Isaksson, and H. Skrifvars. Different transformations for Solving Nonconvex Trim Loss Problems by MINLP. *European Journal of Operational Research*, 105(3):594–603, 1998.
63. J. Wei and J. Realff. Sample Average Approximation Methods for Stochastic MINLPs. *Computers and Chemical Engineering*, 28(3):333–346, 2004.
64. J. W. Yen and J. R. Birge. A stochastic programming approach to the airline crew scheduling problem. Technical report, Industrial Engineering and Management Sciences, Northwestern University. URL: http://users.iems.nwu.edu/~jrbirge//Public/html/new.html .

Application of Quasi Monte Carlo Methods in Global Optimization

Sergei Kucherenko

CPSE, Imperial College London, SW7 2AZ, UK s.kucherenko@imperial.ac.uk

Summary. It has been recognized through theory and practice that uniformly distributed deterministic sequences provide more accurate results than purely random sequences. A quasi Monte Carlo (QMC) variant of a multi level single linkage[1] (MLSL) algorithm for global optimization is compared with an original stochastic MLSL algorithm for a number of test problems of various complexities. An emphasis is made on high dimensional problems. Two different low-discrepancy sequences (LDS) are used and their efficiency is analysed. It is shown that application of LDS can significantly increase the efficiency of MLSL. The dependence of the sample size required for locating global minima on the number of variables is examined. It is found that higher confidence in the obtained solution and possibly a reduction in the computational time can be achieved by the increase of the total sample size N. N should also be increased as the dimensionality of problems grows. For high dimensional problems clustering methods become inefficient. For such problems a multistart method can be more computationally expedient.

Key words: stochastic methods, low-discrepancy sequences, multi level single linkage method

1 Introduction

The motivation for this paper is to develop further efficient and robust optimization methods. Let $f(x) : R^n \to R$ be a continuous real valued objective function. A nonlinear global optimization problem is defined as follows:

$$\min f(x), \quad x \in \mathbb{R}^n \tag{1}$$

subject to

[1] Also see Chapter 8, Sections 2.1, 7.1. In particular, the SobolOpt solver within the *ooOPS* software framework shares the same code as the software implementation proposed in this paper.

$$g(x) = 0, \text{ where } g = \{g_i\}, i = 1, \ldots m_e, \tag{2}$$

$$h(x) \geq 0, \text{ where } h = \{h_i\}, i = m_e + 1, \ldots, m, \tag{3}$$

where x is a vector of bounded continuous variables. No restrictions are imposed on the functional form of the objective function, $f(x)$ or the constraints $g(x)$ and $h(x)$.

There are two kinds of commonly used techniques for solving Eq. (1-3): deterministic and stochastic. Deterministic methods guarantee convergence to a global solution within a specified tolerance (a tolerance is defined as the maximum difference between the objective function value of the numerical solution and the true global optimal solution). For most deterministic methods the complexity of the problem grows exponentially as a function of the number of variables. For high dimensional problems the computational time is usually prohibitively large. Although some efficient methods have been designed for various forms of an objective function and/or the constraints, these methods are tailored to very specific problem structures and cannot be applied in the general high dimensional case. Good surveys of advances in global optimization are given in papers [6, 7, 10].

The present study is confined to stochastic methods and their variants based on deterministic sampling of points. A stochastic approach for global optimization in its simplest form consists only of a random search and it is called Pure Random Search (PRS). In PRS, an objective function $f(x)$ is evaluated at N randomly chosen points and the smallest value of $f(x)$ is taken as an approximation to the global minimum.

For stochastic methods the following result holds: if N points are drawn from a uniform random distribution over the n-dimensional hypercube $H^n = \{x_i \mid 0 \leq x_i \leq 1, i = 1, \ldots, n\}$ and if $f(x)$ is a continuous function defined in the feasible domain $B = H^n$, then the sample point with lowest function value converges to the global minimum. Stochastic search methods yield an asymptotic (in a limit $N \to \infty$) guarantee of convergence. This convergence is with probability 1 (or almost surely).

The PRS approach is not very efficient because the expected number of iterations for reaching a specified tolerance grows exponentially in the dimension n of the problem. Advanced stochastic techniques use stochastic methods to search for the location of local minima and then utilize deterministic methods to solve a local minimization problem. Two phases are considered: global and local. In the global phase, the function is evaluated in a number of randomly sampled points from a uniform distribution over H^n. In the local phase the sample points are used as starting points for a local minimization search. Thus the information obtained on the global phase is refined. For continuous differentiable objective functions classical gradient-based methods are used for local minimization. For non-differentiable functions or functions whose derivatives are difficult to evaluate the local search can be obtained through further sampling in a small vicinity around a starting point. The efficiency of the multistage methods depends both on the performance of the global stochastic and the local minimization phases.

In the simplest form of the multistage approach a local search is applied to every sample point. Inevitably, some local minima would be found many times. The local search is the most computationally intensive stage and ideally it should start just once in every region of attraction. The region of attraction of a local minimum x^* is defined as the set of points starting from which a given local search procedure converges to x^*. This is the motivation behind various versions of clustering methods. An extensive review on this subject can be found in papers [17, 18, 28].

The objective of the global stage is to obtain as much information as possible about the underlying problem with a minimum number of sampled points. To achieve this objective, sampled points should satisfy certain criteria. First, they should be distributed as evenly as possible. Second, on successive iterations new sampled points should fill the gaps left previously. If new points are added randomly, they do not necessarily fill the gaps between the points sampled on previous iterations. As a result, there are always empty areas and regions where the sampled points are wasted due to clustering. No information can be obtained on the behavior of the underlying problem in empty areas.

It has been recognized through theory and practice that a variety of uniformly distributed deterministic sequences provide more accurate results than purely random samples of points. Low-discrepancy sequences (LDS) are designed specifically to place sample points as uniformly as possible. Unlike random numbers, successive low discrepancy points "know" about the position of their predecessors and fill the gaps left previously.

LDS have been used instead of random numbers in evaluating multi-dimensional integrals and simulation of stochastic processes - in the areas where traditionally Monte Carlo (MC) methods were used [9, 25]. It has been found that methods based on LDS, known as quasi Monte Carlo (QMC) methods, always have performance superior to that of MC methods. Improvement in time-to-accuracy using QMC can be as large as several orders of magnitude.

LDS are a natural substitute for random numbers in stochastic optimization methods. As in other areas of applied mathematics, QMC methods provide higher accuracy with fewer evaluations of the objective function. The improvement in accuracy depends on the number of dimensions, the discrepancy of the sequence both of which are known, and the variation of the function, which is generally not known.

Central to the QMC approach is the choice of LDS. Different principles were used for constructing LDS by Holton, Faure, Sobol', Niederreiter and others (good surveys of LDS are given in [4, 14, 15]. Niederreiter's LDS have the best theoretical asymptotic properties. However, many practical studies have proven that Sobol' LDS in many aspects are superior to other LDS [16, 21]. For this reason they were used in the present study. In a classification developed by Niederreiter, the Sobol' LDS are known as (t, s) sequences in base 2 [15, 21]. The Holton LDS [9] were also used for comparison.

There had been a lack of a representative set of test problems for comparing global optimization methods. To remedy this a classification of essentially unconstrained global optimization problems into unimodal, easy, moderately difficult and difficult problems was proposed in [29]. The problem features giving this classification are the chance to miss the region of attraction of the global minimum, "embeddedness" of the global minimum, and the number of minimizers.

The purpose of this paper is the further development of optimization methods with an emphasis on comprehensive testing and a comparison of various techniques on a set of test problems of various complexity in accordance with the classification developed in the paper [29]. In particular: a comparison was made between:

- QMC and stochastic variants of a well known multi level single linkage (MLSL) algorithm [17, 18];
- different implementations of MLSL;
- two different types of LDS;

A number of problems used for testing belong to the category of the difficult multidimensional problems.

The remainder of this paper is organized as follows. A brief analysis of a Quasirandom Search (QRS) method is given in Section 2. Descriptions of MLSL and SL methods are presented in Section 3. Results of a comparison between stochastic MLSL and LDS based MLSL methods are presented in Section 4. Finally, the performance of different techniques is discussed in Section 5.

2 Analysis of Quasirandom Search methods

A general scheme of a QRS method is similar to that of PRS: an objective function $f(\boldsymbol{x})$ is evaluated at N LDS points and then the smallest value of $f(\boldsymbol{x})$ is taken as the global minimum. Generally QRS lacks the efficiency of more advanced methods. However, in some cases QRS has the following advantages over other methods of global optimization:

1. In its most general form it does not use any assumptions about the problem structure. In particular it can be used for any class of objective function (i.e. non-differentiable functions).
2. It can explicitly account for inequality constraints. The feasible region can be non-convex and even disconnected. However, it is not possible to account explicitly for equality constraints and such an optimization problem should be transformed into an unconstrained one.
3. It belongs to the so-called nonadaptive algorithms [30], in which the numerical process depends only on the current state and not on previously

calculated states. In contrast, in adaptive algorithms information is obtained sequentially. Nonadaptive algorithms are superior to adaptive ones in multi-processor parallel computations.

These advantages become more apparent as the number of variables grows. Analysis of QRS is important for understanding the advantages that the use of LDS brings to the multistage approach.

In this section it is assumed for simplicity that the problem is unconstrained and the feasible region is a closed set $K \subset \mathbb{R}^n$, where

$$K = \{x_i \mid x_i^L \leq x_i \leq x_i^U, i = 1, \ldots, n\}.$$

By linear transformation of coordinates K can be mapped into the n-dimensional hypercube H^n, so that the problem is formulated as:

$$\min_{\boldsymbol{x} \in H^n} f(\boldsymbol{x}). \tag{4}$$

Let f^* be an optimal value. Consider a sequence of sets of vectors $\boldsymbol{x}_{(N)} = \{\boldsymbol{x}_j \mid \boldsymbol{x}_j \in H^n, j = 1, \ldots, N\}$ and an approximation f_N^* to f^*:

$$f_N^* = \min_{x_j \in \boldsymbol{x}_{(N)}} f(x_j).$$

On a class of continuous functions $f(\boldsymbol{x})$ and dense sequences in H^n the following result holds [26]:

$$f^* = \lim_{N \to \infty} \min_{M \leq N} f_M^*.$$

For the purposes of error analysis the function $f(\boldsymbol{x})$ is assumed to have piecewise continuous partial derivatives satisfying the conditions:

$$|\partial f / \partial x_i| \leq C_i \qquad i = 1, \ldots, n. \tag{5}$$

From (5) it follows that $f(\boldsymbol{x})$ satisfies a Lipschitz condition:

$$|f(\boldsymbol{x}) - f(\boldsymbol{y})| \leq L\rho(\boldsymbol{x}, \boldsymbol{y}), \tag{6}$$

where L is a Lipschitz constant. The dispersion $d_N(n)$ of the sequence $\boldsymbol{x}_{(N)}$ is defined as [15]:

$$d_N(n) = \sup_{\boldsymbol{x} \in H^n} \min_{1 \leq j \leq N} \rho(\boldsymbol{x}, \boldsymbol{x}_j), \tag{7}$$

where $\rho(\boldsymbol{x}, \boldsymbol{y})$ is the Euclidean distance (metric) between points $\boldsymbol{x}$ and $\boldsymbol{y}$. Using (6) and (7) the approximation error can be written as

$$f_N^* - f^* \leq L d_N(n). \tag{8}$$

As can be seen from (8), $d_N(n)$ defines the "quality" of the sequence. Sequences with small $d_N(n)$ guarantee a small error in a function approximation. For any sequence the following error bounds hold:

$$[1/(N\omega_n)]^{1/n} \leq d_N(n) \leq 2\sqrt{n}(D(n,N)/N)^{1/n}, \tag{9}$$

where $\omega_n = \pi^{n/2}/\Gamma(1+\frac{n}{2})$ is the volume of the n-dimensional unit ball and $D(n,N)$ is the discrepancy of a sequence [15]. Discrepancy is a measure of deviation from uniformity. Apparently, smaller $D(n,N)$ would provide smaller upper estimate of the dispersion $d_N(n)$. LDS are characterized by small $D(n,N)$, therefore every LDS is a low-dispersion sequence (but not conversely).

The best-constructed LDS have $D(n,N) = O(\ln^{n-1} N)$. For such LDS the resulting rate of convergence of QRS as follows from (9) is $O(N^{-1/n} \ln^{(n-1)/n} N)$. This rate is not sufficiently high when n is large. However, it is worth noting that an error bound (8) with $d_N(n)$ given by (9) was obtained in the assumption that function $f(\boldsymbol{x})$ depends equally on all variables: in other words, the constants C_i, i=1,...,n in (5) were assumed to be of the same order of magnitude. This was shown to be "the worst-case scenario" [22, 23]. In practical applications, the function $f(\boldsymbol{x})$ normally strongly depends on a subset of variables: $x_{i_1}, x_{i_2}, \ldots, x_{i_s}, 1 \leq i_1 < i_2 < \ldots < i_s, s < n$ and dependence on other variables can be weak. In this case inequality (6) becomes

$$|f(\boldsymbol{x}) - f(\boldsymbol{y})| \leq L\rho(\boldsymbol{x}', \boldsymbol{y}'),$$

where $\boldsymbol{x}', \boldsymbol{y}'$ are projections of the points $\boldsymbol{x}, \boldsymbol{y}$ on the s-dimensional face $H_{i_1, i_2, \ldots, i_s}$ of H^n. One very useful property of LDS is that the projection of n-dimensional LDS on s-dimensional subspace forms s-dimensional LDS. Then (9) becomes

$$[1/(N\omega_s)]^{1/s} \leq d_N(s) \leq 2\sqrt{s}(D(s,N)/N)^{1/s} \tag{10}$$

and for practical applications n should be substituted by "an effective dimension number" s, which can be much less than n [23]. It can result in a much higher rate of convergence than that predicted by (9). This correction is very important for understanding the advantages of using LDS in QRS. For comparison, a cubic grid provides a better discrepancy measure than (9). At first glance such a grid search may be seen as more efficient than QRS. However, a projection of an n-dimensional cubic grid LDS on s-dimensional subspace does not form an s-dimensional cubic grid because of "the shadow effect" (projections of some points on the coordinate axis would coincide). This means that the correction similar to (10) is not applicable for the cubic grid and its discrepancy measure does not improve as s gets smaller.

Many well-known LDS were constructed mainly upon asymptotic considerations, as a result they do not perform well in real practical tests. The Sobol' LDS were constructed by following three main requirements [24]:

1. Best uniformity of distribution as N goes to infinity.
2. Good distribution for fairly small initial sets.
3. A very fast computational algorithm.

Points generated by the Sobol' LDS produce a very uniform filling of the space even for a rather small number of points N, which is a very important case in practice.

In some cases, it is convenient to employ the dispersion in the maximum (or infinite) metric $\rho'(\boldsymbol{x}, \boldsymbol{y})$ defined by

$$\rho'(\boldsymbol{x}, \boldsymbol{y}) = \max_{1 \le i \le n} |x_i - y_i|,$$

where $\boldsymbol{x} = (x_1, x_2, ..., x_n)$and $\boldsymbol{y} = (y_1, y_2, \ldots, y_n)$. The dispersion $d'_N(n)$ of the sequence $\boldsymbol{x}_{(N)}$ in the maximum metric

$$d'_N(n) = \sup_{x \in H^n} \min_{1 \le j \le N} \rho'(\boldsymbol{x}, \boldsymbol{x}_j)$$

has the following error bounds:

$$\frac{1}{2N^{1/n}} \le d'_N(n) \le \frac{\alpha(n, N)}{N^{1/n}}, \tag{11}$$

where the parameter $\alpha(n, N)$ generally is a weak function of N. For (t, s) sequences this parameter does not depend on N and an improved error bounds has the form

$$\frac{1}{2N^{1/n}} \le d'_N(n) \le \frac{b^{(n+t)/n}}{N^{1/n}}. \tag{12}$$

In particular for the Sobol' LDS (12) becomes

$$\frac{1}{2N^{1/n}} \le d'_N(n) \le \frac{2^{1+T_2(n)/n}}{N^{1/n}}, \tag{13}$$

where $T_2(n)$ is a function with an upper bound

$$T_2(n) < n(\log_2 n + \log_2 \log_2 n + 1). \tag{14}$$

These results were used in the frameworks of quasi random linkage methods presented in papers [12, 19].

QRS was applied to solve global optimization problems in papers [2, 27] as early as 1970 (see also [25, 26]). However, as stated above, with the development of more advanced multistage methods the application of pure QRS is limited mainly to cases of non-differentiable objective functions and to problems in which high accuracy in finding a global solution is not required. In the framework of MLSL, QRS can be seen as a global phase of MLSL. A description of MLSL is given in the next section.

3 Single linkage and multilevel single linkage methods

In the simplest variant of a multistage method, a small number of random points are sampled and then a deterministic local search procedure (LS) is applied to all of these points. All located stationary points are sorted and the one with the lowest value of the objective function is taken as a global minimum. The general scheme of a Multistart (MS) algorithm is as follows:

1. Sample a point from a uniform distribution over H^n.
2. Apply LS to the new sample point.
3. If a termination criterion is not met, then return to Step 1.

One problem with the Multistart technique is that the same local minimum may be located several times. Ideally, LS should be started only once in every region of attraction. A few algorithms had been developed with such a property. Only those sample points whose function values are small enough are chosen as starting points. Points are grouped into clusters, which are initiated by a seed point. The seed point is normally a previously found local minimum $\boldsymbol{x}^* \in \boldsymbol{X}^*$, where $\boldsymbol{X}^*$ is a set of all local minima. All sample points within a critical distance are assigned to the same cluster.

Efficiency can be improved by reducing the number of local searches, namely by discarding some of the sampled points. If $\{f_i\}$ is an ordered set such that $\{f(\boldsymbol{x}_i) \mid f(\boldsymbol{x}_i) < f(\boldsymbol{x}_{i+1}), i = 1, \ldots, N\}$ and $\boldsymbol{X}$ is a corresponding ordered set of all sampled points, then the reduced sample set is taken as:

$$\boldsymbol{X}_r = \{\boldsymbol{x}_i \in \boldsymbol{X} \mid i = 1, \ldots, N_r, N_r = \alpha N\}, \tag{15}$$

where $0 < \alpha < 1$. In this case, some local minima can be discarded without affecting the global minimum search.

An important question in applying any numerical method is when to stop searching for the global minimum. Among various proposed termination criteria of the global stage, one of the most reliable was developed in [3]. It is based on Bayesian estimates for the number of real minima not yet identified and the probability that the next local search will locate a new local minimum. An optimal Bayesian stopping rule is defined as follows: if W different local minima have been found after Nlocal searches started in uniformly distributed points, then the expectation of the number of local minima is

$$W_{exp} = W(N-1)/(N-W-2), \tag{16}$$

provided that $N > W + 2$. The searching procedure is terminated if

$$W_{exp} < W + 0.5. \tag{17}$$

The MLSL method developed by Rinnooy Kan and Timmer [17, 18] is one of the best algorithms among various clustering methods. The general scheme of the MLSL algorithm is outlined below:

1. Set W:=0, k:=0.
2. Set k:=k+1, i:=0.
3. Sample a set $\boldsymbol{x}$ of N points from a uniform distribution over H^n.
4. Evaluate an objective function on set $\boldsymbol{X}$, sort $\{f_i\}$ in order of increasing function values and select a reduced set $\boldsymbol{X}_r$ according to (15).
5. Set $i := i + 1$ and take $\boldsymbol{x}_i \in \boldsymbol{X}_r$.

6. Assign the sample point $\boldsymbol{x}_i$ to some cluster $\boldsymbol{C}_l$ if $\exists \boldsymbol{x}_j \in \boldsymbol{C}_l$ such that $\rho(\boldsymbol{x}_i, \boldsymbol{x}_j) \leq r_k$ and $f(\boldsymbol{x}_j) \leq f(\boldsymbol{x}_i)$, where r_k is a critical distance given by (18). If $\boldsymbol{x}_i$ is not assigned to any cluster yet then start a local search at $\boldsymbol{x}_i$ to yield a local minimum $\boldsymbol{x}^*$. If $\boldsymbol{x}^* \notin \boldsymbol{X}^*$, then add $\boldsymbol{x}^*$ to $\boldsymbol{X}^*$, set $W := W + 1$ and initiate the W-th cluster by $\boldsymbol{x}^*$. Assign $\boldsymbol{x}_i$ to the cluster that is initiated by $\boldsymbol{x}^*$.
7. If $i = N_r$ go to step 8. Else go to Step 5.
8. If $k = Iter_{max}$, where $Iter_{max}$ is the maximum allowed number of iterations, or the stopping rule (15), (16) is satisfied, then stop. Else go to Step 2.

The critical distance r_k is found using cluster analysis on a uniformly distributed sample:

$$r_k = \left(\frac{m(B)}{\omega_n} \frac{\sigma \log(kN_r)}{kN_r} \right)^{1/n}. \tag{18}$$

Here $m(B)$ is the Lebesgue measure (if $B = H^n$ then $m(B) = 1$), k is an iteration index, σ is a known parameter. In our calculations the parameter σ was taken to be 2.0.

Sporadic clustering which is characteristic of relatively small sets of random points would result in inhibiting many LS because such clustered points could be assigned to the same clusters initiated by local minima. A comparison between (10) and (18) shows that the dispersion of LDS and critical distance r_k have a similar asymptotic behavior. It suggests that LDS are better suited for optimization problems than random sets of points. As in other cases of transition from MC to QMC algorithms, a significant improvement in efficiency can be achieved simply by substituting random points with LDS.

Schoen argued that the regularity of LDS can be further exploited [19]. He suggested using instead of (18) a critical distance

$$r_{N,\beta} = \beta N^{-1/n}, \tag{19}$$

where β is a known parameter. It was proved that within the framework of a Simple Linkage (SL) method that if the sampled points are generated according to LDS whose dispersion is limited by (11) then the total number of LS started even if the algorithm is never stopped will remain finite, provided that $\beta > \alpha(n, N)$. A SL method was developed in [12] in order to circumvent some deficiencies of MLSL. A LDS based SL method was presented in the paper [19]. The scheme of the SL method adopted for LDS sampling is the following:

1. Set $N := 0$; choose $\varepsilon > 0$;
2. Let $N := N + 1$;
3. Generate a point $\boldsymbol{x}$ from LDS in H^n;

4. Apply a local search algorithm from $\boldsymbol{x}$ except if $\exists \boldsymbol{x}_j$ in the sample such that:
$$\rho'(\boldsymbol{x}, \boldsymbol{x}_j) \leq r_{N,\beta} \quad \wedge \quad f(\boldsymbol{x}_j) \leq f(\boldsymbol{x}) + \varepsilon$$
5. If stopping criteria is satisfied, then stop. If not, add $\boldsymbol{x}$ to the sample and go to Step 2.

It is important to note that the SL method makes use of the maximum (or infinite) metric $\rho'(\boldsymbol{x}, \boldsymbol{y})$ instead of the Euclidean one which is used in MLSL.

4 Computational experiments

As stated in the paper [29] the choice of test problems should be systematic, so that they represent different types of problems ranging from easy to difficult to solve. Following this strategy, a C++ program called SobolOpt which employs all discussed algorithms, namely stochastic MLSL and its QMC variants with Sobol' and Holton LDS points was applied to a number of test problems of different complexity. All problems presented below are unconstrained, although the techniques used are readily applicable to constrained problems.

A local search was performed using standard nonlinear programming routines from the NAG library [13]. All computational experiments were carried out on an Athlon-800Mhz PC.

In most cases the objective was to find all the local minima that were potentially global. Four criteria for comparing the algorithms were used: (i) success in locating a global minimum; (ii) number of located local minima; (iii) number of calls of a local minimizer; (iv) average CPU time (in seconds).

The results are displayed in the tables. The following notation is used:

- "N" – total number of sampled points in each iteration. For the Sobol' LDS the equidistribution property and improved discrepancy estimates hold for N equal to a power of 2. In all experiments N was taken to be equal to 2^m, where m is an integer number;
- "N_r" – reduced number of sampled points on each iteration;
- "N_{min}" – total number of located minima;
- "*Iter*" – total number of iterations on the global stage;
- "$Iter_{max}$" – maximum number of iterations on the global stage;
- "LM" – number of calls of the local minimizer;
- "GM" – "y" ("n") – global minimum (GM) was found (not found) in a particular run, "Y" – global minimum was found in all four runs, "N" – was not found in any of four runs;
- "LDS Sobol"' – the MLSL method based upon Sobol' LDS sampling;
- "LDS Holton" – the MLSL method based upon Holton LDS sampling;
- "Random" – the MLSL method based upon random sampling;

- "Zakovic" - based upon random sampling implementation of the MLSL algorithm developed in paper [31, 32]
- "LDS Sobol' (NS)", "LDS Holton (NS)", "Random (NS)" – versions of the above mentioned algorithm in which the Bayesian stopping rule is not used, however the maximum number of iterations is limited above by $Iter_{max}$.
- " .... (NC)" – a version of " .... (NS)" algorithm in which clustering is not used (MS method).

Four independent runs for each test problem were performed. For the Random MLSL method all runs were statistically independent. For the LDS Sobol' (Holton) method for each run a different part of the Sobol' (Holton) LDS was used.

4.1 Problem 1: Six-hump camel back function

$f(x,y) = 4\,x^2 - 2.1\,x^4 + 1/3\,x^6 + xy - 4\,y^2 + 4\,y^4$,
$-3.0 \leq x \leq 3.0, -2.0 \leq y \leq 2.0$.
Global Solution:
$f(x,y) = -1.03163$,
$(x,y) = (0.08984, -0.712266)$,
$(x,y) = (-0.08984, 0.712266)$.

This is a well known test for global optimization [8]. There are 6 known solutions, two of which are global. Results for this test are presented in Table 1. According to the classification of problems into the degrees of difficulty suggested in [14] this problem belongs to a class of "easy" (E1) problems.

Algorithm	N/N_r	$Iter$	LM	N_{min}	GM	CPU
LDS Sobol'	256/128	1	6	6	Y	0.1
LDS Holton	256/128	1	6	6	Y	0.12
Random	256/128	1	6	5	Y	0.1
Zakovic	256/128	21	96	6	Y	not available

Table 1. Comparison of various realizations of MLSL for Problem 1.

In all four runs of the LDS Sobol' and Holton algorithms all six local minima were located with just six LM. For the Random MLSL method in one of the four runs only four local minima were found, five - in two runs and six - in one run. For this method, in almost all runs LM was larger than a number of located minima. To compare our results with those of other authors we used a program developed by S. Zakovic [31, 32]. This program was an implementation of the MLSL algorithm, similar to that of Dixon and Jha [5]. In all four runs Zakovic's program located all six minima but at the expense of 21 iterations and 96 calls of the local minimizer. Similar results with LM equal

92 were reported in the paper [5]. It shows that the above mentioned implementations of the MLSL algorithm by other authors are not very efficient. The differences were mainly due to the ways in which clustering and sorting algorithms were implemented. It was not possible to make a straightforward CPU time comparison as Zakovic's program is written in Fortran and makes use of a different local minimizer routine. However, other factors being equal one can expect the CPU time to be proportional to LM.

Other experiments were performed with smaller samples of points (N/N_r = 64/32, N/N_r = 128/64). In these not all local minima were located and in some cases only one global minimum was found. We can conclude that the set of parameters N/N_r = 256/128 were the most efficient settings.

It can be concluded that (i) LDS Sobol' and Holton algorithms are more efficient than other considered methods (ii) our implementation of stochastic MLSL is more efficient than that used in paper [31, 32, 8].

4.2 Problems 2A,B: Griewank function

$$f(x)=1+\sum_{i=1}^{n} x_i^2/d^2 - \prod_{i=1}^{n}(\cos x_i/\sqrt{i})$$

Configurations:

Problem 2A.n=2 , d = 200,

-100 ≤ x_i ≤ 100, i= 1,2.

Problem 2B. n = 10, d = 4000,

-600.0 ≤ x_i ≤ 600.0, i= 1,..,10.

Global Solution:

$f(\boldsymbol{x})$= 0.0,

$\boldsymbol{x} = \{\mathbf{0.0}\}$.

Problem 2A. Both problems belong to the class of "moderate" (M2) [29]. The objective of the test was to evaluate the performance of each method on a problem with a large number of minima. Apart from the global minimum at the origin, this function has some 500 local minima corresponding to the points where the i-th coordinate equals a multiple of $\pi\sqrt{i}$. Because of the very large number of local minima the region of attraction of the global minimum is very small, therefore a very large number of points must be sampled to locate it. In tests with N/N_r = 32768/128 in all four runs all tested algorithms successfully located some 20 minima including the global one (Table 2). LM was equal to the number of located minima. The slightly higher value of the CPU time for LDS Holton is explained by the slower process of generating Holton points compared with that for Sobol' or random points.

The performance of MLSL methods largely depends on the sample size. Other tests were performed with smaller samples, with N ranging from 128 to 16384. None of the methods were able to locate the global minimum in all four runs. This may explain results of similar tests reported in the paper [18]: "for the two-dimensional problem the method never really got started. After

Algorithm	N/N_r	*Iter*	LM	N_{min}	GM	CPU
LDS Sobol'	32768/128	2	23	23	Y	5.0
LDS Holton	32768/128	2	23	23	Y	7.0
Random	32768/128	2	22	22	Y	5.0

Table 2. Comparison of various realizations of MLSL for Problem 2A.

the first sample of 100 points, only one minimum was found in all cases, after which the method terminated. Global minimum was located once, two runs ended with one of the second best minima while seven runs terminated with a minima with a function value close to one". There is a strong dependence of the algorithm efficiency on the size of samples: for this test problem samples of 100 points were not sufficient to locate the global minimum.

It is known that for problems with a very large number of local minima the Bayesian termination criteria do not produce reliable results [18]. Because of this reason a standard MS method which does not use clustering and Bayesian stopping techniques was tested. Table 3 presents results of experiments with $N/N_r = 128/128$.

Algorithm	N/N_r	*Iter*	LM	N_{min}	GM	CPU
LDS Sobol'(NC)	128/128	1	128	107	Y	3.0
LDS Holton (NC)	128/128	1	128	106	Y	3.5
Random (NC)	128/128	1	128	107	n/n/n/y	3.0

Table 3. Comparison of various realizations of MS for Problem 2A.

The solution was limited by a single iteration. Results clearly show the advantages of using LDS points: in all four runs of LDS Sobol' (NC) and LDS Holton (NC) algorithms the global minimum was found in contrast with only one successful run of the Random (NC) algorithm. LM was nearly equal to the number of located minima. It confirms the high efficiency of the MS approach in test problems with a very large number of local minima. It is worth noting that the CPU time was nearly half of that for the MLSL method (with N/N_r = 32768/128, Table 2), while the number of located minima was almost five times higher. The results for the Random (NC) algorithm agree well with the observations made for the same algorithms in the paper [18].

It can be concluded that a reliable detection of the global minimum can be achieved with the MLSL method using large samples or alternatively, with the MS method using small samples of points.

Problem 2B. Problem 2B has an extremely high number of local minima. However, in comparison with the two-dimensional problem 2A it turned out to be much easier to locate the global minimum. This is in line with the results of Törn, Ali and Vjitanen [29]. In tests with the same sample sizes as in various realizations of MLSL for Problem 2A (Table 3) in all four runs all tested algorithms successfully located some 15 minima including the global one (Table 4). Average LM and N_{min} were similar for all methods. The LDS Holton algorithm was the slowest one.

Algorithm	N/N_r	*Iter*	LM	N_{min}	GM	CPU
LDS Sobol'	32768/128	2	18	14	Y	1.2
LDS Holton	32768/128	2	20	16	Y	5.1
Random	32768/128	2	19	14	Y	1.2

Table 4. Comparison of various realizations of MLSL for Problem 2B.

Table 5 illustrates the dependence of the number of LM from the ratio $\gamma = N/N_r$ for the Sobol' algorithm. Reduction in sample size results in increasing LM and the corresponding CPU time. In all four runs the global minimum was found. As sample size decreases CPU time increases super linearly with the number of LM. It is interesting to note that although the number of located N_{min} increased in comparison to previous tests with $N/N_r = 32768/128$, very few second best minima were found. Thus, it can be concluded that the strategy with large samples is more efficient if the objective is to locate only the global minimum.

Algorithm	N/N_r	*Iter*	LM	N_{min}	GM	CPU
LDS Sobol'	4096/128	18	43	33	Y	3.1
LDS Sobol'	2048/128	46	79	59	Y	12.3
LDS Sobol'	1024/128	78	107	81	Y	25.2
LDS Sobol'	512/128	98	150	102	Y	26.4

Table 5. Comparison of various realizations of LDS Sobol' for Problem 2B.

As in the above case of lower dimension (n=2) the MS method performs much better in terms of locating high a number of local minima then the MLSL method. The results of testing with $N/N_r = 512/128$ are presented in Table 2.3B. Calculations were limited to 10 iterations. In addition to locating the global minimum and a large number of local minima all second best minima were located as well. A comparison between MLSL LDS Sobol' $N/N_r = 32768/128$ (Table 4) and MS LDS Sobol' methods (Table 6) shows that the number of located minima increased almost 60 times while the CPU time increased only 30 times. Since in most cases the objective is to locate only the global minimum, in the case of the MS the sample size and maximum number of iterations $Iter_{max}$ can be reduced even further. Other tests showed that a reliable detection of the global minimum can be achieved with N/N_r as small as 32/16 and $Iter_{max}$ =1. The corresponding CPU time for such a case can be reduced to 0.4 s.

Algorithm	N/N_r	$Iter_{max}$	LM	N_{min}	GM	CPU
LDS Sobol' (NC)	512/128	10	1280	796	Y	36.0
LDS Holton (NC)	512/128	10	1280	782	Y	61.0
Random (NC)	512/128	10	1280	776	Y	36.0

Table 6. Comparison of various realizations of MS for Problem 2B.

It can be concluded that the reliable detection of the global minimum can be achieved with the MLSL method and rather large samples. The value of the $\gamma = N/N_r$ has a significant impact on the efficiency of the method: increasing γ can result in a dramatic decrease of the CPU time (Table 5).

For problems with a high number of local minima the MS method can be a good alternative to the MLSL method. Even runs with small sample size can produce a large value of N_{min} (Table 3). A quasi Monte Carlo variant of the MS method is much more efficient than the stochastic one.

4.3 Problems 3A,B: Shubert function

Problem 3A. n=3

$$f(x)=\frac{\pi}{n}\left\{k_1\sin^2(\pi y_1)+\sum_{i=1}^{n-1}(y_i-k_2)^2[1+k_1\sin^2(\pi y_{i+1})]+(y_n-k_2)^2\right\}$$
$$+\sum_{i=1}^{n}u(x_i,10,100,4),$$

$y_i = 1 + 0.25(x_i + 1), k_1 = 10$and $k_2 = 1$,

$$-10 \le x_i \le 10, i = 1, 2, 3.$$

$u(x_i, a, k, m)$ is a penalty function defined by

$$u(x_i,a,k,m)=\begin{cases} k(x_i-a)^m, & x_i > a \\ 0, & -a \le x_i \le a \\ k(-x_i-a)^m, & x_i < -a. \end{cases}$$

Global Solution $f(\boldsymbol{x}) = 0.0$, $\boldsymbol{x}$ = (-1.0, -1.0, -1.0).

Problems 3A and 3B belong to the class of "easy" (E2) problems [29]. Problem 3A has approximately 5^3 local minima. The objective of this test was to test the performance of the LDS Sobol', LDS Holton and Random methods on problems with a large number of minima. In tests with N/N_r = 32/16 and $Iter_{max}$ = 10 in all four runs all algorithms successfully located

Algorithm	N/N_r	$Iter_{max}$	LM	N_{min}	GM	CPU
LDS Sobol'(NS)	32/16	10	13	12	Y	$2.6\ 10^{-2}$
LDS Holton(NS)	32/16	10	13	11	Y	$4.6\ 10^{-2}$
Random(NS)	32/16	10	12	10	Y	$2.6\ 10^{-2}$

Table 7. Comparison of various realizations of MLSL for Problem 3A.

approximately the same number of local minima including the global one (Table 7). LDS Sobol'(NS) showed slightly better performance.

Problem 3B. n=5.

$$f(x) = k_3(\sin^2(\pi k_4 x_1) + \sum_{i=1}^{n-1}(x_i - k_5)^2[1 + k_6 \sin^2(\pi k_4 x_{i+1})] +$$

$$(x_n - k_5)^2[1 + k_6 \sin^2(\pi k_7 x_n)]) + \sum_{i=1}^{n} u(x_i, 5, 100, 4)$$

$$k_3 = 0.1, k_4 = 3, k_5 = 3, k_6 = 1, k_7 = 2.$$

$$-5 \leq x_i \leq 5, i = 1, ..., 5.$$

Global Solution $f(x) = 0.0$, $x = (1.0, 1.0, 1.0, 1.0, 1.0)$.

This problem has approximately 15^5 local minima. The objective of this test was to test the performance of the LDS Sobol', LDS Holton and Random methods on problems with a very large number of minima. In tests with $N/N_r = 1024/512$ and $Iter_{max} = 3$ in all four runs LDS Sobol'(NS) and Random(NS) algorithms successfully located some 60 local minima including the global one (Table 8). There were only three successful runs of the LDS Holton (NS) algorithm. The number of minima located by the LDS Holton (NS) algorithm was also lower than that for other algorithms. This can be explained by the inferior uniformity properties of the Holton LDS even at moderate dimensions.

Algorithm	N/N_r	$Iter_{max}$	LM	N_{min}	GM	CPU
LDS Sobol'(NS)	1024/512	3	68	63	Y	2.1
LDS Holton(NS)	1024/512	3	53	51	y/n/y/y	2.4
Random(NS)	1024/512	3	61	58	Y	1.9

Table 8. Comparison of various realizations of MLSL for Problem 3B.

4.4 Problems 4A,B,C: Schaffler function

$$f(x)=1 + 590\sum_{i=2}^{n} (x_i - x_{i-1})^2 + 6x_1^2 - \cos(12x_1),$$

$-1.05 \leq \quad x_i \quad \leq 2.95$, i =1,. . . ,n.

Configurations:

Problem 4A. n=30.

Problem 4B. n=40.

Problem 4C. n=50.

Global Solution

$f(\boldsymbol{x}) = 0.0$,

$\boldsymbol{x} = \{\mathbf{0.0}\}$.

All three problems belong to the class of "moderate" (M2) [29]. The objective of this test with 5 known solutions was to test the performance of the LDS Sobol', LDS Holton and Random methods on high-dimensional problems. LDS have better uniformity properties than pseudorandom grids. However, this advantage diminishes as the dimensionality n increases. As explained in Section 2, for high-dimensional problems the usage of LDS still can be more efficient than pseudorandom sampling if an objective function $f(\boldsymbol{x})$ strongly depends only on a subset of variables. For such problems an effective dimension number s can be much smaller than n. However, this is not the case for the test problem 4. Apart from variable x_1, all other variables are equally important. This explains why LDS Sobol' and Random methods showed almost the same efficiency (Tables 9, 10, 11). Uniformity properties of Holton LDS rapidly degrade as n grows. Thus for high-dimensional problems the MLSL method based upon Holton LDS sampling becomes less efficient than a stochastic variant of MLSL. Values N and N_r given in Tables 4.1A, 4.1B, 4.1C are the smallest sample sizes for which a global minimum was found in all four runs for LDS Sobol' and Random methods.

Problem 4A. n=30

Algorithm	N/N_r	$Iter$	LM	N_{min}	GM	CPU (s)
LDS Sobol'	8192/256	1	220	3	Y	30.5
LDS Holton	8192/256	1	98	2	y/n/y/n	18.5
Random	8192/256	1	228	3	Y	31.2

Table 9. Comparison of various realizations of MLSL for Problem 4A.

For LDS Holton the global minimum was found in two out of four runs for Problem 4A and Problem 4B, while for Problem 4C this algorithm failed to locate it. A comparison between Problem 4A and Problem 4B for LDS Sobol' and Random shows that for successful location of the global minimum in all four runs it was necessary to increase N in two times and N_r - in 16 times. It resulted in a 30 fold increase of the CPU time.

Problem 4B. n=40

Increasing the dimensionality from n=40 to n=50 resulted in N increasing 64 fold. At the same time N_r increased only 4 fold and the CPU time increased approximately 8 fold.

Algorithm	N/N_r	*Iter*	LM	N_{min}	GM	CPU (s)
LDS Sobol'	16384/4096	1	4028	4	Y	$1.1{\cdot}10^3$
LDS Holton	16384/4096	1	1142	2	n/n/y/y	348.1
Random	16384/4096	1	4031	3	Y	$1.1{\cdot}10^3$

Table 10. Comparison of various realizations of MLSL for Problem 4B.

Problem 4C. n=50

Algorithm	N/N_r	*Iter*	LM	N_{min}	GM	CPU (s)
LDS Sobol'	1048576/16384	1	16206	3	Y	$7.5{\cdot}10^3$
LDS Holton	1048576/16384	1	5665	2	N	$3.3{\cdot}10^3$
Random	1048576/16384	1	16253	3	Y	$7.6{\cdot}10^3$

Table 11. Comparison of various realizations of MLSL for Problem 4C.

For Problem 4C N_r was nearly equal to LM. This is because clustering becomes less efficient as dimensionality grows. Choosing larger σ in (18) may increase the cluster size and thus decrease LM. However, for consistency with other tests experiments σ was kept equal to 2.0.

4.5 Multi-quality Blending Problems

These two blending problems from the petrolchemical industry [1], where they were named example 1 and example 2 respectively. Various types of crude oils of different qualities and coming from different sources are mixed together to produce several end-products subject to certain quality requirements and demands. These are bilinear problems with many local optima. The solution of this type of blending problem is important because of the direct application to the industrial world as well as for the mathematical challenges it poses. Here, we use the general blending problem formulation [1]:

$$
\begin{aligned}
&\min_{y,q,x} \sum_{j=1}^{p} \sum_{i=1}^{n_j} c_{ij} y_{ij} - \sum_{k=1}^{r} d_k \sum_{j=1}^{p} x_{jk}, \\
&\sum_{i=1}^{n_j} y_{ij} - \sum_{k=1}^{r} x_{jk} = 0 \qquad \forall j \le p \\
&q_{jw} \sum_{k=1}^{r} x_{jk} - \sum_{i=1}^{n_j} \lambda_{ijw} y_{ij} = 0 \qquad \forall j \le p, \forall w \le l \\
&\sum_{j=1}^{p} x_{jk} \le S_k \qquad \forall k \le r \\
&\sum_{j=1}^{p} q_{jw} x_{jk} - z_{kw} \sum_{j=1}^{p} x_{jk} = 0 \qquad \forall k \le r^{\forall} w \le l \\
&y^L \le y \le y^U, q^L \le q \le q^U, q^L \le q \le q^U
\end{aligned}
$$

Here y_{ij} is the flow of input stream i into pool j, x_{jk} is the total flow from pool j to product k and q_{jw} is the w-th quality of pool j; p is the number of pools, r the number of products, l the number of qualities, n_j the number of streams; $c_{ij}, d_k, S_k, Z_{kw}, \lambda_{ijw}$ are parameters [1].

The objective function value at the global optimum is -549.8031 in both cases. The values of the variables y, x are also the same in both cases:

y= (7.5443, 19.752, 0, 4.9224, 2.7812)

x= (0, 19.2097, 0, 8.0866, 0, 5.7903, 0, 1.9133)

whereas the values of the quality variables q change:

q' = (3.1708, 2.382, 3.2764, 1.5854, 2.278, 2.8917, 3.361, 1.2166)

q'' = (3.1708, 2.382, 3.2764, 1.5854, 4.2763, 5.382, 2.278, 2.8917, 3.361, 1.2166, 3, 5.083)

where q' are the values of q at the global optimum of example 1, and q'' are the corresponding values for example 2.

Results for this problem and problems 4.6–4.7 are summarized in one table (Table 12).

4.6 A Simple MINLP Example

The developed technique was generalized to account for mixed continuous and discrete variables (MINLP). The solution technique makes use of a continuous reformulation of the problem. Tests presented below have shown a good performance of the multistage methods based on LDS sampling for constrained MINLP problems

The following example was taken from paper [11]. It is an example of a nonlinear mixed-integer problem having one binary variable x_5.

$$\begin{aligned}
&\min x_1^2 + x_1x_2 - x_1x_3 - 2x_1x_4 + x_2^2 + 3x_2x_4 - x_2x_5 + x_3x_4 + \\
&\quad +3x_4 + 2x_4x_5 - x_1 - x_4 - x_6 + e^{-x_2x_3} \\
&x_1 + x_2 - x_3 + x_4 + x_5 = 1, \\
&x_2 - x_4 - x_5 = -1, \\
&x_1 + 2x_2 - 2x_3 \geq 0, \\
&2x_1 + 7x_2 - x_3 \leq 0, \\
&e^{-x_2x_3} - \ln(x_6) - x_2x_6 \leq 1, \\
&x_i \in [0, 10] \ \forall i \leq 4, \\
&x_5 \in \{0, 1\}, \\
&1 \leq x_6 \leq 2
\end{aligned}$$

The global solution is at $\boldsymbol{x}$ = (0, 0, 0, 0, 1, 2), with an objective function value -1.

4.7 Yuan's MINLPs

The following examples come from paper [8]. They are nonlinear mixed-integer problems.

Problem 4.7A

$$\begin{aligned}
&\min_{x,y} 2x_1 + 3x_2 + \tfrac{3}{2}y_1 + 2y_2 - \tfrac{1}{2}y_3 \\
&y_1 + x_1^2 = \tfrac{5}{4}, \\
&\tfrac{3}{2}y_2 + x_2^{3/2} = 3, \\
&y_1 + x_1 \le \tfrac{8}{5}, \\
&y_2 + \tfrac{4}{5}x_5 \le 3 \\
&-y_1 - y_2 + y_3 \le 0, \\
&x_i \in [0, 10] \forall i, \\
&y_i \in \{0, 1\} \forall i
\end{aligned} \tag{20}$$

The global solution of problem (20) is at x = (1.12, 1.31), y = (0, 1, 1) with an objective function value 7.6672.

Problem 4.7B

$$\begin{aligned}
&\min_{x,y}(y_1 - 1)^2 + (y_2 - 2)^2 + (y_3 - 1)^2 + \log(y_4 + 1) + (x_1 - 1)^2 + \\
&\quad +(x_2 - 2) + (x_3 - 3)^2 \\
&y_1 + y_2 + y_3 + x_1 + x_2 + x_3 \le 5, \\
&y_3^2 + x_1^2 + x_2^2 + x_3^2 \le \tfrac{11}{2}, \\
&y_1 + x_1 \le \tfrac{6}{5}, \\
&y_2 + x_2 \le \tfrac{9}{5}, \\
&y_3 + x_3 \le \tfrac{5}{2}, \\
&y_4 + x_1 \le \tfrac{6}{5}, \\
&y_2^2 + x_2^2 \le \tfrac{41}{25}, \\
&y_3^2 + x_3^2 \le \tfrac{17}{4}, \\
&y_2^2 + x_3^2 \le \tfrac{116}{25}, \\
&x_i \in [0, 10] \ \forall i, \\
&y_i \in \{0, 1\} \ \forall i
\end{aligned} \tag{21}$$

The global solution of problem (21) is at $\boldsymbol{x}$ = (0.2, 0.8, 1.908), $\boldsymbol{y}$ = (1, 1, 0, 1) with an objective function value 4.5796.

Algorithm	N/N_r	*Iter*	LM	N_{min}	GM	CPU(s)
Problem 4.5A	64/32	1	32	28	Y	0.01
Problem 4.5B	128/16	1	16	15	Y	0.01
Problem 4.6	8/4	1	4	2	Y	0.001
Problem 4.7A	16/8	1	5	4	Y	0.001
Problem 4.7B	4096/32	1	15	10	Y	0.001

Table 12. Comparison of various realizations of MLSL for Problems 4.5-4.7.

Problems 4.5-4.7 including relatively high dimensional blending problems and all mixed-integer were easy to solve. Problem 4.7B was the most difficult in terms of the required N/N_r, although the presented CPU time does not show the complexity of the problem because of its low dimensionality.

5 Conclusion

In this study QMC and stochastic variants of MLSL were compared. The Program SobolOpt employing the discussed techniques was applied to a number of test problems. When compared with other implementations of MLSL reported in the literature, it showed a superior performance. It was proved that application of LDS results in a significant reduction in computational time for low and moderately dimensional problems. Two different LDS were tested and their efficiency was analyzed. Uniformity properties of Holton LDS degrade as dimensionality grows and for high dimensional problems the MLSL method based on Holton LDS becomes less efficient than the stochastic MLSL method. Sobol' LDS can still be superior to pseudorandom sampling especially for problems in which an objective function strongly depends only on a subset of variables.

To increase the probability of finding the global minimum, the full sample size should be increased with the increase of the dimensionality of a problem. However, it may not be very practical if the objective function is difficult to evaluate. The ratio of the full/reduced sample size γ should be kept high to reduce the computational time. It was shown that the use of a large total number of sampled points is more efficient than that of a small one if the objective is to locate only a global minimum as opposed to locate as many local minima as possible.

Acknowledgments

The financial support of the United Kingdom's Engineering and Physical Sciences Research Council (EPSRC) under Platform Grant GR/N08636 is gratefully acknowledged.

References

1. N. Adhya, M. Tawarmalani, and N. Sahinidis, A lagrangian approach to the pooling problem, *Industrial and Engineering Chemistry Research*, **38**, 1956-1972 (1999).
2. I.I. Artobolevski, M.D. Genkin, V.K. Grinkevich, I.M. Sobol' and R.B. Statnikov, Optimization in the theory of machines by an LP-search, *Dokl. Akad. Nauk SSSR* (in Russian), **200**, 1287-1290 (1971).
3. C.G.E. Boender, The Generalized Multinomial Distribution: *A Bayesian Analysis and Applications,* Ph.D. Dissertation, Erasmus Universiteit Rotterdam, Centrum voor Wiskunde en Informatica (Amsterdam, 1984).
4. P. Bratley, B.L. Fox and H. Niederreiter, Implementation and tests of low-discrepancy sequences, *ACM Trans. Model, Comput. Simulation.,* **2**, 195-213 (1992) .

5. L.C.W. Dixon and M. Jha, Parallel algorithms for global optimization, *Journal of Optimization Theory and Applications*, **79**, no.1, 385-395 (1993).
6. C.A. Floudas and Panos M. Pardalos (eds.), *State of the Art in Global Optimization: Computational Methods and Applications* (Princeton University Press, 1996).
7. C. Floudas and Panos M. Pardalos (eds.), *Encyclopedia of Optimization* (Kluwer Academic Publishers, 2001).
8. C. Floudas, M. Pardalos Panos, C. Adjiman, W.R. Esposito, Z. Gumus, S.T. Harding, J.L. Klepeis, C.A. Meyer and C.A. Schweiger, *Handbook of Test Problems in Local and Global Optimization* (Kluwer Academic Publishers, 1999).
9. J.H. Holton, On the efficiency of certain quasi-random sequences of points in evaluating multi-dimensional integrals, *Numer. Math*, **2**, 84-90 (1960).
10. R. Horst and Panos M. Pardalos (eds.), *Handbook of Global Optimization* (Kluwer Academic Publishers, 1995).
11. L. Liberti, Linearity embedded in nonconvex programs, *Journal of Global Optimization* (to appear).
12. M. Locatelli and F. Schoen, Simple linkage: analysis of a threshold-accepting global optimization method, *Journal of Global Optimization*, **9**, 95-111 (1996).
13. NAG Library (2002); http://www.nag.co.uk.
14. H. Niederreiter, Point sets and sequences with small discrepancy, *Monatsh. Math.*, **104**, 273-337 (1987).
15. H. Niederreiter, *Random Number Generation and Quasi-Monte Carlo Methods* (SIAM, Philadelphia, 1992).
16. S. Paskov and J.F. Traub, Faster evaluation of financial derivatives, *The Journal of Portfolio Management*, **22**, no.1, 113-120 (1995).
17. A.H.G. Rinnooy Kan and G.T. Timmer, Stochastic global optimization methods,part I, *Clustering methods, Mathematical Programming,* **39**, 27-56 (1987).
18. A.H.G. Rinnooy Kan and G.T. Timmer, Stochastic global optimization methods, part II, *Multilevel methods, Mathematical Programming,* **39**, 57-78 (1987).
19. F. Schoen, Random and quasi-random linkage methods in global optimization, *Journal of Global Optimization*, **13**, 445-454 (1998).
20. I.M. Sobol', *Primer for the Monte Carlo Method* (CRC press, Florida, 1994).
21. I.M. Sobol', On quasi-Monte Carlo integrations, *Mathematics and Computers in Simulation*, **47**, 103-112 (1998).
22. I.M. Sobol' On the systematic search in a hypercube, *SIAM J. Numer. Anal.*, **16**, 790-793 (1979).
23. I.M. Sobol', On an estimate of the accuracy of a simple multidimensional search, *Soviet Math. Dokl.*, **26**, 398-401 (1982).
24. I.M. Sobol', On the distribution of points in a cube and the approximate evaluation of integrals, *Comput. Math. Math. Phys.*, **7**, 86-112 (1967).
25. I.M. Sobol', On the search for extreme values of functions of several variables satisfying a general Lipschitz condition, *USSR Comput. Math. Math. Phys.*, **28**, 112-118 (1988).
26. I.M. Sobol', An efficient approach to multicriteria optimum design problems, *Survey Math. Ind.*, **1**, 259-281 (1992).
27. A.G. Sukharev, Optimal strategies of the search for an extremum, *Zh. Vychisl. Mat. i Mat. Fiz* (in Russian), **11**, 910-924 (1971).
28. A. Törn and A. Žilinskas, *Global Optimization, Lecture Notes in Computer Science* , **350** (Springer, Berlin, 1989).

29. A. Törn, M. Afi and S. Vjitanen, Stochastic global optimization, Problem Classes and Solution Techniques, *Journal of Global Optimization*, **14**, 437-447 (1999).
30. J.F. Traub and H. Wozniakowski, *A General Theory of Optimal Algorithms* (Academic Press, New York, 1980).
31. S. Zakovic, Global Optimization *Applied to an Inverse Light Scattering Problem, PhD Thesis*, University of Hertfordshire (Hartfield, 1997).
32. S. Zakovic, Z. Ulanowski, and M C. Bartholomew-Biggs, Application of global optimisation to particle identification using light scattering, *Inverse Problems*, **14**, 1053-1067 (1998).

Part II: Implementations

GLOB – A new VNS-based Software for Global Optimization

M. Dražić[1], V. Kovacevic–Vujcić[2], M. Cangalović[2], and N. Mladenović[34]

[1] Faculty of Mathematics, University of Belgrade, Belgrade mdrazic@sezampro.yu
[2] Faculty of Organizational Sciences, University of Belgrade, Belgrade {verakov,canga}@fon.bg.ac.yu
[3] Mathematical Institute, Serbian Academy of Arts and Sciences, Belgrade nenad@mi.sanu.ac.yu
[4] GERAD and Ecole des Hautes Commerciales, Montreal

Summary. We describe an application of Variable Neighbourhood Search (VNS) methodology to continuous global optimization problems with box constraints. A general VNS algorithm is implemented within the software package GLOB. The tests are performed on some standard test functions and on a class of NP–hard global optimization problems arising in practice. The computational results show the potential of the new software.

Key words: Metaheuristics, variable neighborhood search.

1 Introduction

Global optimization problems have the form

$$\operatorname*{global\,min}_{x \in X} f(x)$$

where $f : R^n \to R$ is a continuous function on an open set containing X and X is a compact set. In most cases of practical interest global optimization is very difficult because of the presence of many local minima, the number of which tends to grow exponentially with the dimension of the problem. Besides, in general it is only possible to design methods that offer an ε-guarantee to find the global minimum. Nevertheless, a number of methods for global optimization problems have been proposed, both deterministic and nondeterministic (for a comprehensive bibliography see [15, 16, 18]).

There are two common approaches to finding the global minimum. The first, so called Multistart[5] Local Search (MS) consists of generating a set of

[5] Also see Chapters 5, 8 (Sections 2.1, 7.1).

random points and using them as starting points of some conventional minimization technique converging to a local minimum. In order to achieve a sufficiently large probability of finding the global minimum, a large number of starting points must be tried. This strategy is time consuming and soon becomes intractable as the dimensionality of the problem increases. The second approach is to design methods which avoid entrapments in local minima and continue search to give near-optimal solutions. The methods of this type are Simulated Annealing (SA) and Genetic Algorithms (GA). The concept of SA is inspired by statistical physics and is in essence a numerical simulation of solids, where by slowly decreasing the temperature the system is being transformed to a state of minimum energy. The idea of GA relies on the Darwinian principle of evolution. GA algorithms crossbreed different trial solutions and allow only the best to survive after several iterations. SA has been applied to a wide range of chemical problems ([4, 8, 20, 21, 22]). GA has been used for finding low–energy conformations of small molecules, predicting the structure of clusters, modeling polymers and proteins [9, 19] and for molecular mechanics calculations [12]. Both Tabu Search (TS) and Variable Neighborhood Search[6] (VNS) belong to intelligent problem–solving algorithms. TS for continuous global optimization has been proposed by Glover in 1994 [5] and has been successfully applied to various real–world problems giving often better results than SA and GA [1, 2, 10]. The VNS approach to continuous global optimization is the most recent [13, 14]. Our work has been motivated by the fact that VNS, [6, 7] for discrete optimization, is conceptually very simple and depends basically on one parameter which determines the number of different neighborhoods in the VNS strategy. The simplicity allows very efficient and flexible implementation.

It is the purpose of this contribution to present the software package GLOB which is based on VNS methodology for global optimization. GLOB is designed primarily as a test platform for comparing VNS strategies with different neighborhood structures and local optimizers. The current version of GLOB handles only problems with box constraints. Future work[7] will focus on extensions to general global optimization problems.

2 VNS methodology

The basic idea of VNS metaheuristic is to use more than one neighborhood structure and to proceed to a systematic change of them within a local search. The algorithm remains in the same solution until another solution better than the incumbent is found and then jumps there. Neighborhoods are usually ranked in such a way that intensification of the search around the current

[6] Also see Chapters 11, 8 (Sections 2.2, 7.2).

[7] Part of this undertaking has already been carried out in the VNS solver implementation of the *ooOPS* software framework, see Chapter 8, Section 7.2.

solution is followed naturally by diversification. The level of intensification or diversification can be controlled by a few easy to set parameters. We may view the VNS as a "shaking" process, where a movement to a neighborhood further from the current solution corresponds to a harder shake. Unlike random restart, the VNS allows a controlled increase in the level of the shake. Let us denote by N_k, $k = 1, ..., k_{\max}$ a finite sequence of pre-selected neighborhood structures, and by $N_k(x)$ the set of feasible solutions corresponding to neighborhood structure N_k at the point x, where x is an initial solution. Let us note that most local search metaheuristics use one neighborhood structure, i.e. $k_{\max} = 1$. The following algorithm presents steps of the basic VNS heuristic.

Repeat until the stopping criterion is met:

(1) Set $k \leftarrow 1$
(2) Until $k > k_{\max}$ repeat the following steps:
  (a) *Shaking:* Generate a point x' at random from $N_k(x)$
  (b) *Local search:* Apply some local search method with x' as the initial solution; denote by x'' the so obtained local minimum.
  (c) *Move or not:* If x'' is better than the incumbent move there ($x \leftarrow x''$) and set $k \leftarrow 1$; otherwise set $k \leftarrow k + 1$

The stopping criterion may be e.g. the predetermined maximal allowed CPU time, the maximal number of iterations, or the maximal number of iterations between two improvements. Let us note that the point x' is generated in Step 2(a) at random in order to avoid cycling which might occur if any deterministic rule was used.

3 Software package GLOB

The software package GLOB is a stand–alone solver for minimization of a continuous function subject to box constraints. The code is written in ANSI C programming language and consists of approximately 7000 lines. The current version of the package implements Variable Neighborhood Search, Multistart Local Search and Random Search (Monte Carlo Method) with possibility to add other global optimization heuristics. The package offers various statistical reporting facilities convenient for research purpose. Thus, GLOB is not primarily designed to be a commercial software with speed as the main objective, but rather as a tool for better understanding of heuristics in global optimization problems.

3.1 Overview of GLOB

GLOB is designed to seek for the global minimum of a continuous (or smooth) function in a finite dimensional box-constrained region:

$$\min f(x_1, x_2, ..., x_n), \quad a_i \leq x_i \leq b_i, \quad i = 1, ..., n$$

Number of space variables is set to maximum of 200 in the current version, but it can be easily increased.

User defines the function f by a C function

```
double user_function ( double *xcoordinates )
```

and, optionally, its gradient by

```
void user_gradient ( double *xcoordinates,
                     double *gradient )
```

If the function to be minimized depends on some parameters with fixed values in minimization process, as in some test functions (e.g. Baluja), their values can be passed to functions via `fun_params` option in job parameter file which defines the values in a global array `Fun_Params[]` that can be used in the user function.

There is a number of well known test functions already built in the package and recognized by their name. In the present time they are: Rosenbrock, Shekel, Hartman, Rastrigin, Griewank, Shubert, Branin, Goldstein and Price, B2, Martin and Gaddy, Baluja f_1, f_2 and f_3. More test functions can be added easily.

The package is designed to be run in a batch mode, so all parameters are defined in a couple of parameter files: main parameter file and job parameter file. Example of those files are given in Appendix 1.

In main parameter file (default name `glob.cfg`) user specifies which job parameter file will be used, the name of the output file and the working directory. The job parameter file contains test function details, various options, limits, parameters as well as reporting options for that job. This approach makes it possible to submit several optimization jobs in batch mode.

3.2 Built–in heuristics and local optimizers

There are three heuristics built in the current version of the package: Random Search (Monte-Carlo), Multistart Local Search (MS), and VNS. In the last two new random points are initial points for a chosen local minimizer. In MS a new random point is generated by the uniform distribution in the whole box region. In VNS new random point is chosen from a series of neighborhoods of the best found optimal point. There are several parameters which define the type of neighborhoods and random distributions (metrics) used for getting the next random point. Local minimizers used in VNS and MS are well known methods Nelder-Mead (NM), Hooke-Jeeves (HJ), Rosenbrock (RO), Steepest Descent (SD), Fletcher-Powell (FP) and Fletcher-Reeves (FR). The first three do not require gradients and can be used for nonsmooth objective functions. The other three methods use information on the gradient, which either can be user supplied or approximately calculated by finite difference method. Many local minimizers use one dimensional optimizer which can be set to Golden Section Search (GS) or Quadratic Approximation method (QA).

3.3 Summary of GLOB options

A typical job parameter file is presented in Appendix 1.

The first group of parameters describe the optimization problem. Setting `function_name` user can choose one of the built–in test functions or define a name for a user supplied function. If a function depends on on some fixed parameters (Baluja for example), the number of parameters can be set by `no_fun_params` and their values in `fun_params`. These values are stored in a global array `Fun_Params[]` and can be used in the user or test function. Box boundary values are set by `left_boundaries` and `right_boundaries` following by an array of values. Initial point is set by `initial_point`. If all the boundary or initial point values are the same, that value is simple entered in `left_boundaries_all`, `right_boundaries_all` or `initial_point_all`. This is very useful for test problems with large space dimension.

If the user knows the exact (or putative) global minimum, he can set it by `glob_known_minimum` and percentage of the difference between current and the optimal function values will be reported. Parameter `boundary_tol` is used for identification of hitting the boundary in a local optimization method. Gradient methods switch to projected gradient method when the point reaches the boundary. The value `significant_fun_difference` is used to distinguish minor function value improvements (mostly around the same local minimum) from more significant ones and is used only in statistical reporting.

The program will in one optimization job try to find a better minimum until `glob_time_limit` seconds are elapsed or `glob_max_iteration` of metaiterations are done. For finding average and best results in number of repeated job runs, user can set `glob_job_repetitions` job runs to be executed automatically with related statistics reported. For portability reasons, the program uses its own uniform random number generator. The seed for this generator can be set in `random_seed`. If set to 0, the seed will be initiated by the system clock. Seed choices not equal to 0 enable the same random point trajectory to be repeated in numerical experiments. For VNS metaheuristics the package can use three different types of neighborhood structures. The first type is L_∞ ball with the uniform distribution in it. The second and the third are L_1 balls with the uniform and a specially designed distribution, respectively. Setting `random_distributions = 1 1 1` means that the three types are changing cyclically, while e.g. `1 0 0` means that only the first neighborhood structure type is used.

The next group of parameters defines the VNS neighborhood structures: `vns_kmax` is the number of balls of different sizes centered at the current point. The sizes of balls can be defined by the user or automatically generated by the software. Two parameters `random_from_disk`, `reject_on_return` and `reject_on_return_k_diff` control random point placement and prevention of returning to the current local minimum. E.g. if the user sets `reject_on_return` to 1, any local optimization started from N_k neighborhood will be terminated if the point returns to N_{k-m} neighborhood, where m is set by `reject_on_return_k_diff`. If `random_from_disk = 1`, then random point is chosen from disk $N_k \setminus N_{k-1}$.

Eight parameters prefixed with `rep_` control the reporting which is described in more details in the next section.

The rest of parameters control the local minimizer used in each metaiteration. The local minimizer terminates if one of the following occurs: number of iterations exceeds some maximal value `ls_nelder_mead_maxiter` in this example), two consecutive iteration points are closer than `ls_eps`, two consecutive function values are closer than `ls_fun_eps` or the norm of the gradient (if used) is less than `ls_grad_eps`.

3.4 Statistical reporting facilities

In one program execution, one or more jobs can be run. If more jobs are run, the best and average optimal results are reported. The user can choose the level of reporting by enabling or disabling certain information to be presented in output file and console screen. The user can get information on every metaiteration or only on successful ones. For each metaiteration, the user can get random point coordinates, coordinates of the best point in local optimization, values of VNS auto recalculated radius values, and some other statistics. At the end of a job (terminated by reaching maximum metaiterations or time limit), statistics on: number of metaiterations, elapsed time, computing effort (number of function and gradient calls) is presented for: overall metaiterations, until the last successful metaiteration and until the last significant metaiteration. The last significant metaiteration is the one after which no significant (defined by an input parameter) function improvement is obtained. This concept recognizes small improvements in the same best found local minimum after the moment when that local minimum is practically found. For the VNS metaheuristic, the number of successful steps is presented for every neighborhood from where that better function value was reached. Also, statistics on number of successful steps for various random distributions (metrics) used for obtaining new starting point is reported.

As has been mentioned already, if the user knows the exact value of global minimum (or putative global minimum) he can set it in a parameter file and % deviation of the current best and the best known objective function value will be calculated as $(f - f_{best})/f_{best} \times 100$ for each reported iteration.

3.5 GLOB Output

For the job parameter file listed in Appendix 1, the output file is presented in Appendix 2.

From this output it could be seen that the better optimum (0.12% better) was found than it was known up to that moment. Also, the best local minimum was found in iteration 1523 (of total 2747 in this 30s time limited job run) but it is not significantly better than the minimum obtained already in iteration 381 for less than 5s. Also one can find that random distribution options 1 and 3 performed better than 2 in this example.

VNS statistics at the end of the report shows that there were 12 successful metaiterations from the first neighborhood (close to current local minimum, mostly slight improvements), and another 12 were successful from more distant starting points mostly leading to other local minima.

4 Numerical experiments

The power of the package GLOB described in Section 2 was tested on three classes of global optimization problems. Experiments were performed on Intel Pentium III processor, 930MHz, 1.00 GB of RAM. Only the first type of neighborhood structures (L_∞ with uniform distribution) was used, i.e. neighborhoods were hypercubes of different sizes centered at the current point, where the sizes were automatically generated. The number of cubes `vns_kmax` was varied in order to investigate its influence on the efficiency of VNS. Note that for `vns_kmax = 1` VNS can be viewed as Multistart Local Search.

4.1 Standard test functions

The VNS heuristic was primarily designed for global optimization problems with large number of local minima, where Multistart Local Search suffers of central limit catastrophy. Nevertheless GLOB was also applied to some standard test functions, which is the usual procedure for newly proposed optimization methods.

These functions are [2, 18], :

BR – Branin ($n = 2$):

$$f(x) = \left(x_2 - \frac{5.1}{4\pi^2}x_1^2 + \frac{5}{\pi}x_1 - 6\right)^2 + 10\left(1 - \frac{1}{8\pi}\right)\cos x_1 + 10,$$
$$-5 \leq x_1 \leq 10, \quad 0 \leq x_2 \leq 15.$$

The global minimum is approximately 0.3979.

GP – Goldstein–Price ($n = 2$):

$$\begin{aligned} f(x) = {} & [1 + (x_1 + x_2 + 1)^2(19 - 14x_1 + 3x_1^2 - 14x_2 + 6x_1x_2 + 3x_2^2)] \times \\ & \times [30 + (2x_1 - 3x_2)^2(18 - 32x_1 + 12x_1^2 + 48x_2 - 36x_1x_2 + 27x_2^2)] \\ & -2 \leq x_i \leq 2, \quad i = 1, 2. \end{aligned}$$

The global minimum is equal to 3.

HTn – Hartman ($n = 3, 6$):

$$f(x) = -\sum_{i=1}^{4} c_i \exp\left[-\sum_{j=1}^{n} a_{ij}(x_j - p_{ij})^2\right],$$
$$0 \le x_i \le 1, \quad i = \overline{1,n}.$$

The values of parameters c_i, a_{ij}, p_{ij} can be found in [16]. For $n = 3$ the global minimum is equal to -3.8628, while for $n = 6$ the minimum is -3.3224.

RS – Rastrigin ($n = 2$):

$$f(x) = x_1^2 + x_2^2 - \cos 18x_1 - \cos 18x_2,$$
$$-1 \le x_i \le 1, \quad i = 1, 2.$$

The global minimum is equal to -2.

SB – Shubert ($n = 2$):

$$f(x) = \left[\sum_{i=1}^{5} i\cos((i+1)x_1 + i)\right]\left[\sum_{i=1}^{5} i\cos((i+1)x_2 + i)\right],$$
$$-10 \le x_i \le 10, \quad i = 1, 2.$$

The global minimum is -186.7309.

ROn – Rosenbrock ($n = 2, 10$):

$$f(x) = \sum_{i=1}^{n-1} \left(100(x_{i+1} - x_i^2)^2 + (1 - x_i)^2\right)$$
$$-10 \le x_i \le 10.$$

The global minimum is 0.

GRn – Griewank ($n = 2, 10$):

$$f(x) = \sum_{i=1}^{n} x_i^2/d - \prod_{i=1}^{n} \cos(x_i/\sqrt{i}) + 1.$$

For $n = 2$, $d = 200$ and $-100 \le x_i \le 100$, $i = \overline{1,n}$, the global minimum is 0. For $n = 10$, $d = 4000$, $-600 \le x_i \le 600$, $i = \overline{1,n}$, the global minimum is also 0.

SHm – Shekel ($n = 4; m = 5, 10$):

$$f(x) = -\sum_{i=1}^{m} \frac{1}{(x-a_i)^T(x-a_i)+c_i}$$

$$x = (x_1, x_2, x_3, x_4)^T, \quad a_i = (a_i^1, a_i^2, a_i^3, a_i^4)^T, \quad 1 \le m \le 10.$$

i	a_i^T				c_i
1	4	4	4	4	0.1
2	1	1	1	1	0.2
3	8	8	8	8	0.2
4	6	6	6	6	0.4
5	3	7	3	7	0.4
6	2	9	2	9	0.6
7	5	5	3	3	0.3
8	8	1	8	1	0.7
9	6	2	6	2	0.5
10	7	3.6	7	3.6	0.5

The global minimum is -10.1532 for $m = 5$ and -10.53641 for $m = 10$.

MG – Martin and Gaddy ($n = 2$):

$$f(x) = (x_1 - x_2)^2 + \left(\frac{x_1 + x_2 - 10}{3}\right)^2$$

$$-20 \le x_i \le 20, \quad i = 1, 2.$$

The global minimum is 0.

B2 ($n = 2$):

$$f(x) = x_1^2 + 2x_2^2 - 0.3\cos(3\pi x_1) - 0.4\cos(4\pi x_2) + 0.7$$

$$-100 \le x_i \le 100, \quad i = 1, 2.$$

The global minimum is 0.

Bf1 – Baluja B_{f_1} ($n = 100, \epsilon = 10^{-5}$):

$$f(x) = -\left(\epsilon + \left| |y_1| + \sum_{i=1}^{n} |y_i| \right| \right)^{-1}, \quad y_1 = x_1, \quad y_i = x_i + y_{i-1},$$

$$-2.56 \le x_i \le 2.56, \quad i = 1, 2, ..., n.$$

The global minimum is $-1/\epsilon = -100000$.

The results of GLOB are summarized in Table 1. The first and the second column contain the test function code and the local minimizer code, respectively. For each of the values `vns_kmax = 1, 2, 3` the average and the best objective function value in 100 runs are given. The computational effort is measured by the average number of function evaluations during the search process used to obtain the first global minimum with precision 10^{-4}, i.e. to satisfy the test $|f(x) - f_{\min}| < 10^{-4}$. Here the number of function evaluations is computed as number of objective function evaluations plus $n\times$ number of gradient evaluations (if any). It can be seen that in most cases function values and computational effort are improved when **vns_kmax** is increased.

function	local minimizer	vns_kmax = 1			vns_kmax = 2			vns_kmax = 3		
		average value	best value	comp. effort	average value	best value	comp. effort	average value	best value	comp. effort
BR	FR+QA	0.3979	0.3979	411	0.3979	0.3979	287	0.3979	0.3979	164
GP	NM	3.0000	3.0000	195	3.0000	3.0000	266	3.0000	3.0000	297
HT3	RO	-3.8627	-3.8628	1305	-3.8627	-3.8628	823	-3.8628	-3.8628	648
HT6	RO	-3.3224	-3.3224	1138	-3.3224	-3.3224	619	-3.3224	-3.3224	586
RS	NM	-1.9963	-2.0000	2364	-2.0000	-2.0000	1411	-2.0000	-2.0000	1194
SB	FP+QA	-186.7309	-186.7309	1112	-186.7309	-186.7309	1181	-186.7309	-186.7309	1047
RO2	RO	0.0000	0.0000	711	0.0000	0.0000	599	0.0000	0.0000	552
RO10	RO	0.0328	0.0000	172905	0.0120	0.0000	135216	0.0055	0.0000	113677
GR2	SD+QA	0.0000	0.0000	787	0.0000	0.0000	757	0.0000	0.0000	990
GR10	SD+QA	0.0000	0.0000	1734	0.0000	0.0000	1430	0.0000	0.0000	1338
SH5	RO	-10.1532	-10.1532	3060	-10.1532	-10.1532	1073	-10.1532	-10.1532	955
SH10	RO	-10.5364	-10.5364	3341	-10.5364	-10.5364	1235	-10.5364	-10.5364	1207
MG	NM	0.0000	0.0000	218	0.0000	0.0000	102	0.0000	0.0000	90
B2	RO	0.0000	0.0000	705	0.0000	0.0000	611	0.0000	0.0000	536
Bf1	HJ	-0.0070	-0.0063	20160399	-3311.77	-3076.17	32121892	-26541.98	-21903.75	35030774

Table 1. Standard test functions.

The figures in Table 1 show that GLOB gives satisfactory results for all of the functions except for Bf1 where obtained function values are far from the global minimum although the reached solutions are very close to the optimal one. This is due to the shape of the function with very narrow and very deep minimum. The results for Bf1 were obtained using manually defined sizes of neighborhoods (as application of automatically generated sizes gave substantially worse results). In order to reach better solutions, experiments with other types of neighborhood structure should be performed.

4.2 Some large real-life problems

Problems of optimal design are natural field of application for global optimization algorithms. Such an engineering problem arises in the spread spectrum radar polyphase code design [3]. This problem is modelled as a min–max nonlinear nonconvex optimization problem with box constraints and exponentially growing number of local minima. It can be expressed as follows:

$$\operatorname*{global\,min}_{x \in X} \; f(x) \equiv \max\{\varphi_1(x), ..., \varphi_{2m}(x)\}$$

$$X = \{(x_1, ..., x_n) \in R^n \mid 0 \le x_j \le 2\pi, \; j = 1, ..., n\}$$

where $m = 2n - 1$ and

$$\varphi_{2i-1} = \sum_{j=i}^{n} \cos\left(\sum_{k=|2i-j-1|+1}^{j} x_k \right), \quad i = 1, ..., n$$

$$\varphi_{2i}(x) = 0.5 + \sum_{j=i+1}^{n} \cos\left(\sum_{k=|2i-j|+1}^{j} x_k \right), \quad i = 1, ..., n-1$$

$$\varphi_{m+i}(x) = -\varphi_i(x), \; i = 1, ..., m.$$

It is proved in [9] that this problem is NP-hard.

In our tests Nelder–Mead method is used as a local minimizer. The dimension n varies from 7 to 20, while $k_{\max}$ takes values 1, 5, 10 and 15. Stopping criterion is CPU time $t_{\max}$, which varies with n from 180 sec for $n = 7$ to 960 sec for $n = 20$.

The results of experiments are summarized in Table 2. For each dimension % deviation of the average and the best objective function value from f_{best} in 10 runs are calculated, where % deviation of objective function value f from f_{best} is defined as $(f - f_{best})/f_{best} \times 100$. Here f_{best} is the best objective function value obtained in [14] by variants of VNS and TS based heuristics.

The results in Table 2 show that the perfomance of VNS depends on the number of neighborhood structures. For all dimensions both the average and the best solutions are improved when `vns_kmax` is increased. This experimentally verifies the conjecture that adding more neighborhood structures

n	t_{max} (sec)	xvs_kmax = 1		vns_kmax = 5		vns_kmax = 10		vns_kmax = 15		f_{best}
		average %	best %	average %	best %	average %	best %	average %	best %	
7	180	-0.07	-0.18	0.35	-0.17	-0.13	-0.17	0.84	-0.15	0.4972
8	240	17.29	7.82	9.25	-0.09	6.31	0.04	5.04	-0.02	0.3871
9	300	55.35	16.41	24.40	2.65	23.41	0.18	30.76	-1.71	0.3290
10	360	56.28	26.94	23.95	9.80	22.51	4.85	15.80	1.77	0.4105
11	420	77.87	54.27	30.33	0.47	19.38	-5.53	16.85	-7.86	0.4024
12	480	63.62	36.43	37.30	20.53	33.36	9.80	23.95	3.11	0.4907
13	540	78.42	54.85	61.68	29.09	46.79	0.64	28.45	-2.22	0.4899
14	600	111.35	84.17	83.16	58.75	70.57	29.35	68.86	29.09	0.4746
15	660	132.38	114.69	95.48	66.94	89.66	59.51	82.43	42.27	0.4857
16	720	189.38	147.92	149.76	117.51	125.54	72.23	131.81	67.89	0.4126
17	780	100.72	68.62	80.89	55.64	48.43	31.19	49.47	13.58	0.6334
18	840	118.61	109.28	85.84	59.12	72.84	26.70	60.37	30.46	0.6404
19	900	164.15	134.61	146.66	91.65	91.49	48.94	81.28	24.44	0.5617
20	960	162.69	148.08	115.48	80.40	72.63	40.51	74.10	34.68	0.6909

Table 2. Radar polyphase code design.

improves perfomance of the search. It should be noted that GLOB has found better solutions than f_{best} in 5 cases ($n = 7, 8, 9, 11, 13$). In all other cases the VNS and TS implementations described in [14] have given better results. This can be explained by the fact that these implementations explicitly use min–max structure of the radar polyphase code design problem, while GLOB is a general purpose software. Namely, results in [14] were obtained with a specially designed local search minimizer which uses information on gradients of functions φ_i active at the current point, while results in Table 2 were obtained with the classical Nelder–Mead method.

The results in Table 2 give an impression that dependence of both the average and the best solutions on `vns_kmax` is monotone and that further increase of this parameter should provide further improvements. Table 3 gives deeper insight in that matter. It contains the average and the best solutions for the radar polyphase code design problem of dimension $n = 20$ and `vns_kmax` $= \overline{1, 25}$. It can be seen that dependence of the average solution on `vns_kmax` is strictly monotone for `vns_kmax` =1 (162.69%) to `vns_kmax = 9` (65.54%) and after that it is oscillating between 43.46% (`vns_kmax = 22`) and 87.01% (`vns_kmax = 13`).

The next class of functions is proposed in [10] for testing global minimization of molecular potential energy. It has the form:

$$f(x) = \sum_{i=1}^{n} \left(1 + \cos 3x_i + \frac{(-1)^i}{\sqrt{10.60099896 - 4.141720682 \cos x_i}} \right),$$

$$0 \le x_i \le 5, \quad i = \overline{1, n}.$$

The results of the experiments for $n = 50, 100, 150, 200$ in 10 runs are summarized in Table 4. In all cases the local minimizer was Steepest Descent

$n = 20$	xvs_kmax											
	1	2	3	4	5	6	7	8	9	10	11	12
average %	162.69	134.39	127.53	122.45	115.48	96.68	90.41	84.91	65.54	72.63	58.40	79.92
best %	148.08	93.27	91.69	108.07	80.40	67.14	45.47	22.68	15.83	40.51	17.28	58.19

$n = 20$	xvs_kmax												
	13	14	15	16	17	18	19	20	21	22	23	24	25
average %	87.01	69.99	74.10	50.42	69.82	59.79	45.67	56.08	64.70	43.46	57.54	73.40	62.86
best %	65.17	37.41	34.68	15.23	12.38	35.71	12.58	26.34	30.49	-5.32	24.87	47.16	37.45

Table 3. Dependence on `vns_kmax`.

method with Quadratic Approximation method for one–dimensional optimization. Column f_{best} contains exact values of global minima. It can be seen that parameter **vns_kmax** has again a strong influence on the quality of GLOB results which further confirms the given conjecture.

5 Conclusion

We describe a new VNS–based software package GLOB for minimization of a continuous function subject to box constraints. The power of GLOB was tested of three classes of problems with encouraging results. The first class consists of some standard test functions used in the literature, while the other two are global optimization problems arising in practice. In all cases numerical

n	t_{max} (sec)	xvs_kmax = 1 average %	best %	vns_kmax = 5 average %	best %	vns_kmax = 10 average %	best %	vns_kmax = 15 average %	best %	f_{best}
50	20	84.17	70.64	41.82	27.89	7.15	3.97	0.00	0.00	-2.0559
100	30	116.97	104.85	74.46	65.65	45.80	41.79	1.99	0.00	-4.1118
150	40	130.30	114.93	80.68	75.65	56.41	45.16	3.45	0.00	-6.1677
200	50	137.67	126.94	81.38	69.62	66.79	59.72	7.56	3.99	-8.2237

Table 4. Molecular potential energy function.

experiments verify the conjecture that adding more neighborhood structures of type l_∞ improves performance of the search. Future work will focus on experiments with other types of neighborhood structures. Extensions of GLOB to global optimization problems subject to general nonlinear constraints are also planned.

References

1. Cvijović, D., and Klinowski, J., Taboo search: an approach to the multiple minima problem, *Science*, 267 (1995) 664–666.
2. Cvijović, D., and Klinowski, J., Taboo search: An approach to the multiple–minima problem for continuous functions, in: Pardalos, P. M., and Romeijn, H. E. (eds.), *Handbook of Global Optimization*, Kluwer, 2002, 387–406.
3. Dukić, M. L., and Dobrosavljević, Z. S., A method of a spread–spectrum radar polyphase code design, *IEEE Journal on Selected Areas in Comm*, 5 (1990) 743–749.
4. Dutta, P., Majumdar, D., and Bhattacharyya, S. P., Global optimization of molecular geometry: A new avenue involving the use of Metropolis simulated annealing, *Chemical Physics Letters*, 181 (1991) 293–297.
5. Glover, F., Tabu search for nonlinear and parametric optimization, *Discrete Applied Mathematics*, 1993.
6. Hansen, P., and Mladenović, N., An introduction to VNS, in: S. Voss et al. (eds.), *Metaheuristics, Advances and Trends in Local Search Paradigms for Optimization*, Kluwer, 1998, 433–458.
7. Hansen, P., and Mladenović, N., Variable neighborhood search in: F. Glover and G. Kochenberger, *Handbook of Mathematics*, Kluwer, 2003, 145–184.
8. Hohl, D., Jones, R. O., Car, R., and Parrinello, M., The structure of selenium clusters: Se3 to Se8, *Chemical Physics Letters*, 139 (1987) 540–545.
9. Judson, R. S., Teaching polymers to fold, *Journal of Physical Chemistry*, 96 (1992) 10102–101104.
10. Kovačević–Vujčić, V., Čangalović, M., Ašić, M., Ivanović, L., and Dražić, M., Tabu search methodology in global optimization, *Computers and Mathematics with Applications*, 37 (1999) 125–133.
11. Lavor, C., and Maculan, N., A function to test methods applied to global minimization of potential energy of molecules, to appear in *Numerical Algorithms*.
12. Linert, W., Margl, P., and Lukovits, I., Numerical minimization procedures in molecular mechanics: structural modeling of the solvation of β–cyclo–dextrin, *Computers & Chemistry*, 16 (1992) 61–69.

13. Mladenović, N., Dražić, M., Čangalović, M., and Kovačevic–Vujčić, V., Variable neighborhood search in global optimization, in: N. Mladenovic, D. Dugosija (eds.), *Proceedings of XXX Symposium on Operations Research, SYM-OP-IS 2003*, Herceg Novi, 2003, 327–330.
14. Mladenović, N., Petrović, J., Kovačević–Vujčić, V., and Čangalović, M., Solving spread-spectrum radar polyphase code design problem by tabu search and variable neighborhood search, *European Journal of Operations Research*, 153 (2003) 389–399.
15. Pardalos, P. M., and Romeijn, H. E. (eds.), *Handbook of Global Optimization*, Kluwer, 2002.
16. Pardalos, P.M., and Rosen, J. B., *Constrained Global Optimization: Algorithms and Applications*, Springer-Verlag, 1987.
17. Petrić, J., Zlobec, S., *Nelinearno programiranje*, Naučna knjiga, 1983.
18. Torn, A., and Zilinskas, A., *Global Optimization*, Springer–Verlag, 1987.
19. Unger, R., and Moult, J., Genetic algorithms for protein folding simulations, *Journal of Molecular Biology*, 231 (1993) 75–81.
20. Van Laarhoven, P. J. M., and Aarts, E. H. L., *Simulated Annealing: Theory and Applications*, Kluwer, 1987.
21. Wille, L.T., Minimum energy configurations of atomic clusters: new results obtained by simulated annealing, *Chemical Physics Letters*, 133 (1987) 405–410.
22. Xiao, Y. L., and Williams, D. E., Minimum energy configurations of atomic clusters: New results obtained by simulated annealing, *Chemical Physics Letters*, 215 (1993) 17–24.

Appendix 1 GLOB parameter files

There are two parameter files: main parameter file and job parameter file. The contents of a typical main parameter file (`glob.cfg`) looks as:

```
# Main configuration file for GLOB

# Job directory defines where to save (some) files.
# Directory must have '\' at the end.

job_name = glob;
job_directory = \projects\glob\data\;

# Output file contains copy of the screen output.
# Parameter file defines file with job parameters.
#
output_file = \projects\glob\data\out.txt;

parameter_file = \projects\glob\data\RadarOdd.cfg;
#parameter_file = \projects\glob\data\Shubert.cfg;
#parameter_file = \projects\glob\data\Branin.cfg;
#parameter_file = \projects\glob\data\Goldstei.cfg;
#parameter_file = \projects\glob\data\Shekel4.cfg;
```

```
#parameter_file = \projects\glob\data\BalujaF1.cfg;
#parameter_file = \projects\glob\data\Rosenbro.cfg;
```

A typical job parameter file (`RadarOdd.cfg`) looks as:

```
# Parameter definition file for GLOB.

function_name = Radar_Odd;
space_dimension = 5;
left_boundaries =   0   0   0   0   0;
right_boundaries =  6.3 6.3 6.3 6.3 6.3;
initial_point =     1   1   1   1   1;
glob_known_minimum = 0.337490;

boundary_tol = 1.0e-10;
significant_fun_diff = 1.0e-4;

glob_job_repetitions = 1;
glob_max_iterations = 100000;
glob_time_limit = 30;
random_seed = 1;

# in array: set to use, 0 not to use certain distribution
#           1. uniform in Linf ball
#           2. uniform in L1 ball
#           3. special in L1 ball
random_distributions = 1 1 1;

#   metaheuristic method:
#       = 1 random point
#       = 2 multistart
#       = 3 VNS
metaheuristic_method = 3;

vns_kmax = 15;
vns_radius = .1 .3 .8 3 10 11 12 13 14 15 16 17 18 19 20 21;
#   VNS radius generation method
#       = 0     user defined
#       = 1     automatic, best point centered
vns_radius_gen_method = 1;

random_from_disk = 1;
reject_on_return = 0;
reject_on_return_k_diff = 1;

rep_job_details = 1;
rep_job_stats = 1;
rep_point_coordinates = 0;
rep_init_point = 0;
```

```
rep_each_meta_iter = 0;
rep_random_point = 0;
rep_new_best_point = 1;
rep_new_vns_radius_gen = 0;

ls_eps = 1.0e-6;
ls_fun_eps = 1.0e-7;
ls_grad_eps = 1.0e-10;

#   local optimizer selection:
#       = 11    steepest descent
#       = 12    Fletcher-Powell
#       = 13    Fletcher-Reeves
#       = 21    Nelder-Mead
#       = 22    Hook-Jeeves
#       = 23    Rosenbrock
local_search_method = 21;

#   gradient calculation method:
#       = 0     user supplied function
#       = 1     finite difference method
grad_calc_method = 1;
grad_calc_step = 1.0e-6;

#   linear (1D) optimizer selection:
#       = 1     golden section method
#       = 2     quadratic method
linear_search_method = 1;

#   for 1D golden section method
ls_1d_gold_eps = 1.0e-6;
ls_1d_gold_maxiter = 500;

#   for Nelder-Mead method:
ls_nelder_mead_init_edge = 0.1;
ls_nelder_mead_maxiter = 500;
```

Appendix 2 GLOB output

For job parameter file listed in Appendix 1, output file would (with unessential reduction) look as

```
        GLOBAL OPTIMIZATION

   Parameters:

     Function name: Radar_Odd
```

```
  Problem dimension: 5
  Left, right boundary and initial point:
    all coordinates:  0.000000  6.300000  1.000000
  Known fun minimum:            0.33749
  Boundary tolerance: 1e-010

  VNS metaheuristic method selected.

  VNS parameters:
    Kmax: 15
    Automatic neighbourhood radius generation selected.
    Random point only from disk ( K-th but not
                                 (K-1)-th neighbourhood )

  Maximum iterations: 100000
   Time limit in sec: 30
         Random seed: 1
  Distributions used: 1 2 3
    Function improvement significance tol: 0.0001

     eps point tolerance for local search: 1e-006
  eps function tolerance for local search: 1e-007
  eps gradient tolerance for local search: 1e-010

  Nelder-Mead local optimizer selected.
  Maximum iterations in Nelder-Mead method: 500
    Init edge length in Nelder-Mead method: 0.1

LS ret codes:
    0 - reject on return back to smaller neighbourhood
    1 - maximum iterations reached
    2 - two points closer than LS_eps
    3 - two function values cloaser than LS_fun_eps
    4 - gradient norm smaller than LS_grad_eps
    5 - null direction vector
    6 - null direction vector after rotation
    7 - maximum step reductions reached


** clock:    0.010s    0.010s  New best point found
   iteration: 1  VNS_k: 1 d1 ( 1 -> 5 )  LS ret: 2  fun: 625
   point dist:     2.27833,  fun diff:    1.29501
Fun = 1.40650133983789  ( 316.75 %)   New best point
** clock:    0.020s    0.010s  New best point found
   iteration: 2  VNS_k: 1 d1 ( 1 -> 3 )  LS ret: 2  fun: 1020
   point dist:    1.24833,  fun diff:   0.164813
Fun = 1.24168798465437  ( 267.92 %)   New best point
** clock:    0.040s    0.020s  New best point found
   iteration: 3  VNS_k: 1 d1 ( 1 -> 4 )  LS ret: 1  fun: 1704
```

```
   point dist:     1.82473,  fun diff:   0.274424
Fun = 0.967263995859989  ( 186.61 %)   New best point

( --> 15 new best points omitted from this listing <-- )

** clock:    2.033s    0.010s  New best point found
   iteration: 171  VNS_k: 1 d3 ( 1 -> 1 )  LS ret: 2 fun: 103507
   point dist: 0.00266369,  fun diff: 4.46617e-005
Fun = 0.358356981348043  (    6.18 %)   New best point
** clock:    3.815s    1.782s  New best point found
   iteration: 332  VNS_k: 11 d1 ( 11 -> 13 )  LS ret: 3 fun: 201826
   point dist:     6.47077,  fun diff: 0.00976652
Fun = 0.348590461358697  (    3.29 %)   New best point
** clock:    4.006s    0.191s  New best point found
   iteration: 349  VNS_k: 2 d2 ( 2 -> 1 )  LS ret: 3 fun: 211927
   point dist:    0.125512,  fun diff: 0.00932393
Fun = 0.339266532745671  (    0.53 %)   New best point
** clock:    4.366s    0.360s  New best point found
   iteration: 381  VNS_k: 2 d1 ( 2 -> 1 )  LS ret: 3 fun: 231608
   point dist:  0.0090049,  fun diff: 0.00212567
Fun = 0.337140866498095  (   -0.10 %)   New best point
** clock:    6.499s    2.133s  New best point found
   iteration: 577  VNS_k: 1 d2 ( 1 -> 1 )  LS ret: 3 fun: 347902
   point dist: 0.000200047,  fun diff: 4.24404e-005
Fun = 0.337098426114104  (   -0.12 %)   New best point
** clock:   16.734s   10.235s  New best point found
   iteration: 1523  VNS_k: 1 d2 ( 1 -> 1 )  LS ret: 3 fun: 912072
   point dist: 6.72353e-005,  fun diff: 1.03975e-005
Fun = 0.337088028580521  (   -0.12 %)   New best point
Timeout of 30 seconds reached.
** clock:   30.003s   13.269s  finish
Best point found:
Fun = 0.337088028580521  (   -0.12 %)       Best point:
  X =     3.45738323045107 1.04894197827272 1.42788440554346
          6.10300822468557 1.91461843833361

Job statistics:
     Random points generated:        2747
      Random points rejected:         510
     Number of improvements for used distributions:
          1:  10
          2:  3
          3:  11
                               overall   last succ. last signif.
        Meta iteration(s):        2747         1523          381
          Time in seconds:      30.003       16.734        4.366
 Comp. effort (fun+N*grad):    1634872       912072       231608
           Function calls:     1634872       912072       231608
           Gradient calls:           0            0            0
```

```
VNS\index{VNS} statistics:
Level,      steps,     succ steps,  rejected steps
   1:         203            12           0
   2:         191             6           0
   3:         185             1           0
   4:         184             0           0
   5:         184             1           0
   6:         183             0           0
   7:         183             1           0
   8:         182             1           0
   9:         181             0           0
  10:         180             1           0
  11:         179             1           0
  12:         178             0           0
  13:         178             0           0
  14:         178             0           0
  15:         178             0           0
 SUM:        2747            24           0
```

Disciplined Convex Programming

Michael Grant[1], Stephen Boyd[1], and Yinyu Ye[12]

[1] Department of Electrical Engineering, Stanford University
{mcgrant,boyd,yyye}@stanford.edu
[2] Department of Management Science and Engineering, Stanford University

Summary. A new methodology for constructing convex optimization models called *disciplined convex programming* is introduced. The methodology enforces a set of conventions upon the models constructed, in turn allowing much of the work required to analyze and solve the models to be automated.

Key words: Convex programming, automatic verification, symbolic computation, modelling language.

1 Introduction

Convex programming is a subclass of nonlinear programming (NLP) that unifies and generalizes least squares (LS), linear programming (LP), and convex quadratic programming (QP). This generalization is achieved while maintaining many of the important, attractive theoretical properties of these predecessors. Numerical algorithms for solving convex programs are maturing rapidly, providing reliability, accuracy, and efficiency. A large number of applications have been discovered for convex programming in a wide variety of scientific and non-scientific fields, and it seems clear that even more remain to be discovered. For these reasons, convex programming arguably has the potential to become a ubiquitous modeling technology alongside LS, LP, and QP. Indeed, efforts are underway to develop and teach it as a distinct discipline [29, 21, 115].

Nevertheless, there remains a significant impediment to the more widespread adoption of convex programming: the high level of expertise required to use it. With mature technologies such as LS, LP, and QP, problems can be specified and solved with relatively little effort, and with at most a very basic understanding of the computations involved. This is not the case with general convex programming. That a user must understand the basics of convex analysis is both reasonable and unavoidable; but in fact, a much deeper understanding is required. Furthermore, a user must find a way to transform his problem into one of the many limited standard forms; or, failing that, develop a custom solver. For potential users whose focus is the application, these requirements can form a formidable "expertise barrier"—especially if it is not yet certain that the outcome will be any better than with other methods. The purpose of the work presented here is to lower this barrier.

In this article, we introduce a new modeling methodology called *disciplined convex programming.* As the term "disciplined" suggests, the methodology imposes a set of conventions that one must follow when constructing convex programs. The conventions are simple and teachable, taken from basic principles of convex analysis, and inspired by the practices of those who regularly study and apply convex optimization today. Conforming problems are called, appropriately, *disciplined convex programs.* The conventions do not limit generality; but they *do* allow much of the manipulation and transformation required to analyze and solve convex programs to be automated. For example, the task of determining if an arbitrary NLP is convex is both theoretically and practically intractable; the task of determining if it is a *disciplined* convex program is straightforward. In addition, the transformations necessary to convert disciplined convex programs into solvable form can be fully automated.

A novel aspect of this work is a new way to define a function in a modeling framework: as the solution of a disciplined convex program. We call such a definition a *graph implementation*, so named because it exploits the properties of epigraphs and hypographs of convex and concave functions, respectively. The benefits of graph implementations to are significant, because they provide a means to support nondifferentiable functions without the loss of reliability or performance typically associated with them.

We have created a modeling framework called cvx that implements the principles of the disciplined convex programming methodology. The system is built around a specification language that allows disciplined convex programs to be specified in a natural mathematical form, and addresses key tasks such as verification, conversion to solvable form, and numerical solution. The development of cvx is ongoing, and an initial version is near release. We will be disseminating cvx freely to encourage its use in coursework, research, and applications.

The remainder of this article begins with some motivation for this work, by examining how current numerical methods can be used to solve a simple norm minimization problem. In §3, we provide a brief overview of convex programming technology; and in §4, we discuss the benefits of modeling frameworks in general, and cvx in particular. Finally, we introduce disciplined convex programming in detail in §5-§10.

2 Motivation

To illustrate the complexities of practical convex optimization, let us consider how one might solve a basic and yet common problem: the unconstrained norm minimization

$$\text{minimize } f(x) \triangleq \|Ax - b\| \qquad A \in \mathbb{R}^{m\times n} \quad b \in \mathbb{R}^m \tag{1}$$

The norm $\|\cdot\|$ has not yet been specified; we will be examining several choices of that norm. For simplicity, we will assume that $n \leq m$ and that A has full rank; and for convenience, we shall partition A and b into rows,

$$A^T \triangleq \begin{bmatrix} a_1 \; a_2 \; \ldots \; a_m \end{bmatrix} \qquad b^T \triangleq \begin{bmatrix} b_1 \; b_2 \; \ldots \; b_m \end{bmatrix} \tag{2}$$

2.1 The norms

The Euclidean norm

The most common choice of the norm in (1) is certainly the ℓ_2 or Euclidean norm,

$$f(x) \triangleq \|Ax - b\|_2 = \sqrt{\textstyle\sum_{i=1}^n (a_i^T x - b_i)^2} \tag{3}$$

In this case, (1) is easily recognized as a least squares problem, which as its name implies is often presented in an equivalent quadratic form,

$$\text{minimize } f(x)^2 = \|Ax - b\|_2^2 = \textstyle\sum_{i=1}^m (a_i^T x - b_i)^2 \tag{4}$$

This problem has an analytical solution $x = (A^T A)^{-1} A^T b$, which can be computed using a Cholesky factorization of $A^T A$, or more accurately using a QR or SVD factorization of A [76]. A number of software packages to solve least squares problems are readily available; *e.g.*, [1, 106].

The Chebyshev norm

For the ℓ_∞ or Chebyshev norm,

$$f(x) \triangleq \|Ax - b\|_\infty = \max_{i=1,2,\ldots,m} |a_i^T x - b_i|, \tag{5}$$

there is no analytical solution. But a solution can be obtained by solving the linear program

$$\begin{array}{ll} \text{minimize} & q \\ \text{subject to} & -q \leq a_i^T x - b_i \leq q, \; i = 1, 2, \ldots, m \end{array} \tag{6}$$

Despite the absence of an analytical solution, numerical solutions for this problem are not difficult to obtain. A number of efficient and reliable LP solvers are readily available; in fact, a basic LP solver is included with virtually every piece of spreadsheet software sold [63].

The Manhattan norm

Similarly, for the ℓ_1 or Manhattan norm

$$f(x) \triangleq \|Ax - b\|_1 = \sum_{i=1}^{m} |a_i^T x - b_i|, \tag{7}$$

a solution can also be determined by solving an appropriate LP:

$$\begin{array}{ll} \text{minimize} \sum_{i=1}^m v_i \\ \text{subject to} -v_i \leq a_i^T x - b_i \leq v_i, \; i = 1, 2, \ldots, m \end{array} \tag{8}$$

So again, this problem can be easily solved with readily available software, even though an analytical solution does not exist.

The Hölder norm, part 1

Now consider the Hölder or ℓ_p norm

$$f(x) \triangleq \|Ax - b\|_p = \left(\sum_{i=1}^m |a_i^T x - b_i|^p\right)^{1/p} \tag{9}$$

for $p \geq 2$. We may consider solving (1) for this norm using Newton's method. For simplicity we will in fact apply the method to the related function

$$g(x) \triangleq f(x)^p = \sum_{i=1}^m |a_i^T x - b_i|^p \tag{10}$$

which yields the same solution x but is twice differentiable everywhere. The iterates produced by Newton's method are

$$\begin{aligned} x_{k+1} &= x_k - \alpha_k \left(\nabla^2 g(x_k)\right)^{-1} \nabla g(x_k) \\ &= x_k - \frac{\alpha_k}{p-1}(A^T W_k A)^{-1} A^T W_k (A x_k - b) \\ &= \frac{p-1-\alpha_k}{p-1} x_k + \frac{\alpha_k}{p-1} \arg\min_{\bar{w}} \left\| W_k^{1/2}(A\bar{w} - b) \right\|_2 \end{aligned} \tag{11}$$

where $x_0 = 0$, $k = 1, 2, 3, \ldots$, and we have defined

$$W_k \triangleq \mathbf{diag}(|a_1^T x_k - b_1|^{p-2}, |a_2^T x_k - b_2|^{p-2}, \ldots, |a_m^T x_k - b_m|^{p-2}). \tag{12}$$

and $\alpha_k \in [0, 1)$ is either fixed or determined at each iteration using a line search technique. Notice how the Newton computation involves a (weighted) least squares problem; In fact, if $p = 2$, then $W_k \equiv I$, and a single Newton iteration produces the correct least squares solution. So the more "complex" ℓ_p case simply involves solving a series of similar least squares problems. This resemblance to least squares turns up quite often in numerical methods for convex programming.

An important technical detail must be mentioned here. When the residual vector $Ax_k - b$ has any zero entries, the matrix W_k (12) will be singular. If $m \gg n$, this will not necessarily render the Hessian $\nabla^2 g(x)$ itself singular, but care must be taken nonetheless to guard for this possibility. A variety of methods can be considered, including the introduction of a slight dampening factor to W_k; *i.e.*, $W_k + \epsilon I$.

Newton's method is itself a relatively straightforward algorithm, and a number of implementations have been developed; *e.g.*, [113]. These methods do require that code be created to compute the computation of the gradient and Hessian of the function involved. This task is eased somewhat by using an automatic differentiation package such ADIC [81] or ADIFOR [10], which can generate derivative code from code that simply computes a function's value.

The Hölder norm, part 2

For $1 < p < 2$, Newton's method cannot reliably be employed, because neither $f(x)$ nor $g(x) \triangleq f(x)^p$ is twice differentiable whenever the residual vector $Ax - b$ has any zero entries. An alternative that works for all $p \in [1, +\infty)$ is to apply a barrier method to the problem. A full introduction to barrier methods is beyond the scope of this text, so we will highlight only key details. The reader is invited to consult [118] for a truly exhaustive development of barrier methods, or [29] for a gentler introduction.

To begin, we note that the solution to (1) can be obtained by solving

$$\begin{array}{ll} \text{minimize} & \mathbf{1}^T v \\ \text{subject to} & |a_i^T x - b_i|^p \leq v_i \;\; i = 1, 2, \ldots, m \end{array} \tag{13}$$

To solve (13), we construct a *barrier function* $\phi : (\mathbb{R}^n \times \mathbb{R}^m) \rightarrow (\mathbb{R} \cup +\infty)$ to represent the inequality constraints [118]:

$$\phi(x,v) \triangleq \begin{cases} \sum_{i=1}^m -\log(v_i^{2/p} - (a_i^T x - b_i)^2) - 2\log v_i & (x,v) \in S \\ +\infty & (x,v) \notin S \end{cases} \tag{14}$$

$$S \triangleq \{\, (x,v) \in \mathbb{R}^n \times \mathbb{R} \mid |a_i^T x - b_i|^p < v_i, \; i = 1, 2, \ldots, m \,\}$$

The barrier function is finite and twice differentiable whenever the inequality constraints in (13) are strictly satisfied, and $+\infty$ otherwise. This barrier is used to create a family of functions g_t parameterized over a quantity $t > 0$:

$$g_t : \mathbb{R}^n \times \mathbb{R}^m \rightarrow \mathbb{R} \cup +\infty, \quad g_t(x,v) \triangleq \mathbf{1}^T v + t\phi(x,v) \tag{15}$$

It can be shown that, as $t \rightarrow 0$, the minimizing values for $g_t(x,v)$ converge to the solution to the original problem. A practical barrier method takes Newton steps to minimize $g_t(x,v)$, decreasing the value of t between iterations in a manner chosen to insure convergence and acceptable performance.

This approach is obviously significantly more challenging than the previous efforts. As with the Newton method for the $p \geq 2$ case, code must be written (or automatic differentiation employed) to compute the gradient and Hessian of $g_t(x, v)$. Furthermore, the authors are unaware of any readily available software implementing a general purpose barrier methods such as this, so one would be forced to write their own.

An uncommon choice

Given a vector $w \in \mathbb{R}^m$, let $w_{[|k|]}$ be the k-th element of the vector after it has been sorted from largest to smallest in absolute value:

$$|w_{[|1|]}| \geq |w_{[|2|]}| \geq \cdots \geq |w_{[|m|]}| \tag{16}$$

Then let us define the *largest-L* norm as follows:

$$\|w\|_{[|L|]} \triangleq \sum_{k=1}^{L} |w_{[|k|]}| \quad (L \in \{1, 2, \ldots, m\}). \tag{17}$$

Solving (1) using this norm produces a vector x that minimizes the sum of the L largest residuals of $Ax - b$. This is equivalent to the ℓ_∞ case for $L = 1$ and the ℓ_1 case for $L = m$, but for $1 < L < m$ this norm produces novel results.

While it may not be obvious that (17) is a norm or even a convex function, it is indeed both. Even less obvious is how to solve this problem—but in fact, it turns out that it can be solved as an LP!

$$\begin{array}{ll} \text{minimize} & \sum_{i=1}^m v_i + Lq \\ \text{subject to} & -v_i - q \leq a_i^T x - b_i \leq v_i + q, \quad i = 1, 2, \ldots, m \\ & v_i \geq 0, \quad i = 1, 2, \ldots, m \end{array} \tag{18}$$

This LP is only slightly larger than the one used for the ℓ_1 case. The result is known—see, for example [124]—but not widely so, even among those who actively study optimization. Thus it is likely that someone wishing to solve this problem would consider a far more difficult approach such as a barrier method or a subgradient method (or not even try).

2.2 The expertise barrier

The conceptual similarity of these problems is obvious, but the methods employed to solve them differ significantly. A variety of numerical algorithms are represented: least squares, linear programming, Newton's method, and a barrier method. In most cases, transformations are required to produce an equivalent problem suitable for numerical solution. These transformations are not likely to be obvious to an applications-oriented user whose primary expertise is not optimization.

As a result of this complexity, those wishing to solve a norm minimization problem may, out of ignorance or practicality, restrict their view to norms for which solution methods are widely known, such as ℓ_2 or ℓ_∞— even if doing so compromises the accuracy of their models. This might be understandable for cases like, say, $\ell_{1.5}$, where the computational method employed is quite complex. But the true computational complexity may be far less severe than it seems on the surface, as is the case with the largest-L norm.

Even this simple example illustrates the high level of expertise needed to solve even basic convex optimization problems. Of course, the situation worsens if more complex problems are considered. For example, adding simple bounds on x (*e.g.*, $l \le x \le u$) eliminates analytical solutions for the ℓ_2 case, and prevents the use of a simple Newton's method for the ℓ_p case.

2.3 Lowering the barrier

In Figure 1, cvx specifications for three of the norm minimization problems presented here are given. In each case, the problem is given in its original form; no transformations described above have been applied in advance to convert them into "solvable" form. Instead, the models utilize the functions `norm_inf`, `norm_p`, and `norm_largest` for the ℓ_∞, ℓ_p, and largest-L norms, respectively. The definitions of these functions have been stored in a separate file `norms.cvx`, and referred to by an `include` command.

```
minimize norm_inf( A x - b );
parameters A[m,n], b[n];
variable x[n];
include "norms.cvx";

minimize norm_p( A x - b, p );
parameters A[m,n], b[n], p >= 1;
variable x[n];
include "norms.cvx";

minimize norm_largest( A x - b, L );
parameters A[m,n], b[n], L in #{1,2,...,n};
variable x[n];
include "norms.cvx";
```

Fig. 1. cvx specifications for the Chebyshev, Hölder, and largest-L cases, respectively.

In most modeling frameworks, function definitions consist of computer code to compute their values and derivatives. This method is not useful for these functions, because they are not differentiable. Graph implementations, which we describe in detail in §10, solve this problem. For now, it is enough to know that they effectively describe the very transformations illustrated in §2.1 above. For example, the definitions for the `norm_inf` and `norm_largest` provide the information necessary to convert their respective problems into

LPs. The definition for norm_p includes a barrier function for its epigraph, which can be used to apply a barrier method to the third problem.

So neither disciplined convex programming nor the cvx framework eliminates the transformations needed to solve any of these convex programs. Rather, they allow the transformations to be *encapsulated*: that is, hidden from the user, and performed without that user's intervention. A function definition can be used in multiple models and shared with many users. A natural, collaborative environment is suggested, where the work of those with advanced expertise in convex programming can share their knowledge with less experienced modelers in a practical way, by creating libraries of function definitions. The task of *solving* convex programs is appropriately returned to experts, freeing applications-oriented users to confidently focus on modeling.

3 Convex programming

A *mathematical program* is an optimization problem of the form

$$\begin{array}{ll} \text{minimize} & f(x) \\ \text{subject to} & g_i(x) \leq 0 \quad i = 1, 2, \dots, n_g \\ & h_j(x) = 0 \quad j = 1, 2, \dots, n_h \end{array} \tag{19}$$

or one that can be readily converted into this form. The vector x is the *problem variable*; the quantity the quantity $f(x)$ is the *objective function*, and the relations $g_i(x) \leq 0$ and $h_j(x) = 0$ are the *inequality* and *equality constraints*, respectively. The study of mathematical programs focuses almost exclusively on special cases of (19). The most popular is certainly the LP, for which the functions f, g_i, h_j are all affine. Least squares problems, QPs, and NLPs can all be represented by this form (19) as well.

A *convex program* (CP) is yet another special case of (19), one in which the objective function f and inequality constraint functions g_i are convex, and the equality constraint functions h_j are affine. The set of CPs is a strict subset of the set of NLPs, and includes all least squares problems, LPs, and convex QPs. Several other classes of CPs have been identified recently as standard forms. These include *semidefinite programs* (SDPs) [155], *second-order cone programs* (SOCPs) [104], and *geometric programs* (GPs) [48, 3, 138, 52, 92]. The work we present her applies to all of these special cases as well as to the general class of CPs.

The practice of modeling, analyzing, and solving CPs is known as *convex programming*. In this section we provide a survey of convex programming, including its theoretical properties, numerical algorithms, and applications.

3.1 Theoretical properties

A number of powerful and practical theoretical conclusions can be drawn once it can be established that a mathematical program is convex. A comprehensive theory of convex analysis was developed by the 1970s [137, 142], and advances have continued since [83, 84, 21, 29].

The most fundamental distinction between CPs and general NLPs is that, for the former, local optima are guaranteed to be global. Put another way, if local optimality can somehow be demonstrated (say, using KKT conditions), then global optimality is assured. Except in certain special cases, no similar guarantees can be made for nonconvex NLPs. Such problems might exhibit multiple local optima, so an exhaustive search would be required to prove global optimality—an intractable task.

Convex programming also has a rich duality theory that is very similar to the duality theory that accompanies linear programming, though it is a bit more complex. The dual of a CP is itself a CP, and its solution often provides interesting and useful information about the original problem. For example, if the dual problem is unbounded, then the original must be infeasible. Under certain conditions, the reverse implication is also true: if a problem is infeasible, then its dual must be unbounded. These and other consequences of duality facilitate the construction of numerical algorithms with definitive stopping criteria for detecting infeasibility, unboundedness, and near-optimality. For a more complete development of convex duality, see [137, 102].

Another important property of CPs is the provable existence of efficient algorithms for solving them. Nesterov and Nemirovsky proved that a polynomial-time barrier method can be constructed for any CP that meets certain technical conditions [117]. Other authors have shown that problems which do not meet those conditions can be embedded into larger problems that do—effectively making barrier methods universal [169, 102, 171].

Finally, we note that the theoretical properties discussed here, including the existence of efficient solution methods, hold even if a CP is nondifferentiable—that is, if one or more of the constraint or objective functions is nondifferentiable. The practical ramifications of this fact are discussed in §3.4.

3.2 Numerical algorithms

The existence of efficient algorithms for solving CPs has been known since the 1970s, but it is only through advances in the last two decades that this promise has been realized in practice. Much of the modern work in numerical algorithms has focused on *interior-point methods* [166, 37, 163]. Initially such work was limited to LPs [88, 133, 73, 89, 109, 105, 62], but was soon extended to encompass other CPs as well [117, 118, 7, 87, 119, 162, 15, 120]. Now a number of excellent solvers are readily available, both commercial and freely distributed.

Below we provide a brief survey of solvers for convex optimization. For the purposes of our discussion, we have separated them into two classes: those that rely on *standard forms*, and those that rely on *custom code*.

Standard forms

Most solvers for convex programming are designed to handle certain prototypical CPs known as *standard forms*. In other words, such solvers handle a limited family of problems with a very specific structure, or obeying certain conventions. The least squares problem and the LP are two common examples of standard forms; we list several others below. These solvers trade generality for ease of use and performance.

It is instructive to think of the collection of standard form solvers as a "toolchest" of sorts. This toolchest is reasonably complete, in that most CPs that one might encounter can be transformed into one (or more) of these standard forms. However, the required transformations are often far from obvious, particularly for an applications-oriented user.

Smooth convex programs

A number of solvers have been designed to solve CPs in standard NLP form (19), under the added condition that the objective function f and inequality constraint functions g_i are *smooth*—that is, twice continuously differentiable—at least over the region that the algorithm wishes to search. We will call such problems *smooth* CPs; conversely, we will label CPs that do not fit this categorization as *nonsmooth* CPs.

Software packages that solve smooth CPs include LOQO [154], which employs a primal/dual method, and the commercial package MOSEK [110], which implements the homogeneous algorithm. These solvers generally perform quite well in practice. Many systems designed for smooth *nonconvex* NLPs will often solve smooth CPs efficiently as well [32, 69, 112, 70, 41, 31, 156, 122, 14, 33, 13, 71, 61, 153]. This is not surprising when one considers that these algorithms typically exploit *local* convexity when computing search directions.

One practical difficulty in the use of smooth CP or NLP solvers is that the solver must be able to calculate the gradient and Hessian of the objective and inequality constraint functions at points of its choosing. In some cases, this may require the writing of custom code to perform these computations. Many modeling frameworks simplify this process greatly in most cases by allowing functions to be expressed in natural mathematical form and compute derivatives automatically (*e.g.*, [56, 18]).

Conic programs

An entire family of standard forms that have become quite common are the primal and dual *conic forms*

$$\begin{array}{lll} \text{minimize } c^T x & & \text{minimize } b^T y \\ \text{subject to } Ax = b & \quad\text{or}\quad & \text{subject to } A^T y + z = c \\ x \in \mathcal{K} & & z \in \mathcal{K}^* \end{array} \qquad (20)$$

in which the sets $\mathcal{K}$ and $\mathcal{K}^*$ are closed, convex *cones* (*i.e.*, they satisfy $\alpha\mathcal{K} \equiv \mathcal{K}$ and $\alpha\mathcal{K}^* \equiv \mathcal{K}^*$ for all $\alpha > 0$). The most common conic form is the LP, for which $\mathcal{K} = \mathcal{K}^*$ is the nonnegative orthant

$$\mathcal{K} = \mathcal{K}^* = \mathbb{R}^n_+ \triangleq \{\, x \in \mathbb{R}^n \mid x_i \geq 0,\ i = 1, \ldots, n \,\} \tag{21}$$

It can be shown that virtually any CP can be represented in conic form, with appropriate choice of $\mathcal{K}$ or $\mathcal{K}^*$ [118]. In practice, two conic forms (besides LP) dominate all recent study and implementation. One is the *semidefinite program* (SDP), for which $\mathcal{K} = \mathcal{K}^*$ is an isomorphism of the cone of positive semidefinite matrices

$$\mathcal{S}^n_+ \triangleq \{\, X = X^T \in \mathbb{R}^{n \times n} \mid \lambda_{\min}(X) \geq 0 \,\}. \tag{22}$$

The second is the *second-order cone program* (SOCP), for which $\mathcal{K} = \mathcal{K}^*$ is the Cartesian product of one or more second-order or Lorentz cones,

$$\mathcal{K} = \mathcal{Q}^{n_1} \times \cdots \times \mathcal{Q}^{n_K}, \quad \mathcal{Q}^n \triangleq \{\, (x, y) \in \mathbb{R}^n \times \mathbb{R} \mid \|x\|_2 \leq y \,\}. \tag{23}$$

SDP and SOCP receive this focused attention because many applications have been discovered for them, and because their geometry admits certain useful algorithmic optimizations [121, 66, 149, 53, 68, 102, 126]. Publicly available solvers for SDP and SOCP include SeDuMi [140], CDSP [22], SDPA [58], SDPT3 [152], and DSDP [12]. These solvers are generally quite efficient, reliable, and are entirely data-driven: that is, they require no external code to perform function calculations.

Geometric programs

Another standard form that has been studied for some time, but which has generated renewed interest recently, is the *geometric program* (GP). The GP is actually a bit of an unique case, in that it is in fact not convex—but a simple transformation produces an equivalent problem that is convex. In convex form, the objective and inequality constraint functions obtain a so-called "log-sum-exp" structure; for example,

$$f(x) \triangleq \log \sum_{k=1}^{M} e^{a_k^T x + b_k} \qquad a_k \in \mathbb{R}^n,\ b_k \in \mathbb{R},\ k = 1, 2, \ldots, M \tag{24}$$

GPs have been used in various fields since the late 1960s [48]. In convex form they are smooth CPs, but recent advances in specialized algorithms have greatly improved the efficiency of their solution [92].

Custom code

There are instances where a CP cannot be transformed into one of the standard forms above—or perhaps the transformations cannot be determined. An alternative is to use one of the methods that we list here, which are *universal* in the sense that they can, in theory, be applied to *any* CP. The cost of this universality is that the user must determine certain mathematical constructions, and write custom code to implement them.

Barrier methods

A barrier method replaces the inequality constraint set

$$\mathcal{S} \triangleq \{ x \in \mathbb{R}^n \mid g_i(x) \leq 0,\ i = 1, \ldots, n_g \} \tag{25}$$

with a twice differentiable convex barrier function $\phi : \mathbb{R}^n \to \mathbb{R}$ satisfying $\mathbf{dom}\, \phi = \operatorname{Int} \mathcal{S}$, producing a modified problem

$$\begin{array}{ll} \text{minimize} & f(x) + t\phi(x) \\ \text{subject to} & h_j(x) = 0 \quad j = 1, 2, \ldots, n_h \end{array} \tag{26}$$

Under mild conditions, the solution to this modified problem converges to that of the original problem as $t \to 0$. Each iteration of a barrier method effectively performs Newton minimization steps on (26) for steadily decreasing values of t. A complete development of barrier methods, including proofs of universality, convergence, and performance, as well as a number of complete algorithms, is given in [118].

There are several practical roadblocks to the use of a barrier method. First of all, this author knows of no publicly-available, *general purpose* barrier solver; someone wishing to use this technique would have to write their own. Even if a barrier solver is found, the user must supply code to compute the value and derivatives of the barrier function. Furthermore, determining a valid barrier function is not always trivial, particularly if the inequality constraints are nondifferentiable.

Cutting plane methods

Localization or cutting-plane methods such as ACCPM [131] require no derivative information for the functions f and g_i, instead relying solely on cutting planes to restrict the search set. The user is expected to supply code to compute subgradients or cutting planes. The performance of these methods is usually inferior to the others mentioned here, but they are ideal for use when second derivative information is not available or difficult to compute. In addition, they often lend themselves to distributed methods for solution.

3.3 Applications

A wide variety of practical applications for convex programming have already been discovered, and the list is steadily growing. Perhaps the field in which the application of convex programming is the most mature and pervasive is control theory; see [44, 8, 11, 40, 114] for a sample of these applications. Other fields where applications for convex optimization are known include, but are not limited to,

- robotics [132, 35, 82];
- pattern analysis and data mining, including support vector machines [141, 172, 134, 164, 91];
- combinatorial optimization and graph theory [101, 5, 28, 77, 57, 55, 90, 170, 174, 111, 42, 60, 59, 85];
- structural optimization [2, 23, 9, 25, 26, 173, 86, 27, 24, 94];
- algebraic geometry [125, 95, 97],[96];
- signal processing [79, 161, 65, 139, 150, 165, 49, 151, 6, 47, 103, 143, 54, 144, 64, 146, 108, 107, 128, 50, 129, 45, 17, 72, 168, 160, 145, 175];
- communications and information theory [19, 98, 43, 135, 75, 148, 4];
- networking [20, 30];
- circuit design [157, 158, 80, 39, 78, 16];
- quantum computation [46];
- neural networks [127];
- chemical engineering [136];
- economics and finance [67, 167].

This list, while large, is certainly incomplete, and excludes applications where only LP or QP is employed. Such a list would be significantly larger; and yet convex programming is of course a generalization of these technologies.

One promising source of new applications for convex programming is the extension and enhancement of existing applications for linear programming. An example of this is *robust linear programming*, which allows uncertainties in the coefficients of an LP model to be accounted for in the solution of the problem, by transforming it into a nonlinear CP [104]. This approach produces robust solutions more quickly, and arguably more reliably, than using Monte Carlo methods. Presumably, robust linear programming would find application anywhere linear programming is currently employed, and where uncertainty in the model poses a significant concern.

Some may argue that our prognosis of the usefulness of convex programming is optimistic, but there is good reason to believe the number of applications is in fact being underestimated. We can appeal to the history of linear programming as precedent. George Dantzig first published his invention of the simplex method for linear programming in the 1947; and while a number of military applications were soon found, it was not until 1955-1960 that the field enjoyed robust growth [38]. Certainly, this delay was in large part due to the dearth of adequate computational resources; but that is the point: the discovery of new applications accelerated only once hardware and software advances made it truly practical to solve LPs.

Similarly, then, there is good reason to believe that the number of known applications for convex programming will rise dramatically if it can be made easier for people to create, analyze, and solve CPs.

3.4 Convexity and differentiability

As mentioned in §3.2, many solvers for smooth (nonconvex) NLPs can be used to effectively solve many smooth CPs. An arguable case can be made that the advance knowledge of convexity is not critical in such cases. Dedicated solvers for smooth CPs do provide some advantages, such as the ability to reliably detect infeasibility and degeneracy; see, for example, [171]. But such advantages may not immediately seem compelling to those accustomed to traditional nonlinear programming.

In the nonsmooth case, the situation is markedly different. Nondifferentiability poses a significant problem for traditional nonlinear programming. The best methods available to solve nondifferentiable NLPs are far less accurate, reliable, or efficient than their smooth counterparts. The documentation for the GAMS modeling framework "strongly discourages" the specification of nonsmooth problems, instead recommending that points of nondifferentiability be eliminated by replacing them with Boolean variables and expressions [18]. But doing so is not always straightforward, and introduces significant practical complexities of a different sort.

In contrast, there is nothing in *theory* that prevents a nonsmooth CP from being solved as efficiently as a smooth CP. For example, as mentioned in §3.1, the proof provided by Nesterov and Nemirovsky of the existence of barrier functions for CPs does not depend on smoothness considerations. And nonsmooth CPs can often be converted to an equivalent smooth problem with a carefully chosen transformation—consider the ℓ_∞, ℓ_1, and largest-L norm minimization problems presented in §2. Of course, neither the construction of a valid barrier function nor the smoothing transformation is always (or even often) obvious.

One might ask: just how often are the CPs encountered in practice nonsmooth? We claim that it is quite often. Most non-trivial SDPs and SOCPs, for example, are nonsmooth. Common convex functions such as the absolute value and most norms are nonsmooth. Examining the current inventory of applications for convex programming, and excluding those that immediately present themselves as LPs and QPs, smoothness is the exception, not the rule.

Thus a convex programming methodology that provides truly practical support for nonsmooth problems is of genuine practical benefit. If such a solution can be achieved, then the *a priori* distinction between convexity and nonconvexity becomes far more important, because the need to avoid nondifferentiability remains only in the nonconvex case.

3.5 Convexity verification

Given the benefits of advance knowledge of convexity, it would be genuinely useful to perform automatic *convexity verification*: that is, to determine whether or not a given mathematical program is convex. Unfortunately, the task of determining whether or not a general mathematical program is convex is at least as difficult as solving nonconvex problems: that is, it is theoretically intractable. Practical attempts have achieved various degrees of success, as we survey here.

Perhaps the most computationally ambitious approach to convexity verification has been independently developed by Crusius [36] and Orban and Fourer [123]. The first of these has been refined and integrated into a commercial offering [116, 63]. These systems combine interval methods, symbolic or automatic differentiation, and other methods to determine if the Hessians of key subexpressions in the objective and constraints are positive semidefinite over an estimate of the feasible region. The efforts are impressive, but these systems do fail to make conclusive determinations in many cases—that is, some problems can neither be proven convex nor nonconvex. Furthermore, these systems are limited to smooth NLPs, due to their reliance on derivative information.

Limiting the scope to one or more standard forms produces more reliable results. For example, many modeling frameworks automatically determine if a model is an LP, enabling specialized algorithms to be selected for them [56, 18]. Similar approaches are employed by modeling tools such as SDPSOL and LMITOOL to automatically verify SDPs [159, 51]. These approaches are effective because these particular standard forms can be recognized entirely through an analysis of their *textual* structure. They are perfectly reliable, making conclusive determinations in every case: *e.g.*, a model is proven to be an LP, or proven otherwise. But of course, generality is significantly compromised. And these systems do not attempt to recognize problems that are *transformable* into the supported standard form. For example, the ℓ_1 norm minimization in §2.1 would have to be manually converted to an LP before it would be recognized as such by these systems.

Yet another alternative is provided by the MPROBE [34] system, which employs numerical sampling to *empirically* determine the shapes of constraint and objective functions. It will often conclusively *disprove* linearity or convexity in many cases, but it can never conclusively prove convexity, because doing so would require an exhaustive search. To be fair, its author makes no claims to that effect, instead promoting MPROBE as a useful tool for interactively assisting the user to make his own decisions.

These practical approaches to automatic convexity verification compromises generality, whether due to limitations of the algorithms or by deliberate restrictions in scope. As we will see below, disciplined convex programming makes a compromise of a different sort, recovering generality by incorporating knowledge provided by its users.

4 Modeling frameworks

The purpose of a *modeling framework* is to enable someone to become a proficient *user* of a particular mathematical technology (*e.g.*, convex programming) without requiring that they may become an expert in it (*e.g.*, interior-point methods). It accomplishes this by providing a convenient interface for specifying problems, and then by automating many of the underlying mathematical and computational steps for analyzing and solving them.

A number of excellent modeling frameworks for LP, QP, and NLP are in widespread use and have had a broad and positive impact on the use of optimization in many application areas, including AMPL [56], GAMS [18], LINGO [99], and Frontline [63]. These frameworks are well-suited for solving smooth CPs as well. More recently, a number of modeling frameworks for semidefinite programming have been developed, including SDPSOL [159], LMITool [51], MATLAB's LMI Control Toolbox [147], YALMIP [100], and SOSTOOLS [130]. These tools are used by thousands in control design, analysis, and research, and in other fields as well.

We are developing a modeling framework called `cvx` to support the disciplined convex programming methodology. The framework addresses a number of the challenges already addressed in this article, including support for nondifferentiable problems, convexity verification, and automatic conversion to solvable form. We have implemented a simple barrier solver for the framework; but in fact, any numerical method currently used to solve CPs can be used to solve DCPs. So we intend to work to create interfaces between `cvx` and other well-known solvers.

```
maximize      entropy( x1, x2, x3, x4 );
subject to    a11 x1 + a12 x2 + a13 x3 + a14 x4 = b1;
              a21 x1 + a22 x2 + a23 x3 + a24 x4 = b2;
                  x1 +     x2 +     x3 +     x4 = 1;
parameters    a11, a12, a13, a14, b1,
              a21, a22, a23, a24, b2;
variables     x1 >= 0, x2 >= 0, x3 >= 0, x4 >= 0;
function entropy( ... ) concave;
```

Fig. 2. An example CP in the `cvx` modeling language.

The `cvx` framework is built around a modeling language that allows optimization problems to be expressed using a relatively obvious mathematical syntax. The language shares a number of basic features with other modeling languages such as AMPL or GAMS, such as parameter and variable declarations, common mathematical operations, and so forth. See Figure 2 for an example of a simple entropy maximization problem expressed in the `cvx` syntax.

Throughout this article, we will illustrate various concepts using examples rendered in the `cvx` modeling language, using a fixed-width font (`example`) to distinguish them. However, because `cvx` is still in development, it is possible that future versions of the language will use a slightly different syntax; and a few examples use features of the language that have not yet been implemented. So the examples should not be treated as definitive references. The reader is invited to visit the Web site `http://www.stanford.edu/~boyd/cvx` to find the most recent information about `cvx`.

5 Disciplined convex programming

Disciplined convex programming is inspired by the practices of those who regularly study and use convex optimization in research and applications. They do not simply construct constraints and objective functions without advance regard for convexity; rather, they draw from a mental library of functions and sets whose convexity properties are already known, and combine and manipulate them in ways which convex analysis insures will produce convex results. When it proves necessary to determine the convexity properties of a new function or set from basic principles, that function or set is added to the mental library to be reused in other models.

Disciplined convex programming formalizes this strategy, and includes two key components:

- An *atom library*: an *extensible* collection of functions and sets, or *atoms*, whose properties of curvature/shape (convex/concave/affine), monotonicity, and range are explicitly declared.
- A *convexity ruleset*, drawn from basic principles of convex analysis, that governs how atoms, variables, parameters, and numeric values can be combined to produce convex results.

A valid *disciplined convex program*, or DCP, is simply a mathematical program built in accordance with the convexity ruleset using elements from the atom library. This methodology provides a teachable conceptual framework for people to use when studying and using convex programming, as well as an effective platform for building software to analyze and solve CPs.

The convexity ruleset, introduced in §6 below, has been designed to be easy to learn and understand. The rules constitute a set of sufficient conditions to guarantee convexity. In other words, any mathematical program constructed in accordance with the convexity ruleset is guaranteed to be convex. The converse, however, is not true: it is possible to construct problems which do not obey the rules, but which are convex nonetheless. Such problems are not valid DCPs, and the methodology does not attempt to accommodate them. This does not mean that they cannot be solved, but it does mean that they will have to be rewritten to comply with the convexity ruleset.

Because the convexity ruleset limits the variety of CPs that can be constructed from a fixed atom library, it follows that the generality of disciplined convex programming depends upon that library being extensible. Each atom must be given a declaration of information about its curvature or shape, monotonicity, and range, information which is referred to when verifying that the atom is used in accordance with the convexity ruleset. We introduce the atom library in detail in §7.

In §9, we examine the consequences of a restricted approach such as this. In particular, we provide some examples of some common and useful CPs that, in their native form, are not *disciplined* convex programs. We show that the these limitations are readily remedied using the extensibility of the atom library. We argue that the remedies are, in fact, consistent with the very thought process that disciplined convex programming is attempting to formalize.

Finally, in §10, we discuss how the elements in the atom library can be *implemented*—that is, how they can be represented in a form usable by numerical algorithms. In addition to some very traditional forms, such as barrier functions, cutting planes, derivatives for Newton's method, and so forth. we introduce the concept of *graph implementations*. Graph implementations allow functions and sets to be defined in terms of other DCPs, and provide such benefits as support for nondifferentiable functions.

Before we proceed, let us address a notational issue. We have chosen to follow the lead of [137] and adopt the extended-valued approach to defining convex and concave functions with limited domains; *e.g.*,

$$f : \mathbb{R} \to (\mathbb{R} \cup +\infty), \qquad f(x) = \begin{cases} +\infty & x < 0 \\ x^{1.5} & x \geq 0 \end{cases} \tag{27}$$

$$g : \mathbb{R} \to (\mathbb{R} \cup -\infty), \qquad g(x) = \begin{cases} -\infty & x \leq 0 \\ \log x & x > 0 \end{cases} \tag{28}$$

Using extended-valued functions simplifies many of the derivations and proofs. Still, we will on occasion use the **dom** operator to refer to the set of domain values that yield finite results:

$$\mathbf{dom}\, f = \{\, x \mid f(x) < +\infty \,\} = [0, +\infty), \tag{29}$$

$$\mathbf{dom}\, g = \{\, x \mid g(x) > -\infty \,\} = (0, +\infty) \tag{30}$$

6 The convexity ruleset

The convexity ruleset governs how variables, parameters, and atoms (functions and sets) may be combined to form DCPs. DCPs are a strict subset of general CPs, so another way to say this is that the ruleset imposes a set of conventions or restrictions on CPs. The ruleset can be separated into four categories: *top-level rules*, *product-free rules*, *sign rules*, and *composition rules*.

6.1 Top-level rules

As the name implies, top-level rules govern the top-level structure of DCPs. These rules are more descriptive than they are restrictive, in the sense that nearly all *general* CPs follow these conventions anyway. But for completeness they must be explicitly stated.

Problem types. A valid DCP can either be:

T1 a *minimization*: a convex objective and zero or more convex constraints;

T2 a *maximization*: a concave objective and zero or more convex constraints; or

T3 a *feasibility problem*: no objective, and one or more convex constraints.

A valid DCP may also include any number of *assertions*; see rule **T9**.

At the moment, support for multiobjective problems and games has not been developed, but both are certainly reasonable choices for future work.

affine = *affine* **(T4)**

convex <= *concave* or *convex* < *concave* **(T5)**

concave >= *convex* or *concave* > *convex* **(T6)**

(*affine*, *affine*, ..., *affine*) in *convex set* **(T7)**

Fig. 3. Valid constraints.

Constraints. See Figure 3. Valid constraints include:

T4 an equality constraint with affine left- and right-hand expressions.

T5 a less than ($<$,$\leq$) inequality, with a convex left-hand expression and a concave right-hand expression;

T6 a greater than ($>$,$\geq$) inequality, with a concave left-hand expression and a convex right-hand expression; or

T7 a set membership constraint $(lexp_1, \ldots, lexp_m) \in cset$, where $m \geq 1$, $lexp_1, \ldots, lexp_m$ are affine expressions, and *cset* is a convex set.

Non-equality ($\neq$) constraints and set non-membership ($\notin$) constraints are not permitted, because they are convex only in exceptional cases—and support for exceptional cases is anathema to the philosophy behind disciplined convex programming.

Constant expressions and assertions.

T8 Any well-posed numeric expression consisting only of numeric values and parameters is a valid constant expression.

T9 Any Boolean expression performing tests or comparisons on valid constant expressions is a valid assertion.

T10 If a function or set is parameterized, then those parameters must be valid constant expressions.

A *constant expression* a numeric expression involving only numeric values and/or parameters; a *non-constant* expression depends on the value of at least one problem variable. Obviously a constant expression is trivially affine, convex, and concave. Constant expressions must be well-posed: which, for our purposes, means that they produce well-defined results for any set of parameter values that satisfy a problem's assertions.

An *assertion* resembles a constraint, but involves only constant expressions. As such, they are not true constraints *per se*, because their truth or falsity is determined entirely by the numerical values supplied for a model's parameters, *before* the commencement of any numerical optimization algorithm. Assertions are not restricted in the manner that true constraints are; for example, non-equality ($\neq$) and set non-membership ($\notin$) operations may be freely employed. Assertions serve as *preconditions*, guaranteeing that a problem is numerically valid or physically meaningful. There are several reasons why an assertion may be wanted or needed; for example:

- to represent physical limits dictated by the model. For example, if a parameter w represents the physical weight of an object, an assertion $w > 0$ enforces the fact that the weight must be positive.
- to insure numeric well-posedness. For example, if x, y, and z are variables and a, b, and c are parameters, then the inequality constraint $ax + by + z/c \leq 1$ is well-posed only if c is nonzero; this can be insured by an assertion such as $c \neq 0$ or $c > 0$.
- to guarantee compliance with the preconditions attached to a function or set in the atom library. For example, a function $f_p(x) = \|x\|_p$ is parameterized by a value $p \geq 1$. If p is supplied as a parameter, then an assertion such as $p \geq 1$ would be required to guarantee that the function is being properly used. See §7.1 for information on how such preconditions are supplied in the atom library.
- to insure compliance with the sign rules §6.3 or composition rules §6.4 below; see those sections and §8 for more details.

The final rule **T10** refers to functions or sets that are parameterized; *e.g.*,

$$f_p : \mathbb{R}^n \to \mathbb{R}, \quad f_p(x) = \|x\|_p = \left(\sum_{i=1}^{n} |x_i|^p \right)^{1/p} \tag{31}$$

$$B_p = \{\, x \in \mathbb{R}^n \mid \|x\|_p \leq 1 \,\} \tag{32}$$

and simply states that parameters such as p above must be constant. Of course this is generally assumed, but we must make it explicit for the purposes of computer implementation.

6.2 The product-free rules

Some of the most basic principles of convex analysis govern the sums and scaling of convex, concave, and affine expressions; for example:

- The sum of two or more convex (concave, affine) expressions is convex (concave, affine).

- The product of a convex (concave) expression and a nonnegative constant expression is convex (concave).
- The product of a convex (concave) expression and a nonpositive constant expression, or the simple negation of the former, is concave (convex).
- The product of an affine expression and any constant is affine.

Conspicuously absent from these principles is any mention of the *product* of convex or concave expressions. The reason for this is simple: there is no simple, general principle that can identify the curvature in such cases. For instance, suppose that x is a scalar variable; then:

- The expression $x \cdot x$, a product of two affine expressions, is convex.
- The expression $x \cdot \log x$, a product between and affine and a concave expression, is convex.
- The expression $x \cdot e^x$, a product between an affine and a convex expression, is neither convex nor concave.

For this reason, the most prominent structural convention enforced by disciplined convex programming is the prohibition of products (and related operations, like exponentiation) between non-constant expressions. The result is a set of rules appropriately called the *product-free rules*:

> *Product-free rule for numeric expressions:* All valid numeric expressions must be product-free; such expressions include:
> PN1 A simple variable reference.
> PN2 A *constant* expression.
> PN3 A call to a function in the atom library. Each argument of the function must be a product-free expression.
> PN4 The sum of two or more product-free expressions.
> PN5 The difference of product-free expressions.
> PN6 The negation of a product-free expression.
> PN7 The product of a product-free expression and a constant expression.
> PN8 The division of a product-free expression by a constant expression.
> We assume in each of these rules that the results are well-posed; for example, that dimensions are compatible.

In the scalar case, a compact way of restating these rules is to say that a valid numeric expression can be reduced to the form

$$a + \sum_{i=1}^{n} b_i x_i + \sum_{j=1}^{L} c_j f_j(arg_{j,1}, arg_{j,2}, \ldots, arg_{j,m_j}) \tag{33}$$

where a, b_i, c_j are constants; x_i are the problem variables; and $f_j : \mathbb{R}^{m_j} \to \mathbb{R}$ are functions from the atom library, and their arguments $arg_{j,k}$ are product-free expressions themselves. Certain special cases of (33) are notable: if $L = 0$, then (33) is a simple affine expression; if, in addition, $b_1 = b_2 = \cdots = b_n = 0$, then (33) is a constant expression.

For an illustration of the use of these product-free rules, suppose that a, b, and c are parameters; x, y, and z are variables; and $f(\cdot)$, $g(\cdot,\cdot)$, and $h(\cdot,\cdot,\cdot)$ are functions from the atom library. Then the expression

$$af(x) + y + (h(x, bg(y, z), c) - z + b)/a \tag{34}$$

satisfies the product-free rule, which can be seen by rewriting it as follows:

$$(b/a) + y + (-1/a)z + af(x) + (1/a)h(x, bg(y, z)c) \tag{35}$$

On the other hand, the following expression does not obey the product-free rule:

$$axy/2 - f(x)g(y, z) + h(x, y^b, z, c) \tag{36}$$

Now certainly, because these rules prohibit *all* products between non-constant expressions, some genuinely useful expressions such as quadratic forms like $x \cdot x$ are prohibited; see §9 for further discussion on this point.

For set expressions, a similar set of product-free rules apply:

Product-free rules for set expressions: All valid set expressions used in constraints must be product-free; such expressions include:

PS1 A call to a convex set in the atom library.

PS2 A call to a function in the atom library. Each argument of the function must be a product-free set expression or a constant (numeric) expression.

PS3 The sum of product-free expressions, or of a product-free expression and a constant expression, or vice versa.

PS4 The difference of two product-free set expressions, or of of a product-free set expression and a constant expression, or vice versa.

PS5 The negation of a product-free set expression.

PS6 The product of a product-free set expression and a constant expression.

PS7 The division of a product-free set expression by a constant expression.

PS8 The intersection of two or more product-free set expressions.

PS9 The Cartesian product of two or more product-free set expressions.

We also assume in each of these rules that the results are well-posed; for example, that dimensions are compatible.

In other words, valid set expressions are reducible to the form

$$a + \sum_{i=1}^{n} b_i S_i + \sum_{j=1}^{L} c_j f_j(arg_{j,1}, arg_{j,2}, \ldots, arg_{j,m_j}) \tag{37}$$

where a, b_i, c_j are constants, and the quantities S_k are either sets from the atom library, or intersections and/or Cartesian products of valid set expressions. The functions $f_j : \mathbb{R}^{m_j} \to \mathbb{R}$ are functions in the atom library, and their arguments $arg_{j,k}$ are product-free set expressions themselves. As we will see in §6.3 below, set expressions are more constrained than numerical expressions, in that the functions f_j must be affine.

It is well understood that the intersection of convex sets is convex, as is the direct product of convex sets; and that unions and set differences generally are not convex. What may not be clear is why **PS8-PS9** are considered "product-free" rules. By examining these rules in terms of indicator functions, the link becomes clear. Consider, for example, the problem

$$\begin{array}{ll} \text{minimize} & ax + by \\ \text{subject to} & (x, y) \in (S_1 \times S_1) \cup S_2 \end{array} \tag{38}$$

If $\phi_1 : \mathbb{R} \to \mathbb{R}$ and $\phi_2 : (\mathbb{R} \times \mathbb{R}) \to \mathbb{R}$ are convex indicator functions for the sets S_1 and S_2, respectively, then the problem can be reduced to

$$\text{minimize } ax + by + (\phi_1(x) + \phi_2(y))\phi_2(x, y) \tag{39}$$

and the objective function now violates the product-free rule. What has occurred, of course, is that the union operation became a forbidden product.

6.3 The sign rules

Once the product-free conventions are established, the sum and scaling principles of convex analysis can be used to construct a simple set of sufficient conditions to establish whether or not expressions are convex, concave, or affine. These conditions form what we call the *sign rules*, so named because their consequence is to govern the signs of the quantities $c_1, \ldots, c_L$ in (33). We can concisely state the sign rules for numeric expressions in the following manner.

Sign rules for numeric expressions. Given a product-free expression, the following must be true of its reduced form (33):

SN1 If the expression is expected to be convex, then each term $c_j f_j(\ldots)$ must be convex; hence of the following must be true:
- $f_j(arg_{j,1}, arg_{j,2}, \ldots, arg_{j,m_j})$ is affine;
- $f_j(arg_{j,1}, arg_{j,2}, \ldots, arg_{j,m_j})$ is convex and $c_j \geq 0$;
- $f_j(arg_{j,1}, arg_{j,2}, \ldots, arg_{j,m_j})$ is concave and $c_j \leq 0$.

SN2 If the expression is expected to be concave, then each term $c_j f_j(\ldots)$ must be concave; hence one of the following must be true:
- $f_j(arg_{j,1}, arg_{j,2}, \ldots, arg_{j,m_j})$ is affine;
- $f_j(arg_{j,1}, arg_{j,2}, \ldots, arg_{j,m_j})$ is concave and $c_j \geq 0$;
- $f_j(arg_{j,1}, arg_{j,2}, \ldots, arg_{j,m_j})$ is convex and $c_j \leq 0$.

SN3 If the expression is expected to be affine, then each function f_j must be affine, as must each of its arguments $arg_{j,1}, arg_{j,2}, \ldots, arg_{j,m_j}$.
SN4 If the expression is expected to be constant, then it must be true that $L = 0$ and $b_1 = b_2 = \cdots = b_n = 0$.
All function arguments must obey these rules as well, with their expected curvature dictated by the composition rules (§6.4).

For example, suppose that that the expression (34) is expected to be convex, and that the atom library indicates that the function $f(\cdot)$ is convex, $g(\cdot,\cdot)$ is concave, and $h(\cdot,\cdot,\cdot)$ is convex. Then the sign rule dictates that

$$af(x) \text{ convex} \implies a \geq 0 \tag{40}$$
$$(1/a)h(x, bg(y,z), c) \text{ convex} \implies 1/a \geq 0 \tag{41}$$

Function arguments must obey the sign rule as well, and their curvature is dictated by the composition rules discussed in the next section. So, for example, if the second argument of h is required to be convex, then

$$bg(y,z) \text{ convex} \implies b \leq 0 \tag{42}$$

It is the responsibility of the modeler to insure that the values of the coefficients $c_1, \ldots, c_L$ obey the sign rule; that is, that conditions such as those generated in (40)-(42) are satisfied. This can be accomplished by adding appropriate assertions to the model; see §8 for an example of this.

There is only one "sign rule" for set expressions:

The sign rule for set expressions. Given a product-free set expression, the following must be true of its reduced form (37):
SS1 Each function f_j, and any functions used in their arguments, must be affine.

Unlike the product-free rule for numerical expressions, functions involved in set expressions are *required* to be affine. To understand why this must be the case, one must understand how an expression of the form (37) is interpreted. A simple example should suffice; for the $L = 0$ case,

$$\begin{aligned} x \in a + \sum_{i=1}^{n} b_i S_i \iff \\ \exists\,(t_1, t_2, \ldots, t_n) \in S_1 \times S_2 \times \cdots \times S_n \quad x = a + \sum_{i=1}^{n} b_i t_i \end{aligned} \tag{43}$$

When $L > 0$, similar substitutions are made recursively in the function arguments as well, producing a similar result: a series of simple set membership constraints of the form $t_k \in S_k$, and a single equality constraint. Thus in order to insure that this implied equality constraint is convex, set expressions (specifically, those used in constraints) must reduce to affine combinations of sets. (Of course, set expressions used in assertions are not constrained in this manner.)

6.4 The composition rules

A basic principle of convex analysis is that the composition of a convex function with an affine mapping remains convex. In fact, under certain conditions, similar guarantees can be made for compositions with nonlinear mappings as well. The ruleset incorporates a number of these conditions, and we have called them the *composition rules.*

Designing the composition rules required a balance between simplicity and expressiveness. In [29], a relatively simple composition rule for convex functions is presented:

Lemma 1. *If* $f : \mathbb{R} \to (\mathbb{R} \cup +\infty)$ *is convex and nondecreasing and* $g : \mathbb{R}^n \to (\mathbb{R} \cup +\infty)$ *is convex, then* $h = f \circ g$ *is convex.*

So, for example, if $f(y) = e^y$ and $g(x) = x^2$, then the conditions of the lemma are satisfied, and $h(x) = f(g(x)) = e^{x^2}$ is convex. Similar composition rules are given for concave and/or nonincreasing functions as well:

- If $f : \mathbb{R} \to (\mathbb{R} \cup +\infty)$ is convex and nonincreasing and $g : \mathbb{R}^n \to (\mathbb{R} \cup -\infty)$ is concave, then $f \circ g$ is convex.
- If $f : \mathbb{R} \to (\mathbb{R} \cup -\infty)$ is concave and nondecreasing and $g : \mathbb{R}^n \to (\mathbb{R} \cup -\infty)$ is concave, then $f \circ g$ is concave.
- If $f : \mathbb{R} \to (\mathbb{R} \cup -\infty)$ is concave and nonincreasing and $g : \mathbb{R}^n \to (\mathbb{R} \cup +\infty)$ is convex, then $f \circ g$ is concave.

In addition, similar rules are described for functions with multiple arguments.

One way to interpret these composition rules is that they only allow those nonlinear compositions that can be to be *separated* or *decomposed.* To explain, consider a nonlinear inequality $f(g(x)) \leq y$, where f is convex and nondecreasing and g is convex, thus satisfying the conditions of Lemma 1. Then it can be shown that

$$f(g(x)) \leq y \quad \Longleftrightarrow \quad \exists z \; f(z) \leq y, \; g(x) \leq z \tag{44}$$

Similar decompositions can be constructed for the other composition rules as well. Decompositions serve as an important component in the conversion of DCPs into solvable form. Thus the composition rules guarantee that equivalence is reserved when these decompositions are performed.

Now the composition rules suggested by Lemma 1 and its related corollaries are a good start. But despite their apparent simplicity, they require a surprising amount of care to apply. In particular, the use of extended-valued functions is a necessary part of the lemma and has subtle impact. For example, consider the functions

$$f(y) = y^2, \quad g(x) = \|x\|_2 \tag{45}$$

Certainly, $h(x) = f(g(x)) = \|x\|_2^2$ is convex; but Lemma 1 would not predict this, because $f(y)$ is not monotonic. A sensible attempt to rectify the problem would be to restrict the domain of the function (in the real-valued sense) to the nonnegative orthant, where it is nondecreasing:

$$\bar{f}(y) = \begin{cases} y^2 & y \geq 0 \\ +\infty & y < 0 \end{cases} \tag{46}$$

But while $\bar{f}$ is nondecreasing over its domain, it is *nonmonotonic* in the extended-valued sense, so the lemma does not apply. The only way to reconcile Lemma 1 with this example is to introduce a far less intuitive version of f which extends it in a nondecreasing fashion:

$$\bar{\bar{f}}(y) = \begin{cases} y^2 & y \geq 0 \\ 0 & y < 0 \end{cases} \tag{47}$$

Figure 4 provides a graph of each function. Forcing users of disciplined con-

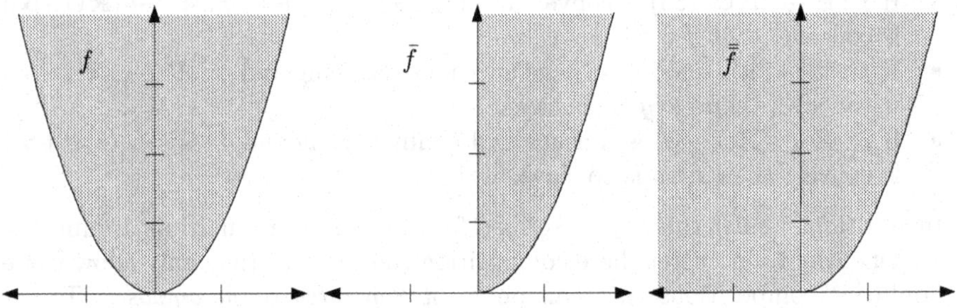

Fig. 4. Three different "versions" of $f(y) = y^2$.

vex programming to consider such technical conditions seems an unnecessary complication, particularly when the goal is to *simplify* the construction of CPs.

To simplify the use of these composition rules, we begin by recognizing something that seems intuitively obvious: f need only be nondecreasing over the range of g. We can formalize this intuition as follows:

Lemma 2. *Let $f : \mathbb{R} \to (\mathbb{R} \cup +\infty)$ and $g : \mathbb{R}^n \to (\mathbb{R} \cup +\infty)$ be two convex functions. If f is nondecreasing over the range of g—i.e., the interval $g(\mathbb{R}^n)$—then $h = f \circ g$ is convex.*

Proof. Let $x_1, x_2 \in \mathbb{R}^n$ and let $\theta \in [0, 1]$. Because g is convex,

$$g(\theta x_1 + (1-\theta)x_2) \leq \theta g(x_1) + (1-\theta)g(x_2) \leq \max\{g(x_1), g(x_2)\} \tag{48}$$

The right-hand inequality has been added to establish that

$$[g(\theta x_1 + (1-\theta)x_2), \theta g(x_1) + (1-\theta)g(x_2)] \subseteq \\ [g(\theta x_1 + (1-\theta)x_2), \max\{g(x_1), g(x_2)\}] \subseteq g(\mathbb{R}^n) \tag{49}$$

f is therefore nondecreasing over this interval; so

$$\begin{aligned} f(g(\theta x_1 + (1-\theta)x_2)) &\leq f(\theta g(x_1) + (1-\theta)g(x_2)) && (f \text{ nondecreasing}) \\ &\leq \theta f(g(x_1)) + (1-\theta)f(g(x_2)) && (f \text{ convex}) \end{aligned} \tag{50}$$

establishing that $h = f \circ g$ is convex.

This lemma does indeed predict the convexity of (45): f is nondecreasing over the interval $[0, +\infty)$, and $g(\mathbb{R}^n) = [0, +\infty)$, which coincide perfectly; hence, $f \circ g$ is convex.

And yet this revised lemma, while more inclusive, presents its own challenge. A critical goal for these composition rules is that adherence can be quickly, reliably, and automatically verified; see §8. The simple composition rules such as Lemma 1 plainly satisfy this condition; but can we be sure to accomplish this with these more complex rules? We claim that it is simpler than it may first appear. Note that our example function $f(x) = x^2$ is nondecreasing over a *half-line*; specifically, for all $x \in [0, +\infty)$. This will actually be true for *any* nonmonotonic scalar function:

Lemma 3. *Let $f : \mathbb{R} \to \mathbb{R} \cup +\infty$ be a convex function which is nondecreasing over some interval $\bar{F} \subset \mathbb{R}$, $\operatorname{Int} \bar{F} \neq \emptyset$. Then it is, in fact, nonincreasing over the entire half-line $F = \bar{F} + [0, +\infty)$; that is,*

$$F = (F_{\min}, +\infty) \quad \text{or} \quad F = [F_{\min}, +\infty). \tag{51}$$

Proof. If $\bar{F}$ already extends to $+\infty$, we are done. Otherwise, select any two points $x_1, x_2 \in \bar{F}$ and a third point $x_3 > x_2$. Then

$$\begin{aligned} & f(x_2) \leq \alpha f(x_1) + (1-\alpha)f(x_3), \quad \alpha \triangleq (x_2 - x_1)/(x_3 - x_1) \\ & \implies \quad f(x_2) \leq \alpha f(x_2) + (1-\alpha)f(x_3) \quad \implies \quad f(x_2) \leq f(x_3). \end{aligned} \tag{52}$$

So $f(x_3) \geq f(x_2)$ for all $x_3 > x_2$. Now consider another point $x_4 > x_3$; then

$$\begin{aligned} & f(x_3) \leq \bar{\alpha} f(x_2) + (1-\bar{\alpha})f(x_4), \quad \bar{\alpha} \triangleq (x_3 - x_2)/(x_4 - x_2) \\ & \implies \quad f(x_3) \leq \bar{\alpha} f(x_3) + (1-\bar{\alpha})f(x_4) \quad \implies \quad f(x_4) \leq f(x_3) \end{aligned} \tag{53}$$

So $f(x_4) \geq f(x_3)$ for all $x_4 > x_3 > x_2$; that is, f is nondecreasing for all $x > x_2 \in \bar{F}$.

In other words, *any* scalar convex function which is nondecreasing between two points is so over an entire half-line. So determining whether f is nondecreasing over $g(\mathbf{dom}\, g)$ reduces to a single comparison between $F_{\min}$ and $\inf_x g(x)$. For concave or nonincreasing functions, similar intervals can be constructed:

- f convex, nonincreasing: $F = (-\infty, F_{\max}]$ or $F = (-\infty, F_{\max})$

- f concave, nondecreasing: $F = (-\infty, F_{\max}]$ or $F = (-\infty, F_{\max})$
- f concave, nonincreasing: $F = [F_{\min}, +\infty)$ or $F = (F_{\min}, +\infty)$

As we show in §7, it is straightforward to include such intervals in the atom library, so that they are readily available to verify compositions.

The task of determining $\inf_x g(x)$ or $\sup_x g(x)$ remains. In §7, we show that function ranges are *included* in the atom library for just this purpose, alongside information about their curvature and monotonicity properties. But often the inner expression $g(x)$ is not a simple function call, but an expression reducible to the form (33):

$$g(x) = \inf_x a + \sum_{i=1}^{n} b_i x_i + \sum_{j=1}^{L} c_j f_j(\ldots) \tag{54}$$

We propose the following heuristic in such cases. Let $X_i \subseteq \mathbf{R}$, $i = 1, 2, \ldots, n$, be simple interval bounds on the variables, retrieved from any simple bounds present in the model. In addition, let $F_j = f_j(\mathbf{R}) \subseteq \mathbf{R}$, $j = 1, 2, \ldots, L$, be the range bounds retrieved from the atom library. Then

$$g(\mathbf{R}) \subseteq a + \sum_{i=1}^{n} b_i X_i + \sum_{j=1}^{L} c_j F_j \tag{55}$$

So (55) provides a conservative bound on $\inf_x g(x)$ or $\sup_x g(x)$, as needed. In practice, this heuristic proves sufficient for supporting the composition rules in nearly all circumstances. In those exceptional circumstances where it the bound is too conservative, and the heuristic fails to detect a valid composition, a model may have to be rewritten slightly—say, by manually performing the decomposition (44) above. It is a small price to pay for added expressiveness in the vast majority of cases.

Generalizing these composition rules to functions with multiple arguments is straightforward, but requires a bit of technical care. The result is as follows:

The composition rules. Consider a numerical expression of the form

$$f(arg_1, arg_2, \ldots, arg_m) \tag{56}$$

where f is a function from the atom library. For each argument arg_k, construct a bound $G_k \subseteq \mathbf{R}$ on the range using the heuristic described above, so that

$$(arg_1, arg_2, \ldots, arg_m) \in G = G_1 \times G_2 \times \cdots \times G_m \tag{57}$$

Given these definitions, (56) must satisfy exactly one of the following rules:

C1-C3 If the expression is expected to be convex, then f must be affine or convex, and one of the following must be true for each $k = 1, \ldots, m$:

C1 f is nondecreasing in argument k over G, and arg_k is convex; or

C2 f is nonincreasing in argument k over G, and arg_k is concave; or

C3 arg_k is affine.

C4-C6 If the expression is expected to be concave, then f must be affine or concave, and one of the following must be true for each $k = 1, \ldots, m$:

C4 f is nondecreasing in argument k over G, and arg_k is concave; or

C5 f is nonincreasing in argument k over G, and arg_k is convex; or

C6 arg_k is affine.

C7 If the expression is expected to be affine, then f must be affine, and each arg_k is affine for all $k = 1, \ldots, m$.

7 The atom library

The second component of disciplined convex programming is the *atom library*. As a concept, the atom library is relatively simple: an extensible list of functions and sets whose properties of curvature/shape, monotonicity, and range are known. The description of the convexity ruleset in §6 shows just how this information is utilized.

As a tangible entity, the atom library requires a bit more explanation. In cvx, the library is a collection of text files containing descriptions of functions and sets. Each entry is divided into two sections: the *declaration* and the *implementation*. The declaration is divided further into two components:

- the *prototype*: the name of the function or set, the number and structure of its inputs, and so forth.
- the *attribute list*: a list of descriptive statements concerning the curvature, monotonicity, and range of the function; or the shape of the set.

The *implementation* is a computer-friendly description of the function or set that enables it to be used in numerical solution algorithms. What is important to note here is that the implementation section is *not used* to determine whether or not a particular problem is a DCP. Instead, it comes into play only *after* a DCP has been verified, and one wishes to compute a numerical solution. For this reason, we will postpone the description of the implementation until §10.

7.1 The prototype

A function prototype models the usage syntax for that function, and in the process lists the number and dimensions of its arguments; *e.g.*,

```
function sq( x );
function max( x, y );
```

$$f(x) = x^2 \tag{58}$$

$$g(x,y) = \max\{x,y\} \tag{59}$$

Some functions are parameterized; *e.g.*, $h_p(x) = \|x\|_p$. In cvx, parameters are included in the argument list along with the arguments; *e.g.*,

```
function norm_p( p, x[n] ) convex( x ) if p >= 1;
```

$$h_p(x) = \|x\|_p \quad (p \geq 1) \tag{60}$$

Parameters are then distinguished from variables through the curvature attribute; see §7.2 below.

The `norm_p` example also illustrates another feature of cvx, which is to allow *conditions* to be placed on the a function's parameters using an `if` construct. In order to use a function with preconditions, they must be enforced somehow; if necessary, by using an assertion. For example, `norm_p( 2.5, x )` would be verified as valid; but if `b` is a parameter, `norm_p( b, x )` would not be, unless the value of `b` could somehow be guaranteed to be greater than 1; for example, unless an assertion like `b > 1` was provided in the model.

Set prototypes look identical to that of functions:

```
set integers( x );
set simplex( x[n] );
set less_than( x, y );
```

$$A = \mathbb{Z} = \{\ldots, -2, -1, 0, 1, 2, \ldots\} \tag{61}$$

$$B = \{\, x \in \mathbb{R}^n \mid x \geq 0,\ \textstyle\sum_i x_i = 1 \,\} \tag{62}$$

$$C = \{\, (x,y) \in \mathbb{R} \times \mathbb{R} \mid x < y \,\} \tag{63}$$

Unlike functions, the actual *usage* of a set differs from its prototype—the arguments are in fact the components of the set, and therefore appear to the left of a set membership expression: *e.g.*,

```
x in integers;
y in simplex;
(x,y) in less_than;
```

$$x \in A \tag{64}$$

$$y \in B \tag{65}$$

$$(x,y) \in C \tag{66}$$

For parameterized sets, there is yet another difference: the parameters are supplied in a *separate* parameter list, preceding the argument list: *e.g.*,

```
set ball_p( p )( x[n] ) if p >= 1;
```

$$B_p(x) = \{\, x \in \mathbb{R}^n \mid \|x\|_p \leq 1 \,\} \quad (p \geq 1) \tag{67}$$

This parameter list remains on the right-hand side of the constraint along with the name of the set:

```
z in ball_p( q );
```

$$z \in B_q \tag{68}$$

7.2 Attributes

As we have seen in §6, the convexity ruleset depends upon one or more of the following pieces of information about each function and set utilized in a DCP. For sets, it utilizes just one piece of information:

- *shape*: specifically, whether or not the set is convex.

For functions, a bit more information is used:

- *curvature*: whether the function is convex, concave, affine, or otherwise.
- *monotonicity*: whether the functions are nonincreasing or nondecreasing; and over what subsets of their domain they are so.
- *range*: the minimum of convex functions and the maximum of concave functions.

The `cvx` framework allows this information to be provided through the use of *attributes*: simple text tags that allow each of the above properties to be identified as appropriate.

Shape

For sets, only one attribute is recognized: `convex`. A set is either convex, in which case this attribute is applied, or it is not. Given the above four examples, only `integers` is not convex:

```
set integers( x );                                  (69)
set simplex( x[n] ) convex;                         (70)
set less_than( x, y ) convex;                       (71)
set ball_p( p )( x[n] ) convex if p >= 1;           (72)
```

Sets which are not convex are obviously of primary interest for DCP, but non-convex sets may be genuinely useful, for example, for restricting the values of parameters to realistic values.

Curvature

Functions can be declared as `convex`, `concave`, or `affine`, or none of the above. Clearly, this last option is the least useful; but such functions can be used in constant expressions or assertions. No more than one curvature keyword can be supplied. For example:

`function max( x, y ) convex;`	$g(x,y) = \max\{x,y\}$	(73)
`function min( x, y ) concave;`	$f(x,y) = \min\{x,y\}$	(74)
`function plus( x, y ) affine;`	$p(x,y) = x + y$	(75)
`function sin( x );`	$g(x) = \sin x$	(76)

By default, a function declared as convex, concave, or affine is assumed to be *jointly* so over all of its arguments. It is possible to specify that it is so only over a subset of its arguments by listing those arguments after the curvature keyword; for example,

```
function norm_p( p, x[n] ) convex( x ) if p >= 1;
```
$$h_p(x) = \|x\|_p \quad (p \geq 1) \tag{77}$$

In effect, this convention allows parameterized functions to be declared: arguments omitted from the list are treated as the parameters of the function and are expected to be constant.

Disciplined convex programming allows only functions which are *globally* convex or concave to be specified as such. Functions which are "sometimes" convex or concave—that is, over a subset of their domains—are not permitted. For example, the simple inverse function

$$f : \mathbb{R} \to \mathbb{R}, \quad f(x) \triangleq 1/x \qquad (x \neq 0) \tag{78}$$

is neither convex nor concave, and so cannot be used to construct a DCP. However, we commonly think of f as convex *if x is known to be positive.* In disciplined convex programming, this understanding must be realized by defining a *different function*

$$f_{\mathsf{cvx}} : \mathbb{R} \to \mathbb{R}, \quad f(x) \triangleq \begin{cases} 1/x & x > 0 \\ +\infty & x \leq 0 \end{cases} \tag{79}$$

which is globally convex, and can therefore be used in DCPs. Similarly, the power function

$$g : \mathbb{R}^2 \to \mathbb{R}, \quad g(x,y) \triangleq x^y \qquad \text{(when defined)} \tag{80}$$

is convex or concave on certain subsets of $\mathbb{R}^2$, such as:

- convex for $x \in [0, \infty)$ and fixed $y \in [1, \infty)$.
- concave for $x \in [0, \infty)$ and fixed $y \in (0, 1]$;
- convex for fixed $x \in (0, \infty)$ and $y \in \mathbb{R}$

In order to introduce nonlinearities such as $x^{2.5}$ or $x^{0.25}$ into a DCP, then, there must be appropriate definitions of these "restricted" versions of the power function:

$$f_y : \mathbb{R} \to \mathbb{R}, \quad f_y(x) = \begin{cases} x^y & x \geq 0 \\ +\infty & x < 0 \end{cases} \quad (y \geq 1) \tag{81}$$

$$g_y : \mathbb{R} \to \mathbb{R}, \quad g_y(x) = \begin{cases} x^y & x \geq 0 \\ -\infty & x < 0 \end{cases} \quad (0 < y < 1) \tag{82}$$

$$h_y : \mathbb{R} \to \mathbb{R}, \quad h_y(x) = y^x \quad (y > 0) \tag{83}$$

Thus the disciplined convex programming approach forces the user to consider convexity more carefully. We consider this added rigor an advantage, not a liability.

Monotonicity

The monotonicity of a function with respect to its arguments proves to be a key property exploited by the ruleset. For this reason, the `cvx` atom library provides the keywords `increasing`, `nondecreasing`, `nonincreasing`, or `decreasing` in each of its arguments. Each argument can be given a separate declaration:

```
function exp( x ) convex increasing;
```

$$f(x) = e^x \tag{84}$$

As far as the convexity ruleset is concerned, strict monotonicity is irrelevant; so, for example, `increasing` and `nondecreasing` are effectively synonymous, as are `decreasing` and `nonincreasing`.

There is one somewhat technical but critical detail that must be adhered to when declaring a function to be monotonic. Specifically, monotonicity must be judged in the *extended-valued* sense. For example, given $p \geq 1$, the function

```
function pow_p( p, x )
    convex( x ) if p >= 1;
```

$$f_p(x) = \begin{cases} x^p & x \geq 0 \\ +\infty & x < 0 \end{cases} \tag{85}$$

is increasing (and, therefore, nondecreasing) over its domain. However, in the extended-valued sense, the function is *nonmonotonic*, so f_p cannot be declared as globally nondecreasing.

As suggested in §6.4, the `cvx` atom library allows *conditions* to be placed on monotonicity. So, for example, $\tilde{f}_p(x)$ is, of course, nondecreasing over $x \in [0, +\infty)$, suggesting the following declaration:

```
function pow_p( p, x ) convex( x ) if p >= 1,
                       increasing( x ) if x >= 0;
```
(86)

Multiple declarations are possible: for example, the function $f(x) = x^2$ is both nonincreasing over $x \in (-\infty, 0]$ and nondecreasing over $x \in [0, +\infty)$:

```
function sq( x ) convex, decreasing if x <= 0,
                   increasing( x ) if x >= 0;
```
(87)

Each argument of a function with multiple inputs can be given a separate, independent monotonicity declaration. For example, $f(x, y) = x - y$ is increasing in x and decreasing in y:

```
function minus( x, y )
        affine increasing( x ) decreasing( y );
```
(88)

Range

Each function definition can include a declaration of its its *range*, using a simple inequality providing a lower or upper bound for the function. For example,

```
function exp( x ) convex, increasing, >= 0;
```
$$f(x) = e^x \tag{89}$$

As with the monotonicity operations, the range must indeed be specified in the extended-valued sense, so it will inevitably be one-sided: that is, all convex functions are unbounded above, and all concave functions are unbounded below.

8 Verification

In order to solve a problem as a DCP, one must first establish that it is indeed a valid DCP—that is, that it involves only functions and sets present in the atom library, and combines them in a manner compliant with the complexity ruleset. A proof of validity is necessarily hierarchical in nature, reflecting the structure of the problem and its expressions. To illustrate the process, consider the simple optimization problem

$$\begin{array}{rl} \text{minimize} & cx \\ \text{subject to} & \exp(y) \le \log(a\sqrt{x} + b) \\ & ax + by = d \end{array} \tag{90}$$

where a, b, c, d are parameters, and x, y are variables. A cvx version of this model is given in Figure 5. Note in particular the explicit declarations of the three atoms exp, log, and sqrt. Usually these declarations will reside in an external file, but we include them here to emphasize that every atom used in a model must be accounted for in the atom library.

```
minimize     c x;
subject to   exp( y ) <= log( a sqrt( x ) + b );
             a x + b y = d;
variables    x, y;
parameters   a, b, c, d;
function     exp( x )  convex   increasing >= 0;
function     sqrt( x ) concave nondecreasing;
function     log( x )  concave increasing;
```

Fig. 5. The cvx specification for (90).

It is helpful to divide the proof into two stages. The first stage verifies that each of expressions involved is product-free. Below is a textual description of this stage. Each line has been indented to represent the hierarchy present in the proof, and includes the rule employed to establish that line of the proof:

cx is product-free, because (**PN6**)
    c is a constant expression(**T8**)
    x is product-free (**PN1**)

$\exp(y)$ is product-free, because (**PN3**)
  y is product-free (**PN2**)
$\log(a\sqrt{x}+b)$ is product-free, because (**PN3**)
  $a\sqrt{x}+b$ is product-free, because (**PN4**)
    $a\sqrt{x}$ is product-free, because (**PN6**)
      a is a constant expression (**PN2**)
      $\sqrt{x}$ is product-free, because (**PN3**)
        x is product-free (**PN2**)
    b is product-free (**PN2,T8**)
$ax+by$ is product-free, because(**PN4**)
  ax is product-free, because (**PN6**)
    a is a constant expression (**T8**)
    x is product-free (**PN1**)
  by is product-free, because (**PN6**)
    b is a constant expression (**T8**)
    y is product-free (**PN1**)
d is product-free (**PN2**)

The second stage proceeds by verifying that the top-level, sign, and composition rules in a similarly hierarchical fashion:

The minimization problem is valid if $a \geq 0$, because (**T1**)
  The objective function is valid, because (**T1**)
    cx is convex (**SN1**)
  The first constraint is valid if $a \geq 0$, because (**T1**)
    $\exp(y) \leq \log(a\sqrt{x}+b)$ is convex if $a \geq 0$, because (**T5**)
      $\exp(y)$ is convex, because (**SN1**)
        $\exp(y)$ is convex, because (**C1**)
          $\exp(\cdot)$ is convex and nondecreasing (atom library)
          y is convex (**SN1**)
      $\log(a\sqrt{x}+b)$ is concave if $a \geq 0$, because (**SN2**)
        $\log(a\sqrt{x}+b)$ is concave if $a \geq 0$, because (**C4**)
          $\log(\cdot)$ is concave and nondecreasing (atom library)
          $a\sqrt{x}+b$ is concave if $a \geq 0$, because (**SN2**)
            $\sqrt{x}$ is concave, because (**C4**)
              $\sqrt{\cdot}$ is concave and nondecreasing (atom library)
              x is concave (**SN2**)
  The second constraint is valid, because (**T1**)
    $ax+by=d$ is convex, because (**T4**)
      $ax+by$ is affine (**SN3**)
      d is affine (**SN3**)

It can be quite helpful to examine the structure of the problem and its validity proof graphically. Figure 6 presents an *expression tree* of the problem (90), annotated with the relevant rules verified at each position in the tree.

The verification process is guaranteed to yield one of three conclusions:

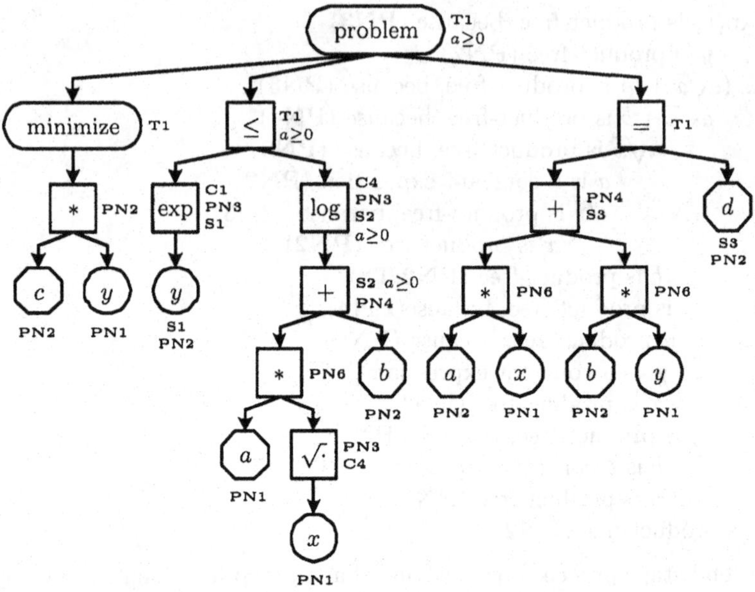

Fig. 6. An expression tree for (90), annotated with applicable convexity rules.

- *valid*: the rules are fully satisfied.
- *conditionally valid*: the rules will be fully satisfied if one or more additional preconditions on the parameters are satisfied.
- *invalid*: one or more of the convexity rules has been violated.

In this case, a conclusion of *conditionally valid* has been reached: the analysis has revealed that an additional condition $a \geq 0$ must satisfied. If this precondition were somehow assured, then the proof would have conclusively determined that the problem is a valid DCP. One simple way to accomplish this would be to add it as an assertion; *i.e.*, by adding the assertion `a >= 0` to the list of constraints. If, on the other hand, we were to do the opposite and add an assertion `a < 0`, the sign rule **SN2** would be violated; in fact, the expression $a\sqrt{x} + b$ would be verifiably convex.

The task of verifying DCPs comprises yet another approach to the challenge of automatic convexity verification described in §3.5. Like the methods used to verify LPs and SDPs, a certain amount of structure is assumed via the convexity rules that enables the verification process to proceed in a reliable and deterministic fashion. However, unlike these more limited methods, disciplined convex programming maintains generality by allowing new functions and sets to be added to the atom library. Thus disciplined convex programming provides a sort of *knowledgebase* environment for convex programming, in which human-supplied information about functions and sets is used to expand the body of problems that can be recognized as convex.

9 Creating disciplined convex programs

As mentioned previously, adherence to the convexity ruleset is sufficient but not necessary to insure convexity. It is possible to construct mathematical programs that are indeed convex, but which fail to be DCPs, because one or more of the expressions involved violates the convexity ruleset.

It is actually quite simple to construct examples of such violations. For example, consider the entropy maximization problem

$$\begin{array}{ll} \text{maximize} & -\sum_{i=1}^{n} x_i \log x_i \\ \text{subject to} & Ax = b \\ & \mathbf{1}^T x = 1 \\ & x \geq 0 \end{array} \tag{91}$$

where $x \in \mathbb{R}^n$ is the problem variable and $A \in \mathbb{R}^{m \times n}$ and $b \in \mathbb{R}^m$ are parameters; and $\log(\cdot)$ is defined in the atom library. The expression $x_i \log x_i$ violates the product-free rule **PS6**—and as a result, (91) is not a DCP, even though it is a well-known CP.

Alternatively, consider the GP in convex form,

$$\begin{array}{lll} \text{minimize} & \log \sum_{k=1}^{K_0} \exp(a_{0k}^T x + b_{0k}) & \\ \text{subject to} & \log \sum_{k=1}^{K_i} \exp(a_{ik}^T x + b_{ik}) \leq 0 & i = 1, 2, \ldots, m \\ & A^{(m+1)} x + b^{(m+1)} = 0 & \end{array} \tag{92}$$

where $x \in \mathbb{R}^n$ is the problem variable,

$$\begin{aligned} A^{(i)} &= \begin{bmatrix} a_{i1} \; a_{i2} \; \ldots \; a_{im_i} \end{bmatrix}^T \in \mathbb{R}^m n m_i n \\ b^{(i)} &= \begin{bmatrix} b_{i1} \; b_{i2} \; \ldots \; b_{im_i} \end{bmatrix}^T \in \mathbb{R}^{m_i} \end{aligned} \qquad i = 1, 2, \ldots, m+1 \tag{93}$$

are parameters; and both $\log(\cdot)$ and $\exp(\cdot)$ are defined in the atom library. This problem satisfies the product-free rules, but the objective function and inequality constraints fail either the sign rule **SN1** or composition rule **C4**, depending on how you verify them. But of course, (92) is a CP.

It is important to note that these violations do not mean that the problems cannot solved in the disciplined convex programming framework; it simply means that they must be rewritten in a compliant manner. In both of these cases, the simplest way to do so is to add new functions to the atom library that encapsulate the offending nonlinearities. By adding the two functions

$$f_{\mathrm{entr}}(x) = \begin{cases} -x \log x & x > 0 \\ 0 & x = 0 \\ -\infty & x < 0 \end{cases} \qquad f_{\mathrm{lse}}(x) = \log \sum_{i=1}^{n} e^{x} \tag{94}$$

to the atom library, both problems can be rewritten as valid DCPs; (91) as

$$\begin{array}{ll} \text{maximize} & \sum_{i=1}^{n} f_{\text{entr}}(x_i) \\ \text{subject to} & Ax = b \\ & \mathbf{1}^T x = 1 \\ & x \geq 0 \end{array} \tag{95}$$

and the GP (92) as

$$\begin{array}{ll} \text{minimize} & f_{\text{lse}}(A^{(0)}x + b^{(0)}) \\ \text{subject to} & f_{\text{lse}}(A^{(i)}x + b^{(i)}) \leq 0, \quad i = 1, 2, \ldots, m \\ & A^{(m+1)}x + b^{(m+1)} = 0 \end{array} \tag{96}$$

The ability to extend the atom library as needed has the potential to be taken to an inelegant extreme. For example, consider the problem

$$\begin{array}{ll} \text{minimize} & f_0(x) \\ \text{subject to} & f_m(x) \leq 0, \quad k = 1, 2, \ldots, m \end{array} \tag{97}$$

where the functions $f_0, f_1, \ldots, f_m$ are all convex. One way to cast this problem as a DCP would simply be to add all $m+1$ of the functions to the atom library. The convexity rules would then be satisfied rather trivially; and yet this would likely require *more*, not *less*, effort than a more traditional NLP modeling method. In practice, however, the functions f_k are rarely monolithic, opaque objects. Rather, they will be constructed from components such as affine forms, norms, and other known functions, combined in ways consistent with the basic principles of convex analysis, as captured in the convexity ruleset. It is *those* functions that are ideal candidates for addition into the atom library.

We should add that once an atom is defined and implemented, it can be freely reused across many DCPs. The atoms can be shared with other users as well. The effort involved in adding a new function to the atom library, then, is significantly amortized. A collaborative hierarchy is naturally suggested, wherein more advanced users can create new atoms for application-specific purposes, while novice users can take them and employ them in their models without regard for how they were constructed.

We argue, therefore, that (91) and (92) are ideal examples of the kinds of problems that disciplined convex programming is intended to support, so that the convexity ruleset poses little practical burden in these cases. While it is true that the term $-x_i \log x_i$ violates the product-free rules, someone interested in entropy maximization does not consider this expression as a product of nonlinearities but rather as a single, encapsulated nonlinearity—as represented by the function f_{expr}. In a similar manner, those studying geometric programming treat the function $f_{\text{expr}}(y) = \log \sum \exp(y_i)$ as a monolithic convex function; it is irrelevant that it happens to be the composition of a concave function and a convex function. Thus the addition of these functions to the atom library coincides with the intuitive understanding of the problems that employ them.

Still, the purity of the convexity rules prevent even the use of obviously convex quadratic forms such as $x^2 + 2xy + y^2$ in a model. It could be argued that this is impractically restrictive, since quadratic forms are so common. And indeed, we are considering extending the relaxation of the product-free rules to include quadratic forms. However, in many cases, a generic quadratic form may in fact represent a quantity with more structure or meaning. For example, traditionally, the square of a Euclidean norm $\|Ax + b\|_2^2$ would be converted to a quadratic form

$$\|Ax + b\|_2 = x^T P x + q^T x + r, \quad P \triangleq A^T A, \quad q \triangleq A^T b, \quad r \triangleq b^T b \tag{98}$$

But within a DCP, this term can instead be expressed as a composition

$$\|Ax + b\|_2 = f(g(Ax + b)), \quad f(y) \triangleq y^2, \quad g(z) \triangleq \|z\|_2 \tag{99}$$

In disciplined convex programming, there is no natural bias against (99), so it should be preferred over the converted form (98) simply because it reflects the original intent of the problem. So we argue that support for generic quadratic forms would be at least somewhat less useful than in a more traditional modeling framework. Furthermore, we can easily support quadratic forms with the judicious addition of functions to the atom library, such as the function $f(y) = y^2$ above, or a more complex quadratic form such as

$$f_Q : \mathbb{R}^n \to \mathbb{R}, \quad f_Q(x) = x^T Q x \qquad (Q \succeq 0). \tag{100}$$

Thus support for quadratic forms is a matter of convenience, not necessity.

10 Implementing atoms

As enumerated in §3.2, there are a variety of methods that can be employed to solve CPs: primal/dual methods, barrier methods, cutting-plane methods, and so forth. All of these methods can be adapted to disciplined convex programming with minimal effort. Limited space prohibits us from examining these methods in detail; please see [74] for a more thorough treatment of the topic. It is sufficient here to say this: that each of these methods will need to perform certain computations involving each of the atoms, each of the functions and sets, employed in the problems they solve. The purpose of the *implementation* of an atom, first introduced in §7 above, is to provide these solvers with the means to perform these calculations.

Disciplined convex programming and the cvx modeling framework distinguish between two different types of implementations:

- a *simple* implementation, which provides traditional calculations such as derivatives, subgradients and supergradients for functions; and indicator functions, barrier functions, and cutting planes for sets; and

- a *graph* implementation, in which the function or set is defined *as the solution to another DCP.*

The computations supported by simple implementations should be quite familiar to anyone who studies the numerical solution of optimization problems. To the best of our knowledge, however, the concept of graph implementations is new, and proves to be an important part of the power and expressiveness of disciplined convex programming.

In `cvx`, an implementation is surrounded by curly braces, and consists of a list of key/value pairs with the syntax *key* := *value*. See Figures 7 and 8 for examples. It is also possible for an implementation to be constructed in a lower-level language like C, but we will not consider that feature here.

10.1 Simple implementations

Any continuous function can have a simple implementation. Simple function implementations use the following *key* := *value* entries:

- `value`: the value of the function.
- `domain_point`: a point on the interior of the domain of the function. If omitted, then the origin is assumed to be in the domain of the function.
- For differentiable functions:
  - `gradient`: the first derivative.
  - `Hessian` (if twice differentiable): the second derivative.
- For nondifferentiable functions:
  - `subgradient` (if convex): a subgradient of a function f at point $x \in \mathbf{dom}\, f$ is any vector $v \in \mathbb{R}^n$ satisfying

$$f(y) \geq f(x) + v^T(y - x) \quad \forall y \in \mathbb{R}^n \tag{101}$$

  - `supergradient` (if concave): a supergradient of a function g at point $x \in \mathbf{dom}\, g$ is any vector $v \in \mathbb{R}^n$ satisfying

$$g(y) \leq g(x) + v^T(y - x) \quad \forall y \in \mathbb{R}^n \tag{102}$$

It is not difficult to see how different algorithms might utilize this information. Most every method would use `value` and `domain_point`, for example. A smooth CP method would depend on the entries `gradient` and `Hessian` to calculate Newton search directions. A localization method would use the entries `gradient`, `subgradient`, and `supergradient` to compute cutting planes.

Any set with a non-empty interior can have a simple implementation. Simple set implementations use the following *key* := *value* entries:

- `interior_point`: a point on the interior of the set.
- `indicator`: an expression that is 0 for points inside the set, and $+\infty$ for points outside the set.
- At least one, but ideally both, of the following:

- **barrier**: a reference to a convex, twice differentiable *barrier function* for the set, declared separately as a function atom with a direct implementation.
- **oracle**: a cutting plane oracle for the set. Given a set $S \subset \mathbb{R}^n$, the cutting plane oracle accepts as input a point $x \in \mathbb{R}^n$; and, if $x \notin S$, returns a separating hyperplane; that is, a pair $(a, b) \in \mathbb{R}^n \times \mathbb{R}$ satisfying

$$a^T x > b, \quad S \subseteq \{\, y \mid a^T y \leq b \,\} \tag{103}$$

If $x \in S$, then the oracle returns $(a, b) = (0, 0)$.

```
function min( x, y ) concave, nondecreasing {
    value := x < y ? x : y;
    supergradient := x < y ? ( 1, 0 ) : ( 0, 1 );
}
set pos( x ) convex {
    interior_point := 1.0;
    indicator := x < 0 ? +Inf : 0;
    oracle := x < 0 ? ( 1, 0 ) : ( 0, 0 );
    barrier := neglog( x );
}
function neglog( x ) convex {
    domain_point := 1.0;
    value := x <= 0 ? +Inf : - log( x );
    gradient := - 1 / x;
    Hessian := 1 / x^2;
}
```

Fig. 7. Simple implementations.

Figure 7 presents several examples of simple implementations. Again, we do not wish to document the cvx syntax here, only illustrate key the feasibility of this approach. Note that the set pos has been given both a barrier function and a cutting plane generator, allowing it to be used in both types of algorithms.

10.2 Graph implementations

A fundamental principle in convex analysis is the very close relationship between convex and concave functions and convex sets. A function $f : \mathbb{R}^n \to \mathbb{R} \cup +\infty$ is convex if and only if its *epigraph*

$$F = \operatorname{epi} f = \{\, (x, y) \in \mathbb{R}^n \times \mathbb{R} \mid f(x) \leq y \,\} \tag{104}$$

is a convex set. Likewise, a function $g : \mathbb{R}^n \to \mathbb{R} \cup +\infty$ is concave if and only if its *hypograph*

$$G = \operatorname{hypo} g = \{\,(x,y) \in \mathbb{R}^n \times \mathbb{R} \mid g(x) \geq y\,\} \tag{105}$$

is a convex set. These relationships can be expressed in reverse fashion as well:

$$f(x) = \inf\{\,y \mid (x,y) \in F\,\} \tag{106}$$
$$g(x) = \sup\{\,y \mid (x,y) \in G\,\} \tag{107}$$

A *graph implementation* of a function is effectively a representation of the epigraph or hypograph of a function, as appropriate, as a disciplined convex feasibility problem. The cvx framework supports this approach using the following *key* := *value* pairs:

- **epigraph** (if convex): the epigraph of the function.
- **hypograph** (if concave): the hypograph of the function.

A simple example is the absolute value function $f(x) = |x|$. The epigraph of this function is

$$\operatorname{epi} f = \{\,(x,y) \mid |x| \leq y\,\} = \{\,(x,y) \mid -y \leq x \leq y\,\} \tag{108}$$

In Figure 8, we show how this epigraph is represented in cvx. Notice that the

```
function abs( x ) convex, >= 0 {
    value := x < 0 ? -x : x;
    epigraph := { -abs <= x <= +abs; }
}
function min( x, y ) concave, nondecreasing {
    value := x < y ? x : y;
    supergradient := x < y ? ( 1, 0 ) : ( 0, 1 );
    hypograph := { min <= x; min <= y; }
}
function entropy( x ) concave {
    value := x < 0 ? -Inf : x = 0 ? 0 : - x log( x );
    hypograph := { ( x, y ) in hypo_entropy; }
}
set simplex( x[n] ) convex {
    constraints := { sum( x ) = 1; x >= 0; }
}
```

Fig. 8. Graph implementations.

name of the function, abs is used to represent the epigraph variable.

The primary benefit of graph implementations is that they provide an elegant means to define nondifferentiable functions. The absolute value function above is one such example; another is the two-argument minimum $g(x,y) = \min\{x,y\}$. This function is concave, and its hypograph is

$$\operatorname{hypo} g = \{\,(x,y,z) \in \mathbb{R} \times \mathbb{R} \times \mathbb{R} \mid z \leq x,\ z \leq y\,\} \tag{109}$$

Figure 8 shows how the function `min` represents this hypograph in `cvx`. Notice that this function has a simple implementation as well—allowing the underlying solver to decide which it prefers to use. Of course, both `abs` and `min` are rather simple, but far more complex functions are possible. For example, graph implementations can be constructed for each of the norms examined in §2.

More subtle but important instances of nondifferentiability occur in functions that are discontinuous at the boundaries of their domains. These functions require special care as well, and graph implementations provide that. For example, consider the scalar entropy function

$$f : \mathbb{R} \to \mathbb{R}, \quad f(x) \triangleq \begin{cases} -x \log x & x > 0 \\ 0 & x = 0 \\ -\infty & x < 0 \end{cases} \tag{110}$$

This function is smooth over the positive interval, but it is discontinuous at the origin, and its derivative is unbounded near the origin. Both of these features cause problems for some numerical methods [93]. Using the hypograph

$$\operatorname{hypo} f = \mathbf{cl}\,\{\, (x, y) \in \mathbb{R} \times \mathbb{R} \mid x > 0, \ -x \log x > y \,\} \tag{111}$$

can solve these problems. In Figure 8, we show the definition of a function `entropy` that refers to a set `hypo_entropy` representing this hypograph. We have chosen to omit the implementation of this set here, but it would certainly contain a definition of the barrier function

$$\begin{aligned} &\phi : \mathbb{R} \times \mathbb{R} \to (\mathbb{R} \cup +\infty), \\ &\phi(x, y) = \begin{cases} -\log(-y - x \log x) - \log x & (x, y) \in \operatorname{Int} \operatorname{epi} f \\ +\infty & \text{otherwise} \end{cases} \end{aligned} \tag{112}$$

[118], as well as an oracle to compute cutting planes $a_1 x + a_2 y \leq b$, where

$$(a_1, a_2, b) \triangleq \begin{cases} (0, 0, 0) & (x, y) \in \operatorname{hypo} f \\ (-1, 0, 0) & x < 0 \\ (\log(y/2) + 1, 1, y/2) & x = 0, \ y > 0 \\ (\log x + 1, 1, x) & x > 0 \end{cases} \tag{113}$$

Graph implementations can also be used to unify traditional, inequality-based nonlinear programming with conic programming. For example, consider the maximum singular value function

$$f : \mathbb{R}^{m \times n} \to \mathbb{R}, \quad f(X) = \sigma_{\max}(X) = \sqrt{\lambda_{\max}(X^T X)} \tag{114}$$

This function is convex, and in theory could be used in a disciplined convex programming framework. The epigraph of f is

$$\begin{aligned} \operatorname{epi} f &= \{\, (X,y) \in \mathbb{R}^{m\times n} \times \mathbb{R} \mid \sigma_{\max}(X) \le y \,\} \\ &= \left\{ (X,y) \;\middle|\; \begin{bmatrix} yI & X \\ X^T & yI \end{bmatrix} \in \mathcal{S}^n_+,\ y \ge 0 \right\} \end{aligned} \tag{115}$$

where $\mathcal{S}^n_+$ is the set of positive semidefinite matrices. For someone who is familiar with semidefinite programming, (115) is likely quite familiar. By burying this construct within the implementation in the atom library, however, it enables people who are *not* comfortable with semidefinite programming to take advantage of its benefits in a traditional NLP-style problem.

Graph implementations are possible for sets as well, through the use of a a single *key* := *value* pair:

- `constraints`: a list of constraints representing the set.

What this means is that a set can be described in terms of a disciplined convex feasibility problem. There are several reasons why this might be used. For example, graph implementations can be used to represent sets with nonempty interiors, such as the set of n-element probability distributions

$$S = \{\, x \in \mathbb{R}^n \mid x \ge 0,\ \mathbf{1}^T x = 1 \,\} \tag{116}$$

A cvx version of this set is given in Figure 8. Graph implementations can also be used to represent sets using a sequence of smooth inequalities so that smooth CP solvers can support them. For example, the second-order cone

$$\mathcal{Q}^n = \{\, (x,y) \in \mathbb{R}^n \times \mathbb{R} \mid \|x\|_2 \le y \,\} \tag{117}$$

can be represented by smooth inequalities as follows:

$$\mathcal{Q}^n = \{\, (x,y) \in \mathbb{R}^n \times \mathbb{R} \mid x^T x / y - y \le 0,\ y \ge 0 \,\} \tag{118}$$

10.3 Using graph implementations

To solve a DCP involving functions or sets with graph implementations, those transformations must be applied through a process we call *graph expansion*, in which the DCP that describes a given atom is incorporated into the problem. To illustrate what this entails, consider the problem

$$\begin{array}{ll} \text{maximize} & \min\{c_1^T x + d_1, c_2^T x + d_2\} \\ \text{subject to} & Ax = b \\ & x \ge 0 \end{array} \tag{119}$$

employing the function $\min\{\cdot,\cdot\}$. The hypograph of this function, presented in (109) above, allows this problem to be rewritten as

$$\begin{array}{ll} \text{maximize} & \sup\{\, y \mid y \le c_1^T x + d_1,\ y \le c_2^T x + d_2 \,\} \\ \text{subject to} & Ax = b \\ & x \ge 0 \end{array} \tag{120}$$

Incorporating the variable y into the model itself yields expanded result

$$\begin{array}{ll} \text{maximize} & y \\ \text{subject to} & y \leq c_1^T x + d_1 \\ & y \leq c_2^T x + d_2 \\ & Ax = b \\ & x \geq 0 \end{array} \tag{121}$$

It is not difficult to see that this problem is equivalent to the original, and yet now it is a simple LP.

Thus, as stated in §2.3, cvx allows the transformations required to convert DCPs into solvable form to be encapsulated—and graph implementations are how this is accomplished. Indeed, consider once again the ℓ_∞, ℓ_1, ℓ_p, and largest-L minimization problems described in §2. The transformations used to solve those problems in that section are, in fact, the very transformations that cvx would use to solve them as well (or at least, very nearly so).

Because graph implementations are expanded before a numerical algorithm is deployed, they require no adjustment on the part of those algorithms to support them. Thus graph implementations are *algorithm agnostic*: any algorithm which can successfully support simply implemented functions and sets—by computing derivatives, sub/supergradients, barrier functions, *etc.*—can solve problems with functions and sets with graph implementations as well. Put another way, algorithms which previously could not support nondifferentiable functions are enabled to do so through cvx.

The concept of graph implementations is based on relatively basic principles of convex analysis; and yet, an applications-oriented user—someone who is not expert in convex optimization—is not likely to be constructing new atoms with graph implementations. Indeed they are not likely to be constructing new simple implementations either. Thankfully, they do not need to *build* them to *use* them. The implementations themselves can be built by those with more expertise in such details, and shared with applications-oriented users. As the development of the cvx framework continues, the authors will build a library of common and useful functions; and we hope that others will do so and share them with the community of users.

11 Conclusion

In this article, we have introduced a new methodology for convex optimization called *disciplined convex programming*. Disciplined convex programming simplifies the specification, analysis, and solution of convex programs by imposing certain restrictions on their construction. These restrictions are simple and teachable; they are inspired by the basic principles of convex analysis; and they formalize the intuitive practices of many who use convex optimization today. Despite the restrictions, generality is preserved through the expandability of the atom library.

We have enumerated a number of the benefits that disciplined convex programing obtains for practical convex optimization. Verifying that a model is a valid DCP is a straightforward and reliable process. Nondifferentiable functions may be freely employed without fear of sacrificing numerical performance. And while we did not explore in detail how DCPs are solved, we did discussed how the implementation of functions and sets enables a variety of numerical methods to be used—methods whose performance and reliability are well-known. We refer the reader to [74] for more development on this topic.

An overarching goal of the development of disciplined convex programming is *unification*. There are no less than *seven* standard forms for convex programming being studied and used today: LS, LP, QP, SDP, SOCP, GP, and smooth CP. Deciding which form best suits a given application is not always obvious; and for many problems, a custom solver is the only appropriate choice. Unification allows modelers to freely consider all of these problem types simultaneously—because they need not think of them as separate types at all.

The principles of disciplined convex programming have been implemented in the `cvx` modeling framework. The current version employs a simple barrier solver, but we intend to develop a more powerful solver in the future, and we hope to convince other developers to provide a link to `cvx` for their own solvers. We will be disseminating `cvx` freely with BSD-like licensing, and it is our hope that it will be used widely in coursework, research, and applications.

The reader is invited to visit the Web site `http://www.stanford.edu/~boyd/cvx` to monitor the development of `cvx`, to download the latest versions, and to read the accompanying documentation.

References

1. E. Anderson, Z. Bai, C. Bischof, J. Demmel, J. Du Croz, A. Greenbaum, S. Hammarling, A. McKenney, S. Ostrouchov, and D. Sorensen. *LAPACK Users' Guide.* SIAM, 1992.
2. W. Achtziger, M. Bendsoe, A. Ben-Tal, and J. Zowe. Equivalent displacement based formulations for maximum strength truss topology design. *Impact of Computing in Science and Engineering*, 4(4):315–45, December 1992.
3. M. Avriel, R. Dembo, and U. Passy. Solution of generalized geometric programs. *International Journal for Numerical Methods in Engineering*, 9:149–168, 1975.
4. M. Abdi, H. El Nahas, A. Jard, and E. Moulines. Semidefinite positive relaxation of the maximum-likelihood criterion applied to multiuser detection in a CDMA context. *IEEE Signal Processing Letters*, 9(6):165–167, June 2002.
5. F. Alizadeh. Interior point methods in semidefinite programming with applications to combinatorial optimization. *SIAM Journal on Optimization*, 5(1):13–51, February 1995.

6. B. Alkire and L. Vandenberghe. Convex optimization problems involving finite autocorrelation sequences. *Mathematical Programming*, Series A, 93:331–359, 2002.
7. E. Andersen and Y. Ye. On a homogeneous algorithm for the monotone complementarity problem. *Mathematical Programming*, 84:375–400, 1999.
8. S. Boyd and C. Barratt. *Linear Controller Design: Limits of Performance.* Prentice-Hall, 1991.
9. M. Bendsoe, A. Ben-Tal, and J. Zowe. Optimization methods for truss geometry and topology design. *Structural Optimization*, 7:141–159, 1994.
10. C. Bischof, A. Carle, G. Corliss, A. Grienwank, and P. Hovland. ADIFOR: Generating derivative codes from Fortran programs. *Scientific Programming*, pages 1–29, December 1991.
11. S. Boyd, L. El Ghaoui, E. Feron, and V. Balakrishnan. *Linear Matrix Inequalities in System and Control Theory.* SIAM, 1994.
12. S. Benson. DSDP 4.5: A daul scaling algorithm for semidefinite programming. Web site: `http://www-unix.mcs.anl.gov/~benson/dsdp/`, March 2002.
13. D. Bertsekas. *Nonlinear Programming.* Athena Scientific, Belmont, Massachusetts, 1995.
14. R. Byrd, N. Gould, J. Norcedal, and R. Waltz. An active-set algorithm for nonlinear programming using linear programming and equality constrained subproblems. Technical Report OTC 2002/4, Optimization Technology Center, Northwestern University, October 2002.
15. O. Bahn, J. Goffin, J. Vial, and O. Du Merle. Implementation and behavior of an interior point cutting plane algorithm for convex programming: An application to geometric programming. Working Paper, University of Geneva, Geneva, Switzerland, 1991.
16. S. Boyd, M. Hershenson, and T. Lee. Optimal analog circuit design via geometric programming, 1997. Preliminary Patent Filing, Stanford Docket S97-122.
17. R. Banavar and A. Kalele. A mixed norm performance measure for the design of multirate filterbanks. *IEEE Transactions on Signal Processing*, 49(2):354–359, February 2001.
18. A. Brooke, D. Kendrick, A. Meeraus, and R. Raman. *GAMS: A User's Guide.* The Scientific Press, South San Francisco, 1998. Web site: `http://www.gams.com/docs/gams/GAMSUsersGuide.pdf`.
19. J. Borwein and A. Lewis. Duality relationships for entropy-like minimization problems. *SIAM J. Control and Optimization*, 29(2):325–338, March 1991.
20. D. Bertsimas and J. Nino-Mora. Optimization of multiclass queuing networks with changeover times via the achievable region approach: part ii, the multi-station case. *Mathematics of Operations Research*, 24(2), May 1999.
21. D. Bertsekas, A. Nedic, and A. Ozdaglar. *Convex Analysis and Optimization.* Athena Scientific, Nashua, New Hampshire, 2004.
22. B. Borchers. CDSP, a C library for semidefinite programming. *Optimization Methods and Software*, 11:613–623, 1999.
23. A. Ben-Tal and M. Bendsoe. A new method for optimal truss topology design. *SIAM J. Optim.*, 13(2), 1993.
24. A. Ben-Tal, M. Kocvara, A. Nemirovski, and J. Zowe. Free material optimization via semidefinite programming: the multiload case with contact conditions. *SIAM Review*, 42(4):695–715, 2000.

25. A. Ben-Tal and A. Nemirovski. Interior point polynomial time method for truss topology design. *SIAM Journal on Optimization*, 4(3):596–612, August 1994.
26. A. Ben-Tal and A. Nemirovski. Robust truss topology design via semidefinite programming. *SIAM J. Optim.*, 7(4):991–1016, 1997.
27. A. Ben-Tal and A. Nemirovski. Structural design via semidefinite programming. In *Handbook on Semidefinite Programming*, pages 443–467. Kluwer, Boston, 2000.
28. S. Boyd and L. Vandenberghe. Semidefinite programming relaxations of non-convex problems in control and combinatorial optimization. In A. Paulraj, V. Roychowdhuri, , and C. Schaper, editors, *Communications, Computation, Control and Signal Processing: a Tribute to Thomas Kailath*, chapter 15, pages 279–288. Kluwer Academic Publishers, 1997.
29. S. Boyd and L. Vandenberghe. *Convex Optimization*. Cambridge University Press, 2004.
30. P. Biswas and Y. Ye. Semidefinite programming for ad hoc wireless sensor network localization. Technical report, Stanford University, April 2004. Web site: `http://www.stanford.edu/~yyye/adhocn4.pdf`.
31. A. Conn, N. Gould, D. Orban, and Ph. Toint. A primal-dual trust-region algorithm for non-convex nonlinear programming. *Mathematical Programming*, 87:215–249, 2000.
32. A. Conn, N. Gould, and Ph. Toint. *LANCELOT: a Fortran Package for Large-Scale Nonlinear Optimization (Release A)*, volume 17 of *Springer Series in Computational Mathematics*. Springer Verlag, 1992.
33. A. Conn, N. Gould, and Ph. Toint. *Trust-Region Methods*. Series on Optimization. SIAM/MPS, Philadelphia, 2000.
34. J. Chinneck. MProbe 5.0 (software package). Web site: `http://www.sce.carleton.ca/faculty/chinneck/mprobe.html`, December 2003.
35. G. Calafiore and M. Indri. Robust calibration and control of robotic manipulators. In *American Control Conference*, pages 2003–2007, 2000.
36. C. Crusius. *A parser/solver for convex optimization problems*. PhD thesis, Stanford University, 2002.
37. T. Terlaky C. Roos and J.-Ph. Vial. *Interior Point Approach to Linear Optimization: Theory and Algorithms*. John Wiley & Sons, New York, NY, 1997.
38. G. B. Dantzig. *Linear Programming and Extensions*. Princeton University Press, 1963.
39. J. Dawson, S. Boyd, M. Hershenson, and T. Lee. Optimal allocation of local feedback in multistage amplifiers via geometric programming. *IEEE Journal of Circuits and Systems I*, 48(1):1–11, January 2001.
40. M. Dahleh and I. Diaz-Bobillo. *Control of Uncertain Systems. A Linear Programming Approach*. Prentice Hall, 1995.
41. S. Dirkse and M. Ferris. The PATH solver: A non-monotone stabilzation scheme for mixed complementarity problems. *Optimization Methods and Software*, 5:123–156, 1995.
42. Y. Doids, V. Guruswami, and S. Khanna. The 2-catalog segmentation problem. In *Proceedings of SODA*, pages 378–380, 1999.
43. T. Davidson, Z. Luo, and K. Wong. Design of orthogonal pulse shapes for communications via semidefinite programming. *IEEE Transactions on Communications*, 48(5):1433–1445, May 2000.

44. G. Dullerud and F. Paganini. *A Course in Robust Control Theory*, volume 36 of *Texts in Applied Mathematics*. Springer-Verlag, 2000.
45. C. de Souza, R. Palhares, and P. Peres. Robust H_∞ filter design for uncertain linear systems with multiple time-varying state delays. *IEEE Transactions on Signal Processing*, 49(3):569–575, March 2001.
46. A. Doherty, P. Parrilo, and F. Spedalieri. Distinguishing separable and entangled states. *Physical Review Letters*, 88(18), 2002.
47. B. Dumitrescu, I. Tabus, and P. Stoica. On the parameterization of positive real sequences and MA parameter estimation. *IEEE Transactions on Signal Processing*, 49(11):2630–2639, November 2001.
48. R. Duffin. Linearizing geometric programs. *SIAM Review*, 12:211–227, 1970.
49. C. Du, L. Xie, and Y. Soh. H_∞ filtering of 2-D discrete systems. *IEEE Transactions on Signal Processing*, 48(6):1760–1768, June 2000.
50. H. Du, L. Xie, and Y. Soh. H_∞ reduced-order approximation of 2-D digital filters. *IEEE Transactions on Circuits and Systems I: Fundamental Theory and Applications*, 48(6):688–698, June 2001.
51. Laurent El Ghaoui, Jean-Luc Commeau, Francois Delebecque, and Ramine Nikoukhah. LMITOOL 2.1 (software package). Web site: `http://robotics.eecs.berkeley.edu/~elghaoui/lmitool/lmitool.html`, March 1999.
52. J. Ecker. Geometric programming: methods, computations and applications. *SIAM Rev.*, 22(3):338–362, 1980.
53. J.-P. A. Haeberly F. Alizadeh and M. Overton. Primal-dual interior-point methods for semidefinite programming: Convergence rates, stability and numerical results. *SIAM J. Optimization*, 8:46–76, 1998.
54. M. Fu, C. de Souza, and Z. Luo. Finite-horizon robust Kalman filter design. *IEEE Transactions on Signal Processing*, 49(9):2103–2112, September 2001.
55. U. Feige and M. Goemans. Approximating the value of two prover proof systems, with applications to max 2sat and max dicut. In *Proceedings of the 3nd Israel Symposium on Theory and Computing Systems*, pages 182–189, 1995.
56. R. Fourer, D. Gay, and B. Kernighan. *AMPL: A Modeling Language for Mathematical Programming*. Duxbury Press, December 1999.
57. A. Frieze and M. Jerrum. Improved approximation algorithms for max k-cut and max bisection. *Algorithmica*, 18:67–81, 1997.
58. K. Fujisawa, M. Kojima, K. Nakata, and M. Yamashita. SDPA (Semi-Definite Programming Algorithm) user's manual—version 6.00. Technical report, Tokyo Insitute of Technology, July 2002.
59. U. Feige and M. Langberg. Approximation algorithms for maximization problems arising in graph partitioning. *Journal of Algorithms*, 41:174–211, 2001.
60. U. Feige and M. Langberg. The rpr^2 rounding technique for semidefinte programs. In *ICALP*, Lecture Notes in Computer Science. Springer, Berlin, 2001.
61. R. Fourer. Nonlinear programming frequently asked questions. Web site: `http://www-unix.mcs.anl.gov/otc/Guide/faq/nonlinear-programming%-faq.html`, 2000.
62. R. Freund. Polynomial-time algorithms for linear programming based only on primal scaling and projected gradients of a potential function. *Mathematical Programming*, 51:203–222, 1991.
63. Frontline Systems, Inc. Premium Solver Platform (software package). Web site: `http://www.solver.com`, September 2004.
64. E. Fridman and U. Shaked. A new H_∞ filter design for linear time delay systems. *IEEE Transactions on Signal Processing*, 49(11):2839–2843, July 2001.

65. J. Geromel. Optimal linear filtering under parameter uncertainty. *IEEE Transactions on Signal Processing*, 47(1):168–175, January 1999.
66. O. Güler and R. Hauser. Self-scaled barrier functions on symmetric cones and their classification. *Foundations of Computational Mathematics*, 2:121–143, 2002.
67. D. Goldfarb and G. Iyengar. Robust portfolio selection problems. Technical report, Computational Optimization Research Center, Columbia University, March 2002. Web site: `http://www.corc.ieor.columbia.edu/reports/techreports/tr-2002-03.pdf`.
68. D. Goldfarb and G. Iyengar. Robust quadratically constrained problems program. Technical Report TR-2002-04, Department of IEOR, Columbia University, New York, NY USA, 2002.
69. P. Gill, W. Murray, and M. Saunders. SNOPT: An sqp algorithm for large-scale constrained optimization. *SIAM Journal on Optimization*, 12:979–1006, 2002.
70. P. Gill, W. Murray, M. Saunders, and M. Wright. User's guide for NPSOL 5.0: A FORTRAN package for nonlinear programming. Technical Report SOL 86-1, Systems Optimization Laboratory, Stanford University, July 1998. Web site: `http://www.sbsi-sol-optimize.com/manuals/NPSOL%205-0%20Manual.p%df`.
71. P. Gill, W. Murray, and M. Wright. *Practical Optimization*. Academic Press, London, 1981.
72. J. Geromel and M. De Oliveira. H_2/H_∞ robust filtering for convex bounded uncertain systems. *IEEE Transactions on Automatic Control*, 46(1):100–107, January 2001.
73. C. Gonzaga. Path following methods for linear programming. *SIAM Review*, 34(2):167–227, 1992.
74. M. Grant. *Disciplined Convex Programming*. PhD thesis, Department of Electrical Engineering, Stanford University, December 2004.
75. D. Guo, L. Rasmussen, S. Sun, and T. Lim. A matrix-algebraic approach to linear parallel interference cancellation in CDMA. *IEEE Transactions on Communications*, 48(1):152–161, January 2000.
76. G. Golub and C. Van Loan. *Matrix Computations*. Johns Hopkins Univ. Press, Baltimore, second edition, 1989.
77. M. Goemans and D. Williamson. Improved approximation algorithms for maximum cut and satisfiability problems using semidefinite programming. *Journal of the ACM*, 42:1115–1145, 1995.
78. M. Hershenson, S. Boyd, and T. Lee. Optimal design of a CMOS op-amp via geometric programming. *IEEE Transactions on Computer-Aided Design*, January 2001.
79. L. Huaizhong and M. Fu. A linear matrix inequality approach to robust H_∞ filtering. *IEEE Transactions on Signal Processing*, 45(9):2338–2350, September 1997.
80. M. Hershenson, S. Mohan, S. Boyd, and T. Lee. Optimization of inductor circuits via geometric programming. In *Proceedings 36th IEEE/ACM Integrated Circuit Design Automation Conference*, 1999.
81. P. Hovland, B. Norris, and C. Bischof. ADIC (software package), November 2003. `http://www-fp.mcs.anl.gov/adic/`.

82. L. Han, J. Trinkle, and Z. Li. Grasp analysis as linear matrix inequality problems. *IEEE Transactions on Robotics and Automation*, 16(6):663–674, December 2000.
83. J.-B. Hiriart-Urruty and C. Lemaréchal. *Convex Analysis and Minimization Algorithms I*, volume 305 of *Grundlehren der mathematischen Wissenschaften*. Springer-Verlag, New York, 1993.
84. J.-B. Hiriart-Urruty and C. Lemaréchal. *Convex Analysis and Minimization Algorithms II: Advanced Theory and Bundle Methods*, volume 306 of *Grundlehren der mathematischen Wissenschaften*. Springer-Verlag, New York, 1993.
85. Q. Han, Y. Ye, and J. Zhang. An improved rounding method and semidefinite programming relaxation for graph partition. *Math. Programming*, 92:509–535, 2002.
86. F. Jarre, M. Kocvara, and J. Zowe. Optimal truss design by interior point methods. *SIAM J. Optim.*, 8(4):1084–1107, 1998.
87. F. Jarre and M. Saunders. A practical interior-point method for convex programming. *SIAM Journal on Optimization*, 5:149–171, 1995.
88. N. Karmarkar. A new polynomial-time algorithm for linear programming. *Combinatorica*, 4(4):373–395, 1984.
89. M. Kojima, S. Mizuno, and A. Yoshise. An $O(\sqrt{n}L)$-iteration potential reduction algorithm for linear complementarity problems. *Mathematical Programming*, 50:331–342, 1991.
90. J. Kleinberg, C. Papadimitriou, and P. Raghavan. Segmentation problems. In *Proceedings of the 30th Symposium on Theory of Computation*, pages 473–482, 1998.
91. J. Keuchel, C. Schnörr, C. Schellewald, and D. Cremers. Binary partitioning, perceptual grouping, and restoration with semidefinite programming. *IEEE Transactions on Pattern Analysis and Machine Intelligence*, 25(11):1364–1379, November 2003.
92. K. Kortanek, X. Xu, and Y. Ye. An infeasible interior-point algorithm for solving primal and dual geometric progams. *Mathematical Programming*, 1:155–181, 1997.
93. K. Kortanek, X. Xu, and Y. Ye. An infeasible interior-point algorithm for solving primal and dual geometric programs. *Mathematical Programming*, 76:155–182, 1997.
94. M. Kocvara, J. Zowe, and A. Nemirovski. Cascading–an approach to robust material optimization. *Computers and Structures*, 76:431–442, 2000.
95. J. Lasserre. Global optimization with polynomials and the problem of moments. *SIAM Journal of Optimization*, 11:796–817, 2001.
96. J. Lasserre. Bounds on measures satisfying moment conditions. *Annals of Applied Probability*, 12:1114–1137, 2002.
97. J. Lasserre. Semidefinite programming vs. LP relaxation for polynomial programming. *Mathematics of Operations Research*, 27(2):347–360, May 2002.
98. H. Lebret and S. Boyd. Antenna array pattern synthesis via convex optimization. *IEEE Transactions on Signal Processing*, 45(3):526–532, March 1997.
99. Lindo Systems, Inc. LINGO version 8.0 (software package). Web site: `http://www.lindo.com`, September 2004.
100. J. Löfberg. YALMIP verison 2.1 (software package). Web site: `http://www.control.isy.liu.se/~johanl/yalmip.html`, September 2001.

101. L. Lovasz. *An Algorithmic Theory of Numbers, Graphs and Convexity*, volume 50 of *CBMS-NSF Regional Conference Series in Applied Mathematics*. SIAM, Philadelphia, 1986.
102. Z.-Q. Luo, J. Sturm, and S. Zhang. Duality and self-duality for conic convex programming. Technical report, Department of Electrical and Computer Engineering, McMaster University, 1996.
103. W. Lu. A unified approach for the design of 2-D digital filters via semidefinite programming. *IEEE Transactions on Circuits and Systems I: Fundamental Theory and Applications*, 49(6):814–826, June 2002.
104. M. Lobo, L. Vandenberghe, S. Boyd, and H. Lebret. Applications of second-order cone programming. *Linear Algebra and its Applications*, 284:193–228, November 1998. Special issue on Signals and Image Processing.
105. R. Monteiro and I. Adler. Interior path following primal-dual algorithms: Part I: Linear programming. *Mathematical Programming*, 44:27–41, 1989.
106. The MathWorks, Inc. *PRO-MATLAB User's Guide*. The MathWorks, Inc., 1990.
107. M. Mahmoud and A. Boujarwah. Robust H_∞ filtering for a class of linear parameter-varying systems. *IEEE Transactions on Circuits and Systems I: Fundamental Theory and Applications*, 48(9):1131–1138, September 2001.
108. G. Millerioux and J. Daafouz. Global chaos synchronization and robust filtering in noisy context. *IEEE Transactions on Circuits and Systems I: Fundamental Theory and Applications*, 48(10):1170–1176, October 2001.
109. N. Megiddo. Pathways to the optimal set in linear programming. In N. Megiddo, editor, *Progress in Mathematical Programming: Interior Point and Related Methods*, pages 131–158. Springer Verlag, New York, 1989. Identical version in: *Proceedings of the 6th Mathematical Programming Symposium of Japan, Nagoya, Japan, 1-35, 1986.*
110. MOSEK ApS. Mosek (software package). Web site: `http://www.mosek.com`, July 2001.
111. S. Mahajan and H. Ramesh. Derandomizing semidefinite programming based approximation algorithms. *SIAM J. of Computing*, 28:1641–1663, 1999.
112. B. Murtaugh and M. Saunders. MINOS 5.5 user's guide. Technical report, Systems Optimizaiton Laboratory, Stanford University, July 1998. Web site: `http://www.sbsi-sol-optimize.com/manuals/Minos%205-5%20Manual.p%df`.
113. J. Moré and D. Sorensen. NMTR (software package), March 2000. Web site: `http://www-unix.mcs.anl.gov/~more/nmtr/`.
114. M. Milanese and A. Vicino. Optimal estimation theory for dynamic systems with set membership uncertainty: An overview. *Automatica*, 27(6):997–1009, November 1991.
115. Y. Nesterov. *Introductory Lectures on Convex Optimization: A Basic Course*, volume 87 of *Applied Optimization*. Kluwer, Boston, 2004.
116. I. Nenov, D. Fylstra, and L. Kolev. Convexity determination in the microsoft excel solver using automatic differentiation techniques. In *The 4th Internation Conference on Automatic Differentiation*, 2004.
117. Yu. Nesterov and A. Nemirovsky. A general approach to polynomial-time algorithms design for convex programming. Technical report, Centr. Econ. & Math. Inst., USSR Acad. Sci., Moscow, USSR, 1988.

118. Yu. Nesterov and A. Nemirovsky. *Interior-Point Polynomial Algorithms in Convex Programming: Theory and Algorithms*, volume 13 of *Studies in Applied Mathematics*. Society of Industrial and Applied Mathematics (SIAM) Publications, Philadelphia, PA 19101, USA, 1993.
119. Yu. Nesterov, O. Pèton, and J.-Ph. Vial. Homogeneous analytic center cutting plane methods with approximate centers. In F. Potra, C. Roos, and T. Terlaky, editors, *Optimization Methods and Software*, pages 243–273, November 1999. Special Issue on Interior Point Methods.
120. S. Nash and A. Sofer. A barrier method for large-scale constrained optimization. *ORSA Journal on Computing*, 5:40–53, 1993.
121. Yu. Nesterov and M. Todd. Self-scaled barriers and interior-point methods for convex programming. *Mathematics of Operations Research*, 22:1–42, 1997.
122. J. Nocedal and S. Wright. *Numerical Optimization*. Springer Series in Operations Research. Springer, New York, 1999.
123. D. Orban and R. Fourer. DrAmpl: a meta-solver for optimization. Technical report, Ecole Poytechnique de Montreal, 2004.
124. M. Overton and R. Womersley. On the sum of the largest eigenvalues of a symmetric matrix. *SIAM Journal on Matrix Analysis and Applications*, 13(1):41–45, January 1992.
125. P. Parrilo. Semidefinite programming relaxations for semialgebraic problems. *Mathematical Programming*, Series B, 96(2):293–320, 2003.
126. G. Pataki. Geometry of cone-optimization problems and semi-definite programs. Technical report, GSIA Carnegie Mellon University, Pittsburgh, PA, 1994.
127. J. Park, H. Cho, and D. Park. Design of GBSB neural associative memories using semidefinite programming. *IEEE Transactions on Neural Networks*, 10(4):946–950, July 1999.
128. R. Palhares, C. de Souza, and P. Dias Peres. Robust H_∞ filtering for uncertain discrete-time state-delayed systems. *IEEE Transactions on Signal Processing*, 48(8):1696–1703, August 2001.
129. R. Palhares and P. Peres. LMI approach to the mixed H_2/H_∞ filtering design for discrete-time uncertain systems. *IEEE Transactions on Aerospace and Electronic Systems*, 37(1):292–296, January 2001.
130. S. Prajna, A. Papachristodoulou, and P. Parrilo. *SOSTOOLS: Sum of squares optimization toolbox for MATLAB*, 2002. Available from `http://www.cds.caltech.edu/sostools` and `http://www.aut.ee.ethz.ch/parrilo/sostools`.
131. O. Pèton and J.-P. Vial. A tutorial on ACCPM: User's guide for version 2.01. Technical Report 2000.5, HEC/Logilab, University of Geneva, March 2001. See also the `http://ecolu-info.unige.ch/~logilab/software/accpm/accpm.html`.
132. E. Rimon and S. Boyd. Obstacle collision detection using best ellipsoid fit. *Journal of Intelligent and Robotic Systems*, 18:105–126, 1997.
133. J. Renegar. A polynomial-time algorithm, based on Newton's method, for linear programming. *Mathematical Programming*, 40:59–93, 1988.
134. B. Radig and S. Florczyk. *Evaluation of Convex Optimization Techniques for the Weighted Graph-Matching Problem in Computer Vision*, pages 361–368. Springer, December 2001.

135. L. Rasmussen, T. Lim, and A. Johansson. A matrix-algebraic approach to successive interference cancellation in CDMA. *IEEE Transactions on Communications*, 48(1):145–151, January 2000.
136. M. Rijckaert and X. Martens. Analysis and optimization of the williams-otto process by geometric programming. *AIChE Journal*, 20(4):742–750, July 1974.
137. R. Rockafellar. *Convex Analysis*. Princeton Univ. Press, Princeton, New Jersey, second edition, 1970.
138. E. Rosenberg. *Globally Convergent Algorithms for Convex Programming with Applications to Geometric Programming*. PhD thesis, Department of Operations Research, Stanford University, 1979.
139. P. Stoica, T. McKelvey, and J. Mari. Ma estimation in polynomial time. *IEEE Transactions on Signal Processing*, 48(7):1999–2012, July 2000.
140. J. Sturm. Using SeDuMi 1.02, a MATLAB toolbox for optimization over symmetric cones. *Optimization Methods and Software*, 11:625–653, 1999.
141. J. A. K. Suykens, T. Van Gestel, J. De Brabanter, B. De Moor, and J. Vandewalle. Least squares support vector machines, 2002.
142. J. Stoer and C. Witzgall. *Convexity and Optimization in Finite Dimensions I*. Springer-Verlag, 1970.
143. U. Shaked, L. Xie, and Y. Soh. New approaches to robust minimum variance filter design. *IEEE Transactions on Signal Processing*, 49(11):2620–2629, November 2001.
144. H. Tuan, P. Apkarian, and T. Nguyen. Robust and reduced-order filtering: new LMI-based characterizations and methods. *IEEE Transactions on Signal Processing*, 49(12):2975–2984, December 2001.
145. H. Tuan, P. Apkarian, T. Nguyen, and T. Narikiyo. Robust mixed H_2/H_∞ filtering of 2-D systems. *IEEE Transactions on Signal Processing*, 50(7):1759–1771, July 2002.
146. C. Tseng and B. Chen. H_∞ fuzzy estimation for a class of nonlinear discrete-time dynamic systems. *IEEE Transactions on Signal Processing*, 49(11):2605–2619, November 2001.
147. The Mathworks, Inc. LMI control toolbox 1.0.8 (software package). Web site: `http://www.mathworks.com/products/lmi`, August 2002.
148. H. Tan and L. Rasmussen. The application of semidefinite programming for detection in CDMA. *IEEE Journal on Selected Areas in Communications*, 19(8):1442–1449, August 2001.
149. T. Tsuchiya. A polynomial primal-dual path-following algorithm for second-order cone programming. Technical report, The Institute of Statistical Mathematics, Tokyo, Japan, October 1997.
150. Z. Tan, Y. Soh, and L. Xie. Envelope-constrained H_∞ filter design: an LMI optimization approach. *IEEE Transactions on Signal Processing*, 48(10):2960–2963, October 2000.
151. Z. Tan, Y. Soh, and L. Xie. Envelope-constrained H_∞ FIR filter design. *IEEE Transactions on Circuits and Systems II: Analog and Digital Signal Processing*, 47(1):79–82, January 2000.
152. R. Tütüncü, K. Toh, and M. Todd. SDPT3—a MATLAB software package for semidefinite-quadratic-linear programming, version 3.0. Technical report, Carnegie Mello University, August 2001.
153. R. Tapia, Y. Zhang, and L. Velazquez. On convergence of minimization methods: Attraction, repulsion and selection. *Journal of Optimization Theory and Applications*, 107:529–546, 2000.

154. R. Vanderbei. LOQO user's manual—version 4.05. Technical report, Operations Research and Financial Engineering, Princeton University, October 2000.
155. L. Vandenberghe and S. Boyd. Semidefinite programming. *SIAM Review*, 38(1):49–95, March 1996.
156. R. Vanderbei and H. Benson. On forumulating semidefinite programming problems as smooth convex nonlinear optimization problems. Technical Report ORFE-99-01, Operations Research and Financial Engineering, Princeton University, January 2000.
157. L. Vandenberghe, S. Boyd, and A. El Gamal. Optimal wire and transistor sizing for circuits with non-tree topology. In *Proceedings of the 1997 IEEE/ACM International Conference on Computer Aided Design*, pages 252–259, 1997.
158. L. Vandenberghe, S. Boyd, and A. El Gamal. Optimizing dominant time constant in RC circuits. *IEEE Transactions on Computer-Aided Design*, 2(2):110–125, February 1998.
159. S.-P. Wu and S. Boyd. SDPSOL: A parser/solver for semidefinite programs with matrix structure. In L. El Ghaoui and S.-I. Niculescu, editors, *Recent Advances in LMI Methods for Control*, chapter 4, pages 79–91. SIAM, 2000.
160. F. Wang and V. Balakrishnan. Robust Kalman filters for linear time-varying systems with stochastic parametric uncertainties. *IEEE Transactions on Signal Processing*, 50(4):803–813, April 2002.
161. S.-P. Wu, S. Boyd, and L. Vandenberghe. FIR filter design via spectral factorization and convex optimization. In B. Datta, editor, *Applied and Computational Control, Signals, and Circuits*, volume 1, pages 215–245. Birkhauser, 1998.
162. M. Wright. Some properties of the Hessian of the logarithmic barrier function. *Mathematical Programming*, 67:265–295, 1994.
163. S. Wright. *Primal Dual Interior Point Methods.* Society of Industrial and Applied Mathematics (SIAM) Publications, Philadelphia, PA 19101, USA, 1999.
164. J. Weickert and Christoph Schnörr. A theoretical framework for convex regularizers in pde-based computation of image motion. *International Journal of Computer Vision, Band 45*, 3:245–264, 2001.
165. S. Wang, L. Xie, and C. Zhang. H_2 optimal inverse of periodic FIR digital filters. *IEEE Transactions on Signal Processing*, 48(9):2696–2700, September 2000.
166. Y. Ye. *Interior-point algorithms: Theory and practice.* John Wiley & Sons, New York, NY, 1997.
167. Y. Ye. A path to the arrow-debreu competitive market equilibrium. Technical report, Stanford University, February 2004. Web site: `http://www.stanford.edu/~yyye/arrow-debreu2.pdf`.
168. F. Yang and Y. Hung. Robust mixed H_2/H_∞ filtering with regional pole assignment for uncertain discrete-time systems. *IEEE Transactions on Circuits and Systems I: Fundamental Theory and Applications*, 49(8):1236–1241, August 2002.
169. Y.Ye, M. Todd, and S. Mizuno. An $O(\sqrt{n}L)$-iteration homogeneous and self-dual linear programming algorithm. *Mathematics of Operations Research*, 19(1):53–67, 1994.
170. Y. Ye and J. Zhang. Approximation for dense-n/2-subgraph and the complement of min-bisection. *Manuscript*, 1999.

171. S. Zhang. A new self-dual embedding method for convex programming. Technical report, Department of Systems Engineering and Engineering Management, The Chinese University of Hong Kong, October 2001.
172. M. Zibulevsky. Pattern recognition via support vector machine with computationally efficient nonlinear transform. Technical report, The University of New Mexico, Computer Science Department, 1998. Web site: `http://iew3.technion.ac.il/~mcib/nipspsvm.ps.gz`.
173. J. Zowe, M. Kocvara, and M. Bendsoe. Free material optimization via mathematical programming. *Mathematical Programming*, 9:445–466, 1997.
174. U. Zwick. Outward rotations: a tool for rounding solutions of semidefinite programming relaxations, with applications to max cut and other problems. In *Proceedings of the 31th Symposium on Theory of Computation*, pages 679–687, 1999.
175. H. Zhou, L. Xie, and C. Zhang. A direct approach to H_2 optimal deconvolution of periodic digital channels. *IEEE Transactions on Signal Processing*, 50(7):1685–1698, July 2002.

Writing Global Optimization Software

Leo Liberti

DEI, Politecnico di Milano, P.zza L. da Vinci 32, 20133 Milano, Italy
leoliberti@yahoo.com

Summary. Global Optimization software packages for solving Mixed-Integer Nonlinear Optimization Problems are usually complex pieces of codes. Some of the difficulties involved in coding a good GO software are: embedding third-party local optimization codes within the main global optimization algorithm; providing efficient memory representations of the optimization problem; making sure that every part of the code is fully re-entrant. Finding good software engineering solutions for these difficulties is not enough to make sure that the outcome will be a GO software that works well. However, starting from a sound software design makes it easy to concentrate on improving the efficiency of the global optimization algorithm implementation. In this paper we discuss the main issues that arise when writing a global optimization software package, namely software architecture and design, symbolic manipulation of mathematical expressions, choice of local solvers and implementation of global solvers.

Key words: MINLP, symbolic computation, multistart, variable neighbourhood search, branch-and-bound, implementation, software design.

1 Introduction

The object of Global Optimization (GO) is to find a solution of a given nonconvex mathematical programming problem. By "solution" we mean here a *global* solution, as opposed to a *local* solution; i.e., a point where the objective function attains the optimal value with respect to the whole search domain. By contrast, a solution is local if it is optimal with respect to a given neighbourhood. We require the objective function and/or the feasible region to be nonconvex because in convex mathematical programming problems every local optimum is also a global one. Consequently, any method solving a convex problem locally also solves it globally.

In this paper we address Mixed-Integer Nonlinear Programming (MINLP) problems in their most general setting:

$$\left.\begin{array}{rrcl} \min\limits_{x\in\mathbb{R}^n} & & c^Tx+f(x) & \\ \text{s.t.} & l\leq & Ax+g(x) & \leq u \\ & x^L\leq & x & \leq x^U \\ & & x_i\in\mathbb{Z} & \forall i\in Z \end{array}\right\} \quad (1)$$

In the above formulation, x are the problem variables; some of them (those indexed by the set $Z \subseteq \{1, \ldots, n\}$) are constrained to take discrete values. The objective function and constraints consist of a linear and nonlinear part: $f : \mathbb{R}^n \to \mathbb{R}$ is a possibly nonlinear function, $g : \mathbb{R}^n \to \mathbb{R}^m$ is a vector of m possibly nonlinear functions, $c \in \mathbb{R}^n$, A is an $m \times n$ matrix, $l, u \in \mathbb{R}^m$ are the constraint bounds (which may be set to $\pm\infty$ if a particular constraint is never active), and $x^L, x^U \in \mathbb{R}^n$ are the variable bounds (again, some of these bounds may be set to $\pm\infty$). We limit the discussion to the case where f, g are continuous functions of their arguments. Formulation (1) encompasses most kinds of mathematical programming problems. For example, if $f = 0$, g is a constant (say $-b$), $l = -\infty$, $u = 0$, $x^L = 0$, $x^U = \infty$, and $Z = \emptyset$ we have the canonical formulation of a Linear Programming problem (LP).

At the time of writing this paper, there is no GO software established as standard. In fact, the GO software market is still rather poor and definitely belonging to the academic world; GO is not being used extensively in the corporate environment yet. Part of the reason for this is that linear modelling is often a sufficient approximation to real-life processes, so GO software is not required. Even when a nonlinear model arises, a lot of effort is put into linearizing it by standard or novel modelling techniques. Finally, most GO algorithms rely on calling a local optimization procedure as a black-box function, and the fastest local optimization algorithms for Nonlinear Programming (NLP) problems are often inherently fragile: they may fail to converge for many different reasons even when the problem is reasonably smooth and well-behaved. This makes general-purpose robust GO software codes virtually non-existent (by contrast, GO software targeted at solving one particular problem can be made rather robust). As there are no standardized guidelines for designing GO software, this paper attempts to fill the gap by discussing a set of methods that should make general-purpose GO software robust and hopefully efficient. This work is based on two different "software frameworks" for GO designed by the author. One of these, *ooOPS*(object-oriented OPtimization System) [39], can be tested via the online interface at `http://liberti.dhs.org/liberti/ooOPS`. The other, MORON (MINLP Optimization and Reformulation Object-oriented Navigator), is still very much work in progress.

Most published papers on GO proposing a novel algorithm or a variant of an existing algorithm also include computational results which have been derived from some kind of software implementation of the method being proposed. This suggests that there should be quite a lot of working GO software available. Unfortunately, this is not the case. Most of these implementations are no more than prototypes designed to produce the computational results.

Methodologically speaking, at least in the academic world, there is nothing wrong about writing software with a view to publishing a paper and nothing else. This is not the approach to software writing that we are proposing to illustrate here, however. We are interested in reviewing software design methods and GO algorithms for general-purpose global optimization software. A clarifying analogy could be that we mean to describe methods to write a GO software akin to what CPLEX [29] is to mixed-integer linear programming. CPLEX is a software that can potentially solve any MILP problem, regardless of problem structure. It would be desirable to have a corresponding software for solving arbitrarily structured MINLP problems. CPLEX owes much of its success to a period of intense research in the field of solution methods for MILPs in their most general form (cutting plane, branch-and-bound, branch-and-price), as well as algorithmic improvements which made the proposed algorithm practically viable (new families of cuts, polyhedral theory, automatic reformulation methods, hierarchies of convex relaxations leading to the convex hull). CPLEX is by no means the only MILP-solving software on the market: but its efficiency is widely recognized, and we believe that it may be called a *de facto* standard software for solving MILPs. In the last two decades, many precise and heuristic GO methods have been proposed for solving MINLPs. The work on algorithmic improvements for these methods, however, is lagging behind if compared to the MILP scene. For instance, symbolic reformulation techniques for MINLPs are still largely an unexplored world, at least as far as computer implementations go: the most common way to proceed seems to be that the reformulation is carried out by hand, and then the reformulated problem is hard-coded in a "single-purpose" solver wrapper. By contrast, automatic symbolic reformulation algorithms are a crucial part of the CPLEX pre-solver. One possible explanation for such a different state of affairs is that software design for MILP solvers is inherently simpler than that required by GO algorithms. The situation is very different in GO algorithms where re-entrancy, good memory management, efficient data passing, and the ability to treat complex pieces of software like an NLP local solver as a black box are all of paramount importance. Yet, all of these issues have hardly been addressed in the existing GO literature. This paper attempts to move a few steps in this direction.

We propose a software design based on a framework that deals with the basic tasks required by any optimization solver code: reading the problem into memory, performing symbolic manipulation, providing and modifying problem structure data, wrapping the solvers into independent callable modules, configuring and running the solvers. Each solver can be called by any other solver on any given problem, effectively allowing the use of any solver (be it local or global) as a black-box. Solvers usually provide a numerical solution as output, but a solver in this framework may even be a specialized symbolic manipulation routine whose job is to change the structure of the problem.

The rest of this paper is organized as follows. Section 2 is a review of three existing algorithms that can be applied to optimization problems in form (1), where Z (the set of integer variables) may or may not be non-empty; namely, MultiStart (MS), Variable Neighbourhood Search (VNS) and spatial Branch-and-Bound (sBB). Section 3 is a review of existing general-purpose GO software packages. Section 4 lays the foundations for the software framework where the GO solver codes are executed. Section 5 is an in-depth treatment of the techniques used in the symbolic manipulation of the mathematical expressions in the objective function and constraints of the problem. Section 6 is a review of some of the existing local LP and NLP solvers. Section 7 contains descriptions of the global optimization solver implementations of MS, VNS and sBB within the described framework.

2 Global Optimization algorithms

This section presents three of the the existing algorithms targeted at the solution of problem (1): MultiStart (MS) (in fact a variant thereof called Multi Level Single Linkage[1] (MLSL)), Variable Neighbourhood Search[2] (VNS) and spatial Branch-and-Bound (sBB). The first two are classified as stochastic algorithms, the latter as deterministic. Note that this section only describes algorithmic structure. The implementation is discussed in Section 7.

Most GO algorithms are two-phase. The solution space S is explored exhaustively in the global phase, which iteratively identifies a promising starting point $\tilde{x}$. In the local phase, a local optimum x^* is found starting from each $\tilde{x}$. The local phase usually consists of a deterministic local descent algorithm which the global phase calls as a black-box function. The global phase can be stochastic or deterministic. Algorithms with a stochastic global phase are usually heuristic algorithms, whereas deterministic global phases often provide a certificate of optimality, making the algorithm precise.

Stochastic global phases identify the starting points $\tilde{x}$ either by some kind of sampling in S (sampling approach), or by trying to escape from the basin of attraction of the local minima x^* found previously (escaping approach), or by implementing a blend of these two approaches. Stochastic global phases do not offer certificates of optimality of the global optima they find, and they usually only converge to the global optimum with probability 1 in infinite time. In practice, though, these algorithms are very efficient, and are, at the time of this writing, the only viable choice for solving reasonably large-scale MINLPs. The efficiency of stochastic GO algorithms usually depends on the proper fine-tuning of the algorithmic parameters controlling intensification of sampling, extent of escaping and verification of termination conditions.

Deterministic global phases usually work by partitioning S into smaller sets $S_1, \ldots, S_p$. The problem is then solved globally in each of the subsets S_j. The global solution of each restriction of the problem to S_j is reached

[1] Also see Chapter 5.

[2] Also see Chapters 6, 11 (Section 1.1).

by recursively applying the global phase to each S_j until a certificate of optimality can be obtained for each S_j. The certificate of optimality is obtained by computing upper and lower bounds u, l to the objective function value. A local optimum x^* in S_j is deemed global when $|u - l| < \varepsilon$, where $\varepsilon > 0$ is a (small) constant. The convergence proofs for these algorithms rely on the analysis of the sequence of upper and lower bounds: it is shown that these sequences contain ε-convergent subsequences. The certificate of optimality for the global optimum of the problem with respect to the whole solution space S is therefore really a certificate of ε-global optimality. Such global phases are called Branch-and-Select (the partitioning of the sets S_j is called branching; the algorithm relies on selection of the most promising S_j for the computation of bounds) [85]. Deterministic algorithms tend to be fairly inefficient on large-scale MINLPs, but they perform well on small and medium-scale problems. The efficiency of Branch-and-Select algorithms seems to depend strongly on the particular instance of the problem at hand, and on the algebraic formulation of the problem.

We note here, in passing, that not all solution methods for GO problems follow the approach of finding candidate solution points $\tilde{x}$ and applying local descent to them to identify the closest local minimum x^*. Where f, g are very expensive to evaluate, the local phase is usually skipped (as it requires many function evaluations) and x^* is set to $\tilde{x}$ (thus, these GO algorithms only consist of the global phase). There exist algebraic methods (based on the computation of Gröbner bases) that solve polynomially constrained polynomial problems, which do not actually present either a global or a local phase [25].

In the rest of this section, we shall give a short presentation of the following stochastic algorithms: Multistart (MS), Variable Neighbourhood Search (VNS); and of the deterministic algorithm called "spatial Branch-and-Bound" (sBB). In fact, there are many other stochastic GO algorithms. To name but a few which are not discussed in this paper: tabu search [36], genetic algorithms [72], simulated annealing [47], differential evolution [74], adaptive Lagrange multiplier methods [86], ant colony simulation [50], ruin and recreate [77], dynamic tunnelling methods [65].

2.1 Multistart

Multistart (MS) algorithms are conceptually the most elementary GO algorithms. Many local descents are performed from different starting points. These are sampled with a rule that is guaranteed to explore the solution space exhaustively (in infinite time), and the local minimum with the best objective function value is deemed the "global optimum". MS algorithms are stochastic GO algorithms with a sampling approach. Their many variants usually differ in sampling and local descent strategies.

One of the main problems that MS algorithms face is that the same local optimum is identified many times when the sampling rule picks starting points in the basin of attraction of the same local optimum. Since the local descent is the most computationally expensive step of MS, it is important to control the extent to which this situation occurs. Obviously, identifying a local optimum many times is wasteful. The most common method used to inhibit multiple local descents to start in the same basin of attraction is called *clustering*. Sampled starting points are grouped together in clusters of nearby points, and only one local descent is performed in each cluster, starting from the most promising (in terms of objective function value) cluster point. One particularly interesting idea for clustering is the Multi Level Single Linkage (MLSL) method [60, 61]: a point x is clustered together with a point y if x is not too far from y and the objective function value at y is better than that at x. The clusters are then represented by a directed tree, the root of which is the designated starting point from where to start the local optimization procedure (see Fig. 1).

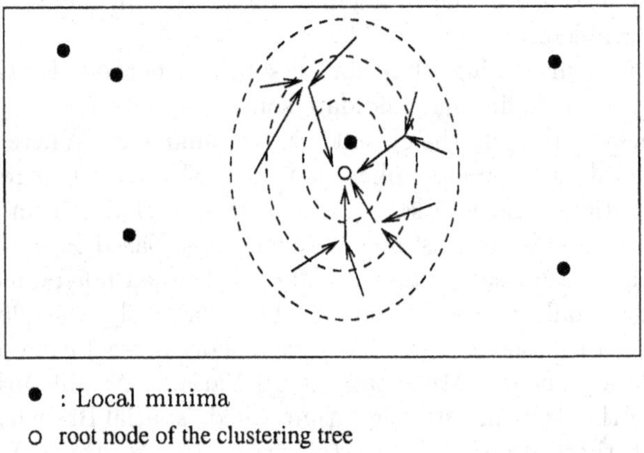

Fig. 1. Linkage clustering in the stochastic global phase. The points in the cluster are those incident to the arcs; each arc (x, y) expresses the relation "x is clustered to y". The arcs point in the direction of objective function descent. The root of the tree is the "best starting point" in the cluster.

One of the main problems with clustering is that as the number of problem variables increases, the sampled points are further apart (unless one is willing to spend an exponentially increasing amount of time on sampling, but this is very rarely acceptable) and cannot be clustered together so easily.

Despite their conceptual simplicity, MS algorithms for GO usually perform rather well on medium to large scale problems. The work of Locatelli and Schoen on MS with random and quasi-random sampling shows that MS is, to date, the most promising approach to solving the Lennard-Jones potential energy problem, arising in the configuration of atoms in a complex molecule [45, 67, 68, 46, 69].

A MS algorithm for GO problems in form (1) (called SobolOpt[3]) was developed by Kucherenko and Sytsko [37]. The innovation of the SobolOpt algorithm is that the sampling rule is not random but deterministic. More precisely, it employs Low-Discrepancy Sequences (LDSs) of starting points called *Sobol' sequences* whose distributions in Euclidean space have very desirable uniformity properties. Uniform random distributions where each point is generated in a time interval (as is the case in practice when generating a sampling on a computer) are guaranteed to be uniformly distributed in space in infinite time with probability 1. In fact, these conditions are very far from the normal operating conditions. LDSs, and in particular Sobol' sequences, are guaranteed to be distributed in space as uniformly as possible even in finite time. In other words, for any integer $N > 0$, the first N terms of a Sobol' sequence do a very good job of filling the space evenly. One further very desirable property of Sobol' sequences is that any projection on any coordinate hyperplane of the Euclidean space $\mathbb{R}^n$ containing N n-dimensional points from a Sobol' sequence will still contain N projected $(n-1)$-dimensional Sobol' points. This clearly does not hold with the uniform grid distribution where each point is located at a coordinate lattice point (in this case the number of projected points on any coordinate hyperplanes is $O(N^{\frac{n-1}{n}})$, as shown in Fig. 2). The comparison between grid and Sobol' points in $\mathbb{R}^2$ is shown in Fig. 3.

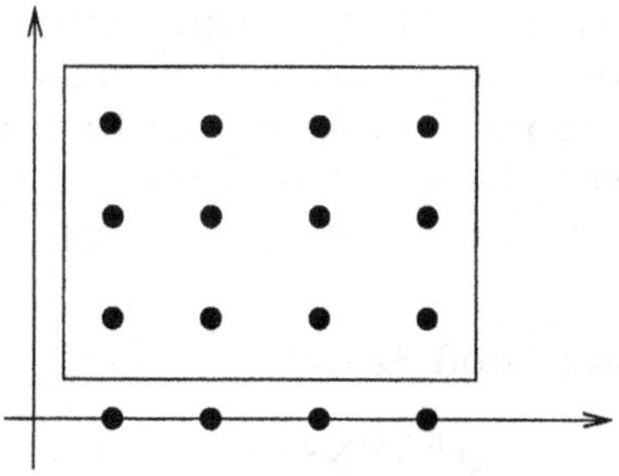

Fig. 2. Projecting a grid distribution in $\mathbb{R}^2$ on the coordinate axes reduces the number of projected points. In this picture, $N = 12$ but the projected points are just 4.

The SobolOpt algorithm has been used to successfully solve the Kissing Number Problem (KNP — determining the maximum number of non-overlapping spheres of radius 1 that can be arranged adjacent to a central sphere of radius 1) up to 4 dimensions, using a GO formulation proposed in [43]. The kissing number in 3 dimensions was first conjectured by Newton to be equal to 12. Newton was proven to be right only 250 years later by Leech.

[3] Also see Chapter 5, which is an in-depth analysis of the SobolOpt algorithm.

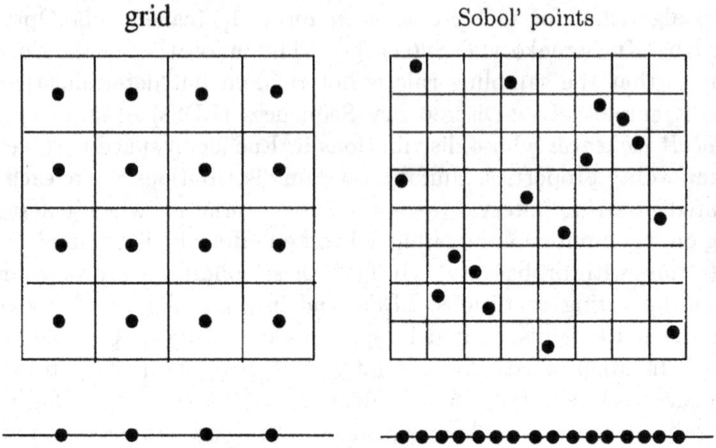

Fig. 3. Comparison between projected distribution of grid points and Sobol' points in $\mathbb{R}^2$.

The 4D case was only very recently proved to be equal to 24 (a result by O. Musin, still unpublished). The 5D case is still open. Unfortunately the problem formulation for the 5D case is too large to be solved by the SobolOpt algorithm. Research in this field is ongoing.

A computational comparison of SobolOpt versus a spatial Branch-and-Bound algorithm has been carried out and discussed in [44], showing that SobolOpt performs well in box-constrained as well as equation and inequality-constrained NLP problems. Some positive results have been obtained even for modestly-sized MINLPs with a few integer variables[4].

2.2 Variable Neighbourhood Search

Variable Neighbourhood Search[5] (VNS) is a relatively recent metaheuristic which relies on iteratively exploring neighbourhoods of growing size to identify better local optima [28, 27, 26]. More precisely, VNS escapes from the current local minimum x^* by initiating other local searches from starting points sampled from a neighbourhood of x^* which increases its size iteratively until a local minimum better than the current one is found. These steps are repeated until a given termination condition is met. VNS is a combination of both the sampling and the escaping approaches, and has been applied to a wide variety of problems both from combinatorial and continuous optimization. Its early applications to continuous problems were based on a particular problem structure. In the continuous location-allocation problem the neighbourhoods

[4] Also see Chapter 5, Section 4 for computational experiments with the SobolOpt algorithm.

[5] Also see Chapters 6, 11 (Section 1.1).

were defined according to the meaning of problem variables (assignments of facilities to customers, positioning of yet unassigned facilities and so on) [12]. In the bilinearly constrained bilinear problem the neighbourhoods took advantage of a kind of successive linear programming approach, where the problem variables can be partitioned so that fixing the variables in either set yields a linear problem; the neighbourhoods of size k were then defined as the vertices of the LP polyhedra that were k pivots away from the current vertex. In summary, none of the early applications of VNS was a general-purpose one.

The first VNS algorithm targeted at problems with fewer structural requirements, namely, box-constrained NLPs, was given in [51][6] (the paper focuses on a particular class of box-constrained NLPs, but the proposed approach is general). Since the problem is assumed to be box-constrained, the neighbourhoods arise naturally as hyperrectangles of growing size centered at the current local minimum x^*.

1. Set $k \leftarrow 1$, pick random point $\tilde{x}$, perform local descent to find a local minimum x^*.
2. Until $k > k_{\max}$ repeat the following steps:
   a) define a neighbourhood $N_k(x^*)$;
   b) sample a random point $\tilde{x}$ from $N_k(x^*)$;
   c) perform local descent from $\tilde{x}$ to find a local minimum x';
   d) if x' is better than x^* set $x^* \leftarrow x'$ and $k \leftarrow 1$; go to step 2;
   e) set $k \leftarrow k + 1$

In the pseudocode algorithm above, the termination condition is taken to be $k > k_{\max}$. This is the most common behaviour, but not the only one (the termination condition can be based on CPU time or other algorithmic parameters). The definition of the neighourhoods may vary. If $N_k(x)$ is taken to be a hyperrectangle $H_k(x)$ of size. k centered at x, sampling becomes easy; there is a danger, though, that sampled points will actually be inside a smaller hyperrectangular neighbourhood. A way to deal with this problem is to take $N_k(x) = H_k(x) \backslash H_{k-1}(x)$, although this makes it harder to sample a point inside the neighbourhood.

Some work is ongoing to implement a modification of the VNS for GO so that it works on problems in the more general form (1). This is obtained by replacing the box-constrained local descent algorithm in step (2c) with an SQP algorithm capable of locally solving constrained NLPs. Integer variables are dealt with in the global phase by employing a suitable neighbourhood structure.

[6] Also see Chapter 6, which is an in-depth analysis of the implementation of the VNS algorithm for box-constrained global optimization in [51], together with a presentation of computational results.

2.3 Spatial Branch-and-Bound

Spatial Branch-and-Bound (sBB) algorithms are the extension of traditional Branch-and-Bound (BB) algorithms to continuous solution spaces. They are termed "spatial" because they successively partition the Euclidean space where the problem is defined into smaller and smaller regions where the problem is solved recursively by generating converging sequences of upper and lower bounds to the objective function value. Traditional BB algorithms are used for finding the optimal solution of MILP problems. They work by generating subproblems where some of the integer variables are fixed and the others are relaxed, thus yielding an LP, which is easier to solve. Eventually, the solution space is explored exhaustively and the best local optimum found is shown to be the optimal solution.

Central to each sBB algorithm is the concept of a *convex relaxation* of the original nonconvex problem; this is a convex problem whose solution is guaranteed to provide an underestimation for the objective function optimal value in the original problem. At each iteration of the algorithm, restrictions of the original problem and its convex relaxations to particular sub-regions of space are solved, so that a lower and an upper bound to the optimal value of the objective function can be assigned to each sub-region; if the bounds are very close together, a global optimum relative to the subregion has been identified. The particular selection rule of the sub-regions to examine makes it possible to exhaustively explore the search space rather efficiently.

Most sBB algorithms for the global optimization of nonconvex NLPs conform to the following general framework:

1. (Initialization) Initialize a list of regions to a single region comprising the entire set of variable ranges. Set the convergence tolerance $\varepsilon > 0$, the best current objective function value as $U := \infty$ and the corresponding solution point as $x^* := (\infty, \ldots, \infty)$. Optionally, perform optimization-based bounds tightening (see Section 7.3) to try to reduce the variable ranges.
2. (Choice of Region) If the list of regions is empty, terminate the algorithm with solution x^* and objective function value U. Otherwise, choose a region R (the "current region") from the list according to some rule (a popular choice is: choose the region with lowest associated lower bound). Delete R from the list. Optionally, perform feasibility-based bounds tightening on R (see Section 7.3) to attempt further reduction of the variable ranges.
3. (Lower Bound) Generate a convex relaxation of the original problem in the selected region R and solve it to obtain an underestimation l of the objective function with corresponding solution $\bar{x}$. If $l > U$ or the relaxed problem is infeasible, go back to step 2.
4. (Upper Bound) Solve the original problem in the selected region with a local minimization algorithm to obtain a locally optimal solution x' with objective function value u.

5. (Pruning) If $U > u$, set $x^* = x'$ and $U := u$. Delete all regions in the list that have lower bounds bigger than U as they cannot possibly contain the global minimum.
6. (Check Region) If $u - l \leq \varepsilon$, accept u as the global minimum for this region and return to step 2. Otherwise, we may not yet have located the region global minimum, so we proceed to the next step.
7. (Branching) Apply a branching rule to the current region to split it into sub-regions. Add these to the list of regions, assigning to them an (initial) lower bound of l. Go back to step 2.

The first paper concerning continuous global optimization with a BB algorithm dates from 1969 [21]. In the 1970s and 1980s work on continuous or mixed-integer deterministic global optimization was scarce. Most of the papers published in this period dealt either with applications of special-purpose techniques to very specific cases, or with theoretical results concerning convergence proofs of BB algorithms applied to problems with a particular structure. In the last decade three sBB algorithms for GO appeared, targeted at constrained NLPs in form (1).

- The Branch-and-Reduce algorithm, by Sahinidis and co-workers [62, 63], which was then developed into the BARON software (see below), possibly the best sBB implementations around to date (see Section 3.1).
- The αBB algorithm, by Floudas and co-workers [8, 2, 1, 3, 4, 5], that addressed problems of a slightly less general form than (1). This algorithm was also implemented in a software which was never widely distributed (see Section 3.2).
- The sBB algorithm with symbolic reformulation, by Smith and Pantelides [71, 75]. There are two implementations of this algorithm: an earlier one which was never widely distributed, called GLOP (see Section 3.3), and a recent one which is part of the *ooOPS* system, and for which development is still active.

All three algorithms derive lower bounds to the objective function by solving a convex relaxation of the problem. The main algorithmic difference among them (by no means the only one) is the way the convex relaxation is derived, although all three rely on the symbolic analysis of the problem expressions.

The list above does not exhaust all of the sBB variants that appeared in the literature fairly recently. As far as we know, however, these were the main contributions targeted at general-purpose GO, and whose corresponding implementations were undertaken with a software-design driven attitude aimed at producing a working software package. Other existing approaches for which we have no information regarding the implementation are Pistikopoulos' Reduced Space Branch-and-Bound approach [16] (which only applies to continuous NLPs), Grossmann's Branch-and-Contract algorithm [87] (which also only applies to continuous NLPs) and Barton's Branch-and-Cut framework [34]. We do not include interval-based sBB techniques here because their performance is rather weak compared to the algorithms cited above, where the lower bound is obtained with a convex relaxation of the problem. Interval-based sBB algorithms are mostly used with problems where the objective function and constraints are difficult to evaluate or inherently very badly scaled.

Convex relaxation

It is very difficult to devise an automatic method[7] for generating tight convex relaxations. Thus, many algorithms targeted at a particular class of problems employ a convex relaxation provided directly by the user — in other words, part of the research effort is to generate a tight convex relaxation for the problem at hand. The standard automatic way to generate a convex relaxation consists in linearizing all nonconvex terms in the objective function and constraints and then replacing each nonconvex definition constraint with the respective upper concave and lower convex envelopes. More precisely, each nonconvex term is replaced by a linearization variable w (also called added variable) and a defining constraint $w =$ nonconvex term. The linearized NLP is said to be in *standard form*. The standard form is useful for all kinds of symbolic manipulation algorithms, as the nonconvex terms are all conveniently listed in a sequence of "small" constraints which do not require complex tree-like data structures to be stored [71, 41].

The defining constraint is then replaced by a constraint

$$\text{lower convex envelope} \leq w \leq \text{upper concave envelope}.$$

Since it is not always easy to find the envelopes of a given nonconvex term, slacker convex (or linear) relaxations are often employed. This approach to linearization was first formalized as an automatic algorithm for generating convex relaxations in [75], and implemented in the the GLOP software (see Section 3.3). The approach used by BARON is similar (see Section 3.1). The αBB code avoids this step but is limited to solving problems in a given form (which, although very general, is not as general as (1)).

The downside to using linearization for generating the convex relaxation is that the standard form is a lifting reformulation, that is, the number of variables in the problem is increased (one for each nonconvex term). Furthermore, even if envelopes (i.e. tightest relaxations) are employed to over- and under-estimate all the nonconvex terms in a term by term fashion, the resulting problem relaxation is not guaranteed to be the tightest possible relaxation of the problem. Quite on the contrary, relaxations obtained automatically in this way may in fact be very slack.

Common underestimators for bilinear [49, 7], trilinear, fractional, fractional trilinear [4], convex univariate, concave univariate [75] and piecewise convex and concave univariate terms [40] are all found in the literature.

[7] In fact, the work presented in Chapter 7 can also be seen as a first step in this direction, providing automatic symbolic techniques to verify the convexity properties of a given optimization problem and to generate new convex problems.

3 Global Optimization software

This section is a literature review on the existing software for GO. The review does not include software targeted at particular classes of problems, focusing on general-purpose implementations instead.

3.1 BARON

The BARON software (BARON stands for "Branch And Reduce Optimization Navigator"), written by Sahinidis and co-workers, implements a sBB-type algorithm called Branch-and-Reduce (because it makes extensive use of range reduction techniques both as a preprocessing step and at each algorithmic iteration), first described in [62, 63]. BARON aims at solving factorable nonconvex MINLPs. At the outset (1991), BARON was first written in the GAMS modelling language [11]. It was then re-coded in Fortran in 1994 and again in 1996 in a combination of Fortran and C for a more efficient memory management. The code was enriched in the number of local solvers during the years, and put online until around 2002, when it was decided that it would be distributed commercially as a MINLP solver for the GAMS system. Nowadays it can be purchased from GAMS (`www.gams.com`); for evaluation purposes, it is possible to download the whole GAMS modelling language, together with all the solvers, and run it in demo mode without purchasing a license. Unfortunately, the demo mode for global optimization solvers is limited to solving problems with 10 variables at most.

BARON and the Branch-and-Reduce algorithm it implements are further described in [66, 82, 83]. The documentation of the GAMS software contains an article about the usage of the BARON solver within GAMS. BARON is currently regarded as the state of the art implementation for a sBB solver, and in this author's experience the praise is wholly deserved at the time of writing this paper.

The main feature in this Branch-and-Bound implementation is the range reduction technique employed before and after solving the lower and the upper bounding problems. Because range reduction allows for tighter convex underestimators, the algorithm has the unusual feature that a subproblem can be solved many times in the same node. As long as the range reduction techniques manage to find reduced ranges for at least one variable, the convexification on the same node becomes tighter and the variable bounds in both lower and upper bounding problems change. The subproblems are then solved repeatedly until (a) the range reduction techniques fail to change the variable bounds or (b) the number of times a subproblem is allowed to be solved in the same node reaches a pre-set limit. Let P be the original problem and R its convex relaxation. Let L be a lower bound for the objective function of R and U an upper bound for the objective function of P. If the constraint $x_j - x_j^U \leq 0$ is active at the solution (i.e. the solution has $x_j = x_j^U$ with Lagrange multiplier $\lambda_j^* > 0$) and if $U - L < \lambda_j^*(x_j^U - x_j^L)$, then increase the variable lower bound:

$$x_j^L := x_j^U - \frac{U-L}{\lambda_j^*}.$$

Similarly, if the constraint $x_j^L - x_j \leq 0$ is active at the solution with Lagrange multiplier $\mu_j^* > 0$ and if $U - L < \mu_j^*(x_j^U - x_j^L)$, then decrease the variable upper bound:

$$x_j^U := x_j^L + \frac{U-L}{\mu_j^*}.$$

The geometrical interpretation of these range reduction tests is illustrated in Figure 4. It can be seen that the changes of the bounds effected by these rules do not exclude feasible solutions of P with objective function values which are lower than U. Even if the variable bounds are not active at the

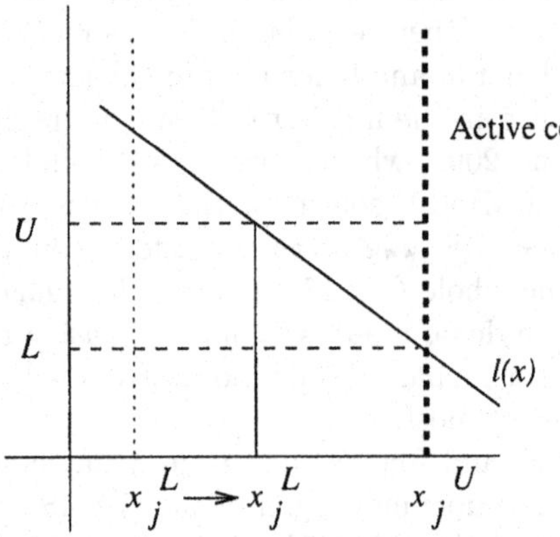

Fig. 4. Range reduction test. $l(x)$ is the straight line $\lambda_j^* x_j + L$.

solution, it is possible to "probe" the solution by fixing the variable value at one of the bounds, solving the partially restricted relaxed problem and checking whether the corresponding Lagrange multiplier is strictly positive. If it is, the same rules as above apply and the variable bounds can be tightened.

Another range reduction test is as follows: suppose that the constraint $\bar{g}_i(x) \leq 0$ (where $\bar{g}$ is the relaxed convex underestimator for the original problem constraint g) is active at the solution with a Lagrange multiplier $\lambda_i^* > 0$. Let U be an upper bound for the original problem P. Then the constraint

$$\bar{g}_i(x) \geq -\frac{U-L}{\lambda_i^*}$$

does not exclude any solutions with objective function values better than U and can be added as a "cut" to the current formulation of R to tighten it further.

As explained in Section 2.3, the lower bound to the objective function in each region is obtained by solving a convex relaxation of the problem. The techniques used by BARON to form the nonlinear convex relaxation of factorable problems are based on a symbolic analysis of the form of the factorable function. Each nonconvex term is analysed iteratively and then the convexification procedure is called recursively on each nonconvex sub-term. This approach to an automatic construction of the convex relaxation was first proposed in [75] (also see the implementation notes for the sBB solver in Section 7.3), and makes use of a tree-like data structure for representing mathematical expressions (see Section 5). The novelty of the BARON approach, and one of the main reasons why BARON works so well, is that it employs very tight linear relaxations for most nonconvex terms.

In particular, convex and concave envelopes are suggested for various types of fractional terms, based on the theory of convex extensions [80, 81]. The proposed convex underestimator for the term $\frac{x}{y}$, where $x \in [x^L, x^U]$ and $y \in [y^L, y^U]$ are strictly positive, is as follows:

$$\left.\begin{array}{c} z \geq \frac{x^L}{y_a}(1-\lambda) + \frac{x^U}{y_b}\lambda \\ y^L \leq y_a \leq y^U \\ y^L \leq y_b \leq y^U \\ y = (1-\lambda)y_a + \lambda y_b \\ x = x^L + (x^U - x^L)\lambda \\ 0 \leq \lambda \leq 1 \end{array}\right\} \tag{2}$$

The underestimator is modified slightly when $0 \in [x^L, x^U]$:

$$\left.\begin{array}{c} z \geq \frac{x^L(y^L+y^U-y_a)}{y^L y^U}(1-\lambda) + \frac{x^U}{y_b}\lambda \\ y^L \leq y_a \leq y^U \\ y^L \leq y_b \leq y^U \\ y = (1-\lambda)y_a + \lambda y_b \\ x = x^L + (x^U - x^L)\lambda \\ 0 \leq \lambda \leq 1 \end{array}\right\} \tag{3}$$

It is shown that these underestimators are tighter than all previously proposed convex underestimators for fractional terms, in particular:

- the bilinear envelope:

$$\max\left\{\frac{xy^U - yx^L + x^L y^U}{(y^U)^2}, \frac{xy^L - yx^U + x^U y^L}{(y^L)^2}\right\}$$

- the nonlinear envelope:

$$\frac{1}{y}\left(\frac{x+\sqrt{x^L x^U}}{\sqrt{x^L}+\sqrt{x^U}}\right)^2.$$

The above convex nonlinear underestimators are then linearized using an outer approximation approach.

Furthermore, there is a specific mention of piecewise convex and piecewise concave univariate terms (called *concavoconvex* by the authors) and their respective convex and concave envelopes [83]. The convexification of this type of nonconvex term — an example of which is the term x^3 when the range of x includes 0 — presents various difficulties, and it is usually not catered for explicitly (see [40] for a detailed study). An alternative to this envelope is suggested which circumvents the issue: by branching on the concavoconvex variable at the point where the curvature changes (i.e. the point where the concavoconvex term changes from concave to convex or vice versa) at a successive Branch-and-Bound iteration, the term becomes completely concave and completely convex in each region.

Other notable features of the BARON software include: generating valid cuts during pre-processing and and during execution; improving the branching scheme by using implication lists which for each nonconvex term point out the variable which most contributes to its nonconvexity [66, 83, 82, 64]; and most importantly, targeting particular problem formulations with specialized solvers [83]. These are available for:

- mixed-integer linear programming;
- separable concave quadratic programming;
- indefinite quadratic programming;
- separable concave programming;
- linear multiplicative programming;
- general linear multiplicative programming;
- univariate polynomial programming;
- 0-1 hyperbolic programming;
- integer fractional programming;
- fixed charge programming;
- problems with power economies of scale;

besides the "default" solver for general nonconvex factorable problems. BARON runs as a solver of GAMS; therefore, it runs on all architectures where GAMS can run.

3.2 αBB

The αBB algorithm [8, 6, 1, 5] solves problems where the expressions in the objective function and constraints are in factorable form. The convex relaxation of general twice-differentiable nonconvex terms is carried out by using a quadratic underestimation (based on the α parameter, which gives the name to the algorithm). Quadratic underestimations work for any twice-differentiable nonconvex term, but are usually very slack. In order to relax the original problem to a tight convex underestimator, the "usual" convex underestimators are proposed for bilinear, trilinear, fractional, fractional trilinear and concave univariate terms [4].

A function $f(x)$ (where $x \in \mathbb{R}^n$) is underestimated over the entire domain $[x^L, x^U] \subseteq \mathbb{R}^n$ by the function $L(x)$ defined as follows:

$$L(x) = f(x) + \sum_{i=1}^{n} \alpha_i (x_i^L - x_i)(x_i^U - x_i)$$

where the α_i are positive scalars that are sufficiently large to render the underestimating function convex. A good feature of this kind of underestimator is that, unlike other underestimators, it does not introduce any new variable or constraint, so that the size of the relaxed problem is the same as the size of the original problem regardless of how many nonconvex terms it involves. Since the sum $\sum_{i=1}^{n} \alpha_i (x_i^L - x_i)(x_i^U - x_i)$ is always negative, $L(x)$ is an underestimator for $f(x)$. Furthermore, since the quadratic term is convex, all nonconvexities in $f(x)$ can be overpowered by using sufficiently large values of the α_i parameters. From basic convexity analysis, it follows that $L(x)$ is convex if and only if its Hessian matrix $H_L(x)$ is positive semi-definite. Notice that:

$$H_L(x) = H_f(x) + 2\Delta$$

where $\Delta \equiv \mathrm{Diag}_{i=1}^{n}(\alpha_i)$ is the matrix with α_i as diagonal entries and all zeroes elsewhere (diagonal shift matrix). Thus the main focus of the theoretical studies concerning all αBB variants is on the determination of the α_i parameters. Some methods are based on the simplifying requirement that the α_i are chosen to be all equal (uniform diagonal shift matrix), others reject this simplification (non-uniform diagonal shift matrix). Under the first condition, the problem is reduced to finding the parameter α that makes $H_L(x)$ positive semi-definite. It has been shown that $H_L(x)$ is positive semi-definite if and only if:

$$\alpha \geq \max\{0, -\frac{1}{2} \min_{i, x^L \leq x \leq x^U} \lambda_i(x)\}$$

where $\lambda_i(x)$ are the eigenvalues of $H_f(x)$. Thus the problem is now of finding a lower bound on the minimum eigenvalue of $H_f(x)$. The most promising method to this end seems to be Interval Matrix Analysis. Various $O(n^2)$ and $O(n^3)$ methods have been proposed to solve both the uniform and the non-uniform diagonal shift matrix problem [19].

The αBB code itself is unfortunately not publicly distributed; refer to `http://titan.princeton.edu` for further details. The αBB code is provided with a front-end parser module which is accepts mathematical expressions and generates corresponding "code lists". The parser is capable of reading certain types of quantifications in enumeration, summations and products, but is not equivalent to a full-fledged modelling language like AMPL or GAMS. Following parsing, automatic differentiation is applied to the code lists to generate the first and second order derivatives. Code lists for the lower bounding problem are also automatically generated. The main sBB iteration loop can then be started. αBB runs on Unix architectures.

3.3 GLOP

GLOP is the implementation of the spatial Branch-and-Bound with symbolic reformulation proposed in [71, 73, 75]. This was the first sBB algorithm which generated convex relaxations automatically for optimization problems in general form (1), using the linearization method explained in Section 2.3. Other distinctive features of this algorithm are optimization and feasibility-based range reduction procedures. The sBB solver in *ooOPS* is based around a sBB derived directly from this algorithm. More information can be found in Section 7.3.

The GLOP code is not distributed. Extensive information about this software can be found in [71]. A brief description of this software is given here for the following reasons:

- Some of the ideas for *ooOPS* were borrowed from GLOP. More precisely, the software design and architecture are different, but the sBB algorithm implemented in GLOP is the basis of the sBB algorithm in *ooOPS*.
- GLOP makes an interesting case study for an advanced software design, with solvers calling each other, implemented with programming techniques of about a decade ago. The first implementation was carried out mostly in C, with some of the local solvers being coded in Fortran.
- GLOP was experimentally inserted in the integrated software environment gPROMS for process synthesis [15]; in particular, a parallel version (which can be called from gPROMS) was coded in Modula-2. No other sBB algorithm targeted at problems in form (1) ever had a working parallel implementation, to the best of our knowledge.
- GLOP is one of the few general-purpose sBB codes which can generate nonlinear, as well as linear, convex relaxation. Usually, the trade-off between tightness of convex relaxation and speed of solution is won by the latter; thus, most sBB implementations only produce linear relaxation. There are cases, however, where having a very tight convex relaxation (which may be nonlinear) is advantageous.

GLOP uses the binary tree data structure to manipulate mathematical expressions symbolically (see Section 5.1). Initially, GLOP reads a problem definition file in a pre-parsed proprietary format which is basically a text description of the binary trees representing the mathematical expressions in the problem, as well as the other numerical data defining the problem. Derivatives are computed symbolically, and some degree of symbolic simplification is enforced. The user can select whether a nonlinear or a linear convex relaxation is desired, as well as the local upper and lower bounding solvers to use. The selection includes CPLEX and MINOS for the lower bounding solver, and CONOPT, together with various experimental local NLP codes, for the upper bounding solver. E. Smith, the author of the software, also designed and implemented a local NLP solver based on successive linear programming.

GLOP was not written according to object-oriented software design principles. Some of the overall algorithmic control is carried out using global variables. Some care has been paid to the global variables not interfering with parallel execution. However, this is not fully re-entrant software design. As a stand-alone software, GLOP runs on Unix architectures. It was tested on Solaris and Linux. As part of gPROMS, it runs on all architectures where gPROMS runs.

3.4 *ooOPS*

ooOPS stands for object-oriented OPtimization System. It is a software framework for doing optimization. As such, it contains a parser module, various reformulator modules, and various global and local solver modules; it has full symbolic manipulation capabilities (with mathematical expressions represented by binary trees) and is designed to tackle large scale global optimization problems. Its software design is such that the code is fully re-entrant. The architecture makes it possible for a number of modules to configure the solver parameters, even at different stages of the solution process. Its parser is interfaced with AMPL so that a full-fledged modelling language can be used to input problems. The parser and the rest of the systems are separated, so that *ooOPS* can actually be used as an external library (with a well documented API [39]).

ooOPS was designed as the environment where an advanced sBB solver code based on the spatial Branch-and-Bound with symbolic reformulation (see Section 3.3) should have been executed. As such, solvers are largely interchangeable, in the sense that they all bind to the same set of library calls. Technically, it is even possible to call sBB itself as the local solver of another sBB instance. With time, more global solvers were added to *ooOPS*, so that now it can be considered as a general-purpose global optimization software of good quality, implementing sBB, MLSL and VNS algorithms targeted at solving MINLPs in form (1). *ooOPS* is fairly reliable, and was tested on a number of different problem classes (bilinear pooling and blending problems, Euclidean location-allocation problems, the Kissing Number problem, molecular distance geometry problems, and various other problem classes and instances) with considerable success. *ooOPS* consists of over 40000 lines of C++, in addition to the code of several local solvers (SNOPT [23], UCF from the NAG library [52], `lp_solve` [9]). The distribution package of *ooOPS* contains all the source code for compiling the system. It was decided, at this stage, to interface only to solvers whose source code is available (hence the prominent exclusion of CPLEX from the list of local solvers). *ooOPS* can be linked as a static executable including all the solvers, or as a dynamically-linked executable which loads the available solver modules as needed.

The public distribution of *ooOPS* is at the moment still being discussed, but a binary release under the form of an AMPL solver is planned. *ooOPS* runs on most Unix architectures. It has been tested on Linux and Solaris; earlier versions had been produced to run on Windows, both under the CYGWIN environment and under the MS Visual C++ compiler, but maintenance of the Windows-based versions has been discontinued. *ooOPS* was mostly written by the author of this paper, but C. Pantelides, B. Keeping, P. Tsiakis, S. Kucherenko of CPSE, Imperial College, London all contributed to the software.

Section 3.4 gives some details about the inner working of *ooOPS*. Please note that not all of the *ooOPS* system conforms to the guidelines given in Section 4 for writing a good optimization software; some of the ideas given in Section 4 were developed by considering the inefficiencies of the existing *ooOPS* implementation.

Object classes

ooOPS consists of 4 major classes of objects, each with its own interface.

1. The `ops` object.
   An `ops` object is a software representation of a programming problem. The corresponding interface provides the following functionality.
   - It allows problem objects to be constructed and modified in a structured manner.
   - It allows access to all numerical and symbolic information pertaining to the problem in structured, flat (unstructured) and standard form.
2. The `opssystem` object.
   This is formed by the combination of an `ops` object with a solver code. The corresponding interface provides the following functionality.
   - It allows the behaviour of the solver to be configured via the specification of any algorithmic parameters that the solver may support.
   - It permits the solution of the problem.
3. The `opssolvermanager` object.
   This corresponds to a particular solver and allows the creation of an `opssystem` object combining this solver with a given `ops` object. The corresponding interface provides the following functionality.
   - It allows the creation of many different `opssystem` objects, all of which have the same solver parameter configuration.
4. The `convexifiermanager` object.
   This embeds the convexification code. The corresponding interface provides the following functionality.
   - It allows the creation of the convex relaxation of the problem.
   - It allows the on-the-fly update of the convex relaxation when the variable ranges change.

Control flow

In this section, we describe how the user code (also called the *client*) calls the objects described above to formulate and solve an optimization problem. This involves a number of steps.

1. Construction of the problem `ops` object.
2. Creation of an `opssolvermanager` for the solver code to be used.
3. Creation of the necessary `opssystem` (by passing the `ops` object created at step 1 to the `opssolvermanager` created at step 2).
4. Solution of the problem (via the `opssystem`'s `Solve()` method).
5. Solution data query (via the `ops` object's interface).

Note that the `opssystem`'s `Solve()` method places the solution values back in the `ops` object containing the original problem, so they can be recovered later by the user code (also called the *client* code) by using a variable query method.

3.5 Other GO software packages

There are a few other GO codes targeted at fairly large classes of NLPs and MINLPs, which we do not discuss in detail either because of availability issues, lack of stability or simply because they are numerical solvers rather than complete software frameworks.

- The LGO (Lipschitz Global Optimizer) solver, coded and marketed by János Pintér [57, 58], only requires function evaluation (no derivatives) as external user-defined routines. Bounds on the objective function value are obtained through Lipschitz analysis, and then employed in a spatial Branch-and-Bound framework. Currently, several versions of this solver exist: the core library object (for both Windows and Unix operating systems), an integrated development environment for Windows, and solver engines for Microsoft Excel, GAMS, Mathematica and Matlab. See the website `http://www.pinterconsulting.com/l_s_d.html` for more information.
- The GlobSol solver, by Robert Kearfott [31], is based on interval arithmetics to compute upper and lower bounds for a spatial Branch-and-Bound framework. There is no integrated environment for this solver.
- The Coconut Environment. This is a large global optimization project which is ongoing at Universität Wien, headed by A. Neumaier. One of the products of this project is a software framework for global optimization, which offers an API [70] for external linking, and includes local and global solvers. Unfortunately, the code is still too unstable to be used productively. See the project website at `http://www.mat.univie.ac.at/coconut-environment/`.

- Lindo API [79]. This is an optimization framework for programmers. It offers an API and several solvers, among which a global one. A limited version can be downloaded for evaluation purposes from the website `http://www.lindo.com`.
- MORON stands for MINLP Optimization and Reformulation Object-oriented Navigator. This project was started by the author of this paper in order to produce a GO code that could be released to the public. MORON is still very much work in progress, but it will offer substantial improvements over $oo\mathcal{OPS}$, the main one being much more advanced symbolic manipulation capabilities. MORON uses an n-ary tree representation for mathematical expressions (see Section 5.3), which makes it possible to simplify algebraic expressions much more effectively. The software architecture ideas are mostly borrowed from $oo\mathcal{OPS}$, but MORON shares no code with $oo\mathcal{OPS}$.

4 Optimization software framework design

As has been remarked, a general-purpose GO software is a complex program requiring interactions among many software modules. One of the major difficulties, requiring a high degree of programming abstraction, is to be able to replace a given software module with another one of the same type but different mathematical properties. For example, as most GO algorithm call a local optimization procedure as a sub-algorithm, we may wish to replace the default local solver embedded in the GO solver with a sparse local solver when solving large-scale problems to global optimality. Thus the local solver should not be "hard-wired" in the GO solver, but rather be a pluggable software module. Suppose further that we know the problem to be a bilinear one: then we might wish to replace the standard pre-solver (*reformulator* module) with one which is more apt to the task. With this philosophy, nearly every step in the global solution of a problem is implemented as a pluggable software module. This calls for a *core* software module where every other module can be plugged in. The core module supplies the necessary Application Programming Interfaces (APIs) so that the different modules can be called using the same standard function calls.

This type of software design may, with a bit of tweaking, be implemented in almost every programming language. Its proper semantic domain, however, is within the object-oriented programming paradigm. We chose C++ [78] to write implementations of the proposed software design, as it is widely available on almost every hardware and operating system platform.

4.1 Control Flow

In this software description we follow a top-down approach, so that the reader may find higher level requirements before reading about the consequent implementation choices at lower level. The first thing that is necessary to know about how a software works is the control flow.

At first, a file representation of problem (1) is read into memory, parsed, and transformed into a memory representation of the problem. This task is carried out by the *parser* module, and the memory representation of the problem is stored within the data fields of the core module, which exposes a common API for each solver module to retrieve the problem data. Subsequent to parsing, the first task is to apply a symbolic reformulator to the problem. The reformulator tries a number of standard symbolic manipulations to reduce the number of variables and constraints, as well as reformulating the problem to a simpler, or more convenient form, if required. Next, based on user request, the control is passed to the *solver* module. Each solver is embedded into a *wrapper* module which allows solver software written by different people with different ad-hoc APIs to be interfaced with the core module uniformly. Solver modules written natively for the core module may dispense with the wrapper and directly use the core module's API. Each solver may call other modules (reformulators or solvers) as required by the algorithm being implemented. Before the solver terminates its execution, it stores the optimal solution, the optimal objective function value and other information regarding algorithmic performance and solution reliability within the core module, so that this information can be accessed by other solvers. When the first-level solver (i.e. the solver that was called first) terminates, the solution is output to the user and program execution stops.

Fig. 5 shows the control flow of a first-level solver module requiring two different second-level solvers, one for the main problem and the other to be applied to an auxiliary problem derived from the main one by means of symbolic reformulation (for example, consider a sBB algorithm requiring local solutions to a nonconvex NLP problem and to its convex relaxation).

4.2 Data Flow

Since we are envisaging an object-oriented architecture and re-entrant coding, global variables may not be used. The data relating to the mathematical formulation of the problem (i.e. the equations in (1)) are stored in the core module, as has been remarked. In practice, these data are fields of a class called `Problem` (more or less equivalent to what the class `ops` is in the *ooOPS* software, see Section 3.4). Since the parser module builds the `Problem` object corresponding to the main problem, it must have read/write permissions on `Problem` objects. Consider also that some GO algorithms need to solve auxiliary problems at each step as part of the overall solution method (e.g. sBB solves a lower-bounding problem at each step); since the auxiliary problems may in principle be solved by any of the available solvers, it makes sense to embed them in an object of the `Problem` class. The reformulator module which generates the auxiliary problem must therefore have read/write permissions

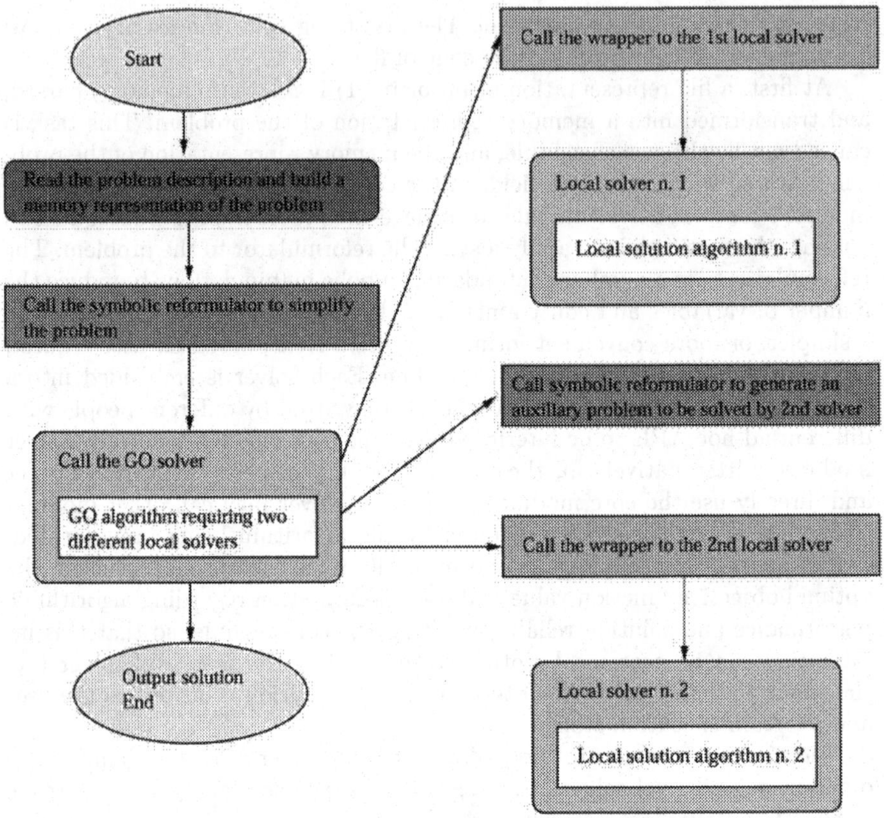

Fig. 5. Example of control flow.

on `Problem` objects. Care must be taken to make the variable numbering consistent between the main problem and the derived auxiliary problems. Solvers need to read problem data rather than modify the problem, so they only need read access to `Problem` objects.

The `Problem` class, implementing the core module, is the fundamental class of the proposed GO framework. It stores information about variables, constraints and objective function. A variable has the following basic properties: integrality (whether it is an integer or a continuous variable), bounds and current value. A constraint has two basic properties: symbolic expression of the constraint and bounds. The objective function also has two basic properties, namely the symbolic expression it consists of, and the optimization direction (min or max). For simplicity, and without loss of generality, we shall thereafter assume that the optimization direction is that of minimization. The `Problem` class offers an API containing methods for performing the following actions:

- creation of problem entities
  1. create a new variable (including integrality, bounds and current value)
  2. create a new constraint (including constraint bounds)
  3. create the objective function (including optimization direction)
- output of problem data
  1. get problem sizes (number of variables, number of constraints, number of integer variables, number of linear constraints)
  2. get variable integrality, linearity, bounds and current values
  3. get constraint symbolic expression and bounds
  4. get the symbolic expression for the first (optionally second) order partial derivative of each constraint with respect to each variable
  5. get the symbolic expression and the optimization direction of the objective function
  6. get the symbolic expression of the first (optionally second) order partial derivative of the objective function with respect to each variable
- modification of problem data
  1. modify a variable (including integrality, bounds and current value)
  2. modify a constraint (including symbolic expression and bounds)
  3. modify the objective function
- output of dynamic information (i.e. depending on the current variable values)
  1. evaluate the constraint at the current variable values
  2. evaluate the constraint derivatives at the current variable values
  3. evaluate the objective function at the current variable values
  4. evaluate the objective function derivatives at the current variable values
  5. test whether current variable values are a feasible solution
  6. test whether a variable only occurs linearly in the problem or not
  7. test whether problem has a particular structure (linear, bilinear, convex)

The part of the API that tests wether the problem has a particular structure is called the symbolic analyser, and can also be considered as a separate module instead of being part of the `Problem` API. Testing whether a problem is linear or bilinear is easy; devising an algorithmic convexity test is far from trivial[8].

Fig. 6 shows an example of data flow in the proposed framework. The arrows represent the relationship "modifies the data of" between modules. It appears clear that certain reformulators change the core modules (`Problem` object), whilst others read the data from the core module to generate another core module for an auxiliary problem. Each solver wrapper has to internally store the data of the problem it is solving in order to pass these data to the solver proper in its required format. Figure 6 describes a sBB solver: the core module for the main problem contains the data relative to the problem.

[8] Some work is being carried out on such a task; see Chapter 7.

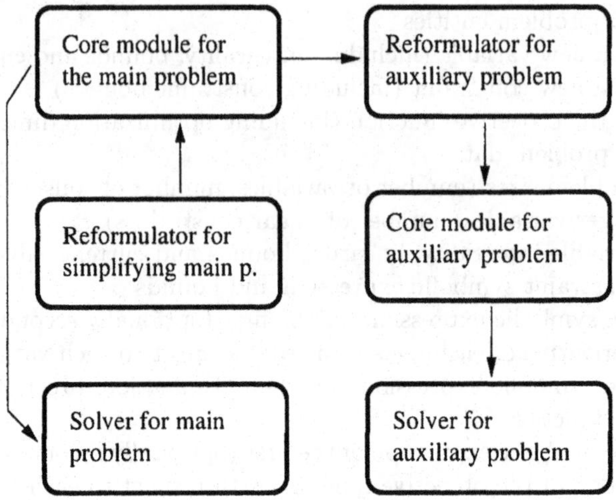

Fig. 6. Example of data flow. An arrow between modules A and B means that A modifies the internal data of B.

This data is modified when the problem is simplified by the simplifying reformulator. The core module then loads its data in the main solver module (the global solver), which solves the problem. During the solution process, the solver module tells the core module to create a reformulator for generating a convex relaxation. The core module loads its data in the reformulator, which generates another core module for the auxiliary problem (the convex relaxation). The main solver then instructs the core module of the auxiliary problem to load its data in the auxiliary solver (the local solver), which solves the convex relaxation. The implementation of this sequence of data exchanges might be carried out so as to reduce the amount of transferred data.

4.3 Parser module

The ideal parser reads the mathematical formulation of an optimization problem, and transforms it in the appropriate data structures. Since optimization problems are often expressed in terms of quantifiers, indices and multidimensional sets of various kinds, to design such a parser is akin to crafting a modelling language, which is an extremely difficult and time-consuming task. Existing modelling languages (e.g. AMPL [18, 11]) usually provide an API for external solvers. If the API goes as far as providing symbolic information for the equations, we can design our software framework as a "super-solver" hooked to the modelling language software. The implementations described in this paper use AMPL as a modelling language. The documented AMPL API is very limited in its capabilities for passing structural problem information to solvers. However, AMPL also offers an undocumented API [20, 22] which makes it possible to read the mathematical expressions in the problem in a tree-like fashion. Both *ooOPS* and MORON work by using the undocumented AMPL API for building the internal data structures.

It is, of course, very easy to write a parser that reads optimization problems in flat form, i.e., where no quantifier appears in any of the problem expressions (for example, a constraint like $\forall i \in N(x_i + x_1 \leq 1)$ would be written in the flat form as a list of $|N|$ constraints where all the indices have been made explicit). Such parsers can be written by using standard programming tools like LEX and YACC [38] or from scratch: one good starting point can be found in the first chapters of [78]. Again, both *ooOPS* and MORON are equipped with such flat form parser modules. Interfacing with AMPL was actually carried out with a small external program which uses the undocumented AMPL API to produce a flat form representation of the problem that the optimization software can read.

A standard MINLP description file format

The widespread adoption of a standard, flat-form problem definition file for MINLPs (akin to what the MPS and LP file formats are to linear programming) would be a very desirable event. AMPL is capable of producing `.nl` ASCII files which are complete descriptions of MINLP problems. However, they are so cryptic to the human eye that suggesting their adoption as standard would pose serious problems. Unfortunately, the same goes for the rather widespread SIF format [84], which has some limitations in the type of function structure it is able to express (the function must be factorable). *ooOPS* and MORON currently read flat-form definition files in the following format.

```
# a comment
variables =
        VLowerBound_1 < VarName_1 < VUpperBound_1/VarType_1,
                              ⋮
        VLowerBound_n < VarName_n < VUpperBound_n/VarType_n;
objfun =
        [c^T x + f(x)];
constraints =
        [CLowerBound_1 < A_1 x + g_1(x) < CUpperBound_1],
                              ⋮
        [CLowerBound_m < A_m x + g_m(x) < CUpperBound_m];
startingpoint =
        x'_1, ..., x'_n;
options =
        ParameterName_1 ParameterValue_1,
                              ⋮
        ParameterName_k ParameterValue_k;
```

The symbols are the same as in (1). Here, $A_1, \ldots, A_m$ are the rows of the matrix A. The symbol "$<$" has been employed instead of "$\leq$" because the text file is less cluttered with just one ASCII symbol (`<`) instead of two (`<=`), but the semantics of the (`<`) symbol is actually "less than or equal to". *VarType* is a string (`Integer` or `Continuous`) describing the type of variable. Although objective function and constraints are separated in linear and nonlinear parts in the format description above, there is no reason why this task should not be performed directly by the software. Indeed, MORON separates linear and nonlinear parts automatically. *ooOPS*, which relies on a less advanced symbolic manipulation library, requires the linear parts to be made explicit in the description file, with the special character "`|`" syntactically separating linear and nonlinear parts (with the semantics of "sum"). The nonlinear functions $f(x), g_1(x), \ldots, g_m(x)$ are strings describing the mathematical expressions. The only really non-trivial piece of code required for reading the above problem description file is therefore a small parser that, given a string containing a mathematical expression (containing just variable names, numbers, and operators, without indices or quantifiers), builds a binary (or n-ary) tree representing the expression. As has been remarked in Section 4.3, this task is actually fairly easy.

Below is an example of a problem expressed in the format described in this section.

```
# problem: Yuan 1988 (MINLP)
variables =
   0 < y1 < 1 / Integer,
```

```
   0 < y2 < 1 / Integer,
   0 < y3 < 1 / Integer,
   0 < x4 < 10 / Continuous,
   0 < x5 < 10 / Continuous;
objfun =
   [ 2*x4 + 3*x5 + 1.5*y1 + 2*y2 - 0.5*y3 ];
constraints =
   [1.25 < y1 + x4^2 < 1.25],
   [3 < 1.5*y2 + x5^1.5 < 3],
   [MinusInfinity < y1 + x4 < 1.6],
   [MinusInfinity < y2 + 1.3333*x5 < 3],
   [MinusInfinity <  -y1 - y2 + y3 < 0];
startingpoint =
   0, 0, 1, 1, 1;
options =
   MainSolver sBB,
   sBBLowerBoundSolver lp_solve,
   sBBUpperBoundSolver snopt;
```

Although a starting point is usually not required for most GO algorithms, it was decided to include the possibility of passing this information to the solver, since this standard might be used as input for a local NLP solver.

We suggest that a suitable extension of this file format (encompassing an arbitrary number of objective functions and other minor adjustments) should be used as a standard MINLP problem description file format.

4.4 Data structures

The most important data structures in the `Problem` class refer to the main problem entities: variables, objective function and constraints. The mathematical expressions defining objective function and constraints are stored symbolically in a tree-like fashion, as explained in Section 5 below. Many of the other properties of the problem entities, like names, bounds and so on, can be implemented as desired, as they do not pose any particular practical problem. *ooOPS* and MORON use the C++ STL `string` class [78] for names and the basic `double` type for real numbers.

The most puzzling issues in designing data structures for problem entities are to do with indexing of variables and constraints, and usually give rise to the worst bugs. Since many global optimization algorithms work by reformulating the problem symbolically, and since each reformulation is conveniently stored in a separate object, it is important to retain the identity of each variable and constraint, even when they undergo symbolic transformations. Variable IDs are the most critical pieces of information: a sBB algorithm has to act alternately on the original problem and on its convex relaxation. Since the convex relaxation may be the result of a complex symbolic manipulation (involving adding and removing problem variables as needed), a mapping between variables in the original and in the convexified problem is of the utmost importance. After much coding experimentation, the most practical solution seems to be the following:

- the original problem is first simplified as much as possible, and as many variables as possible are deleted from the problem;
- the variables in the simplified problem are flagged "unchangeable" to all subsequent reformulator modules;
- reformulators which perform liftings, i.e. which add variables to the problem, flag the added variables as "changeable", so that subsequent reformulators can remove them if required.

Each variable is assigned a unique ID when the problem is first read by the parser. These IDs are stored in lists. Deletion of a variable is equivalent to removing the corresponding ID from the list; no other variable is re-indexed (obviously, deleting a variable can only be done if the variable does not appear in any of the problem expressions). A variable can be added by finding an unassigned variable ID and inserting it into the list. Cycling over all variables is equivalent to traversing the list. Similar lists containing variable names, bounds, values and other properties are easily kept synchronized with the "basic" ID list.

Constraints can be dealt with similarly, although symbolic manipulation of constraint indices is usually much less problematic.

4.5 Configuration of solver parameters

The performance of most solver codes is usually hugely conditioned by proper parameter settings (like tolerances, limits on the number of iterations and so on). If a GO algorithm relies on a cascaded sequence of solvers being called, each solver must be properly configured; this, however, poses some problems, because whilst some configuration parameters can be set by the user, others are better left to the higher-level solver, which will set them based on the current performance of the algorithm run. Thus, solvers are first created in memory by a *solver manager*, which pre-configures them; subsequently, the solver and the `problem` object being solved are bound together in a *solver system*, which offers further configuration capabilities and finally offers the API for starting the solution process. A parameter list is stored both in the solver manager and in the solver system (which inherits a pre-configured parameter list from the solver manager). This treatment of parameter setting follows the guidelines of the Global CAPE-Open consortium for chemical engineering software [33, 54, 55].

5 Symbolic manipulation of mathematical expressions

Symbolic computation is usually something that the mathematical practitioner employs dedicated software for; software like Maple, Mathematica,

Matlab and so on. Numerical methods, and in particular optimization, have always been implemented in the spirit of number crunching, the most notable exception being linear algebra, for the simple reason that doing symbolic manipulation on linear expressions is very easy. It is difficult to even envisage how to write a symbolic computation program on a computer whose basic data types are integer and floating point numbers. The following books provide good introductions to symbolic computation methods [76, 14, 13].

Symbolic computation relies on a machine representation of mathematical operations on some numbers or literal symbols (constants, variables, or expressions involving constants and variables). Usually, one of the following techniques is employed to represent these operations:

- binary trees;
- lists;
- n-ary trees.

5.1 Binary Trees

Binary trees have been proposed as a way of representing mathematical expressions by Knuth [35] and made their way in computational engineering and other fields of scientific computing [10]. This representation is based on the idea that operators, variables and constants are nodes of a digraph; binary operators have two outcoming edges and unary operators only have one; leaf nodes have no outcoming edge. One disadvantage is that binary tree representation makes it cumbersome to implement associativity. For example, the expression $y + x + 2x + 3x$ is represented as $(((y + x) + 2x) + 3x)$, so it would require three recursive steps to lower tree ranks to find out that it is possible to write it as $(y+6x)$. Another disadvantage is that different parsers may have different representations for the same expressions. With the example above, a "left-hand-side-first" parser would create $(y + (x + (2x + 3x)))$ instead of the "right-hand-side-first" $(((y + x) + 2x) + 3x)$.

Where symbolic manipulation is only desired to compute symbolic derivatives and performing little or no symbolic manipulation, this approach may be the best, as it is simpler to implement than the other techniques and generally performs very efficiently [53, 39]. $oo\mathcal{OPS}$ uses binary trees to store expressions.

5.2 Lists

The representation of algebraic expressions by lists dates back to the AI-type languages Prolog and Lisp. Lisp, in particular, was so successful at the task that a lot of CASes, today, are still based on Lisp's list manipulation abilities. Prolog has some interesting features in conjunction with symbolic computation, in particular the "computation-reversing" ability, by which if you compute a symbolic derivative and do not bother to simplify it, Prolog lets you integrate it symbolically performing virtually no calculation at all.

Any symbolic computation library written in Prolog/Lisp faces the hard problem of implementing an API which can be used by procedural languages like Fortran, C or C++. Whilst technically not impossible, the architectures and OSes offering stable and compatible Lisp and C/C++ compilers are few. GNU/Linux actually has object-compatible Lisp and C/C++ compilers; however, the GNU Lisp compiler uses an array of internal data structures which are very difficult to read from a C/C++ program, making data interchange between the different modules hard to implement.

There are two other problems faced by Prolog/Lisp programs: portability (many Prolog/Lisp compilers implement different dialects of the languages) and a reduced user base.

5.3 n-ary Trees

Expression representation by n-ary trees can be seen as a combination of the previous two techniques. MORON makes use of this representation. In order to characterize this representation formally, we need some definitions.

An *operator* is a node in a directed tree-like graph. Let L be the set

$$\{+, -, \times, /, \hat{}, (-1)\times, \log, \exp, \sin, \cos, \tan, \cot, \text{VAR}, \text{CONST}\}$$

of *operator labels*. An operator with label VAR is a *variable*, an operator with label CONST is a *constant*. Operator nodes may generally have any number of outcoming edges; variables and constants have no outcoming edges and are called *leaf nodes*. A variable is also characterized by a non-negative integer index i, and a constant by a value which is an element of a number field F. We shall assume $F = \mathbb{R}$ (or at least, a machine representation of $\mathbb{R}$) in what follows, but this can vary. Let V be the set of all variable-type operator nodes. Let $T_0 = V \cup \mathbb{R}$. This is the set of the terminal (or leaf) nodes, i.e. the variables and constants. Now for each positive integer i, define recursively $T_i = L \times (T_{i-1} \cup T_0)^{<\omega}$. Elements of T_i are operator nodes having *rank* i. Basically, an element of T_i is made up of an operator label $l \in L$ and a finite number of subnodes. A *subnode* s of n is a node s in the digraph so that there is an edge leaving node n and entering s.

The biggest advantage of n-ary tree representation is that it makes it very fast and easy to perform expression simplification. Another advantage is that expression evaluation on n-ary trees is faster than that obtained with a binary tree structure [42].

5.4 Main symbolic manipulation algorithms

In this section we discuss the most important algorithms for symbolic manipulation: allocation and deallocation of memory for recursive data structures like trees, evaluation of an expression at a point, symbolic differentiation and basic simplification rules. Many more symbolic algorithms are actually used in both *ooOPS* and MORON, including a standard representation for mathematical expressions, equality tests, advanced simplification routines and separation of linear and nonlinear parts of an expression. The latter is particularly important as most nonlinear local solvers require the linear parts of the constraints to be input separately from the nonlinear parts.

Allocation and deallocation

One of the biggest challenges in tree handling is memory allocation/deallocation. A tree-node class normally consists of its own semantic data and two (or more, in the case of n-ary trees) references to its children subnodes. In this setup, the following memory-related problems arise.

- When a node is allocated, the subnodes are not automatically allocated, so they have to be allocated manually when the need arises (see fig. 7).

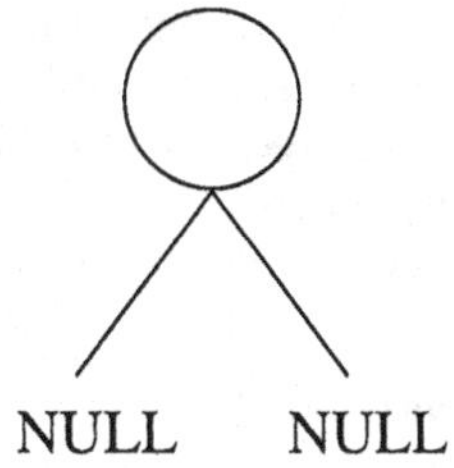

Fig. 7. Allocation of a node does not allocate subnodes.

- When a node is copied, the question arises whether all the subnode hierarchy should be copied or just the references to the immediate subnodes (see fig. 8). The two cases must be treated separately.
- Supposing a tree has been allocated, with some of its nodes copied hierarchically and some others just copied as references, how does one deallocate the tree? Just deleting all the nodes will not work because the nodes copied as references are still in use by other trees. A node cannot hold the information about all the trees it belongs to, as it may belong to a huge number of trees. The standard way to deal with this situation is to store a counter in each node that counts the number of times it is copied as a reference. Each time the deallocation of the node is requested the counter is decreased. The node is truly deallocated only when the counter is zero.

Evaluation

Normal mathematical operations on expressions amount to the manipulation of nodes. These can be copied to form new expressions, replaced by other

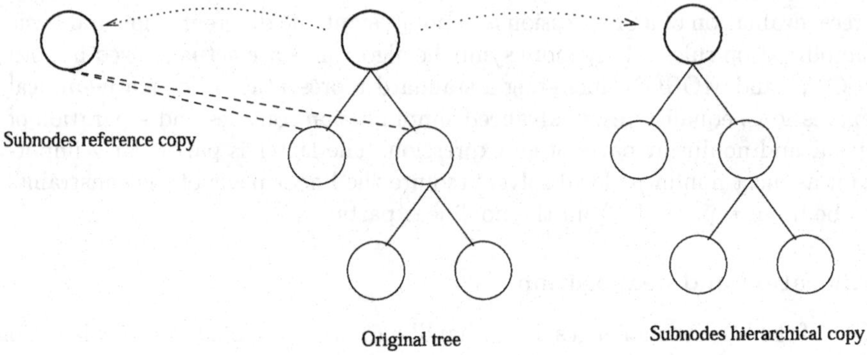

Fig. 8. Copy of subnodes and copy of references.

nodes to achieve symbolic simplification and so on. Two expressions can be summed (or in fact acted on by any other binary operator) just by creating a new top node and setting the top nodes of the two expressions as its two subnodes. Substitution of a variable can be obtained by replacing the relevant variable indices in the terminal nodes; special care must be taken if these terminal nodes are in use in other trees within the program. If they are, then new terminal nodes should be created before the changes occur. Evaluation of an expression tree is obtained by recursing on each node in the following way.

```
evaluate(node, variable_values) {
  if (node is terminal) {
    return variable value corresponding to variable index;
  } else {
    result = 0;
    for each subnode in node {
      partial_result = evaluate(subnode, variable_values);
      result = result (node.operator) partial_result;
    }
    return result;
  }
}
```

Both `result` and `partial_result` in the above pseudocode indicate real numbers. If the same tree has to be evaluated several times, it may be convenient to store the linear order of recursive evaluation of each node and then call a modified evaluation procedure which is linear in nature rather than recursive. This avoids the computational overhead of recursiveness (although optimizing compilers reduce that overhead considerably nowadays). The details of this strategy are analysed in [32].

Differentiation

Derivatives of the objective function and constraints are used (or can be used) by most nonlinear deterministic local solvers, so an efficient way to calculate these derivatives gains good computational savings. Ordinarily, derivatives are computed by finite differences or by hard-coding the symbolic derivative inside the program at compile-time. The first method has huge associated computational costs, whilst the second method only targets programs devised to solve one particular optimization problem, which is not suitable for our purposes.

There are two alternatives: automatic differentiation (AD) [24] and symbolic derivative computation. A. Griewank, in his fascinating 1989 paper on AD, claims that AD is better than symbolic differentiation. However, the limitation of the symbolic differentiation methods he was referring to were twofold: hardware-wise, in lack of RAM (he was using a Sun3 with 16MB ram); and software-wise, in having to pass the output of a symbolic derivative computed by Macsyma 1 or Maple 1 to a Fortran compiler before the evaluation, apparently with the overhead of shell-piping mechanisms. Besides, AD is usually employed for pieces of software where the user supplies the complex evaluation routines as source code. This would make the optimization framework dependent on a compiler and a linker, which is not usually an acceptable choice for a stand-alone software. For this reason, we chose to employ symbolic differentiation techniques, which perform very fast when the evaluation occur on expression trees.

Once the symbolic derivatives are calculated for the objective function and constraints, they only need to be evaluated in order to produce the derivative value at a point. Other advantages of this approach are that (a) by using expression trees and recursive procedures, symbolic derivatives are not computationally expensive to construct; and (b) the derivative values they provide are exact, whereas finite difference methods can only approximate the values at best.

The following pseudocode shows how to construct a node representing the symbolic derivative of `node` with respect to `variable`.

```
diff(node, variable) {
  retnode = 0;
  for each subnode in node {
    if (subnode depends on variable) {
      if (subnode is terminal) {
        retnode = 1;
      } else {
        case of node.operator {
        case '+':
          retnode = retnode + diff(subnode, variable);
        case '-':
          retnode = retnode - diff(subnode, variable);
        case '*':
```

```
            retnode = retnode + diff(subnode, variable) * node / subnode;
          // and all the other derivative rules...
          }
        }
      }
    }
    return retnode;
  }
```

Notice that `retnode` in the above pseudocode indicates a node, so all the operations that act on `retnode` (addition, subtraction, multiplication, division and so on) are to be implemented as procedures which manipulate expression trees.

Derivative rules are the usual ones; the rule for multiplication is expressed in a way that allows for n-ary trees to be correctly derived:

$$\frac{\partial}{\partial x}\prod_{i=1}^{n} f_i = \sum_{i=1}^{n}\left(\frac{\partial f_i}{\partial x}\prod_{j\neq i} f_j\right).$$

All valid algebraic simplifications can be used to simplify symbolic expressions. However, all simplifications have an associated computational cost so it is essential to find a balance between the degree of simplification of each expression and the cost of the simplification itself. Simplifying $\sin^2(f(x)) + \cos^2(f(x)) = 1$, for example, involves a tree search that spans 7 nodes (addition, exponentiation and "2" as a constant in two nodes, sine and cosine operators) and is of limited use, so it is not advisable to employ it unless it is known in advance that most of the expressions will involve sines and cosines.

Simplification

The basic simplifications which should be carried out are the following:

$$constant\ (operator)\ constant = constant(result of operation)$$

$$-(-f(x)) = f(x)$$
$$f(x) + 0 = f(x)$$
$$f(x) - 0 = f(x)$$
$$0 - f(x) = -f(x)$$
$$f(x) \times 1 = f(x)$$
$$\frac{f(x)}{1} = f(x)$$
$$(f(x))^0 = 1$$
$$(f(x))^1 = f(x)$$
$$f(x) + f(x) = 2f(x)$$
$$f(x) - f(x) = 0$$
$$f(x) \times f(x) = (f(x))^2$$
$$\frac{f(x)}{f(x)} = 1$$

Note that binary trees are not commutative, so the commuted simplifications

$$0 + f(x) = 0$$
$$1 \times f(x) = f(x)$$

should also be carried out. Note also that after a tree has been simplified once there is scope for further simplification. For example applying the above rules in succession to $x^{2y-2y} - 1$ would gather $x^0 - 1$, but it is evident that the expression can be simplified even more. A second application of the rules would gather $1 - 1$ and a third application would finally gather 0. Ideally, thus, simplification should be carried out repeatedly until the expression does not change under the simplification rules. Where this is too computationally expensive, a compromise may be enforced.

6 Local solvers

By "local solvers" we mean here those solvers which decide on the local or global optimality of a point by performing an analysis of a point neighbourhood. As such, local solvers may implement local solution algorithms (for NLPs) and global solution algorithms (for LPs). Mostly, local solvers are used within global solvers as black-box calls to solve the main problem, or auxiliary problems, locally. In most GO algorithms, the global phase has a limited numerical knowledge of the problem structure; the "dirty work" is usually performed by the local solvers. In the case of LPs, the local solver needs to know the linear coefficients of all the variables in the objective and the constraints, as well as the variable and constraint bounds. In the case of NLPs, the local solver needs to evaluate objective function, constraints, first derivatives of both, and optionally also second derivatives, for any given point. Large-scale local NLP solvers need to be explicitly told about linear and nonlinear parts of each expression in the problem. As most local solvers are in fact "solver libraries", with varying degrees of user-friendliness, sometimes the problem variables and/or the constraints need to be re-ordered. This is very time-consuming and error-prone, as the inverse re-ordering needs to be applied to the solution vector.

The task of interfacing a local solver with the rest of the optimization system is carried out by the solver wrapper. In this author's experience, solver wrappers for existing local NLP solvers (specially those requiring variable and constraint re-ordering) are the most frustrating source of software bugs. Interfacing with local LP solvers is easier; however, since many local LP solvers do not accept constraints in "double bounded format" ($LowerBound \leq g(x) \leq UpperBound$), preferring the "single bounded format" instead ($g(x) \gtreqless Bound$), some of the constraints might have to be replicated with different directions and bounds.

One word should be spent about the reliability of local solver codes. Whereas LP solvers are next to 100% reliable, some of the most efficient algorithms for the local solution of nonconvex NLPs are inherently unreliable. Sequential Quadratic Programming (SQP), for example, is a standard and widely used technique for locally solving constrained NLPs in general form. SQP might fail (rather spectacularly in certain cases) if a feasible starting point cannot be provided, or if the linearized constraints are infeasible (even though the original nonlinear constraints may be feasible). Both these occurrences are far from rare, so local NLP solvers are rarely reliable. Since most GO algorithms delegate the numerical work to the local solvers, a global solver is only as reliable as its local sub-solver, that is to say, not very reliable at all. Therefore, it is always a good idea for the wrapper to be able to deal properly with all the return messages of the local solver; in our opinion, it is also a good idea to have the wrapper double-check on the feasibility of the solution provided by the local solver.

ooOPS at the moment has three local solvers: NPSOL (or rather, the VCF optimization code in the NAG library), a rather old version of SNOPT (which is a large-scale modification of NPSOL), and `lp_solve`, which is a free LP solver.

MORON is interfaced to a recent version of SNOPT and the GLPK [48] local LP solver.

7 Global solvers

GO algorithms mostly require very high-level steps, like local solution of subproblems, symbolic manipulation of mathematical expressions, and so on.

Embedding global solvers within optimization environment which offer these possibilities makes it possible to implement and test a global solver in a very short time. In this section we shall describe the three global solvers found in *ooOPS*. At this stage, MORON only has a very preliminary version of a sBB solver, which is not discussed here.

7.1 SobolOpt multistart algorithm

SobolOpt[9] (also see Section 2.1) is an implementation of a Multi-Level Single Linkage (MLSL) algorithm; its main strength is that it employs certain Low-Discrepancy Sequences (LDSs) of sampling points called Sobol' sequences whose distributions in Euclidean space have very desirable uniformity properties. Let Q be the set of pairs of sampled points q together with their evaluation $f(q)$ (where f is the objective function). Let S be the list of all local minima found up to now.

1. (Initialization) Let $Q = \emptyset$, $S = \emptyset$, $k = 1$ and set $\varepsilon > 0$.
2. (Termination) If a pre-determined termination condition is verified, stop.
3. (Sampling) Sample a point q_k from a Sobol' sequence; add $(q_k, f(q_k))$ to Q.
4. (Clustering distance) Compute a distance r_k (which is a function of k and n; there are various ways to compute this distance, so this is considered as an "implementation detail" — one possibility is $r_k = \beta k^{-\frac{1}{n}}$, where β is a known parameter and n is the number of variables).
5. (Local phase) If there is no previously sampled point $q_j \in Q$ (with $j < k$) such that $||q_k - q_j|| < r_k$ and $f(q_j) \leq f(q_k) - \varepsilon$, solve problem (1) locally with q_k as a starting point to find a solution y with value $f(y)$. If $y \notin S$, add y to S. Set $k \leftarrow k + 1$ and repeat from step 2.

The algorithm terminates with a list S of all the local minima found. Finding the global minimum is then a trivial matter of identifying the minimum with lowest objective function value $f(y)$. Two of the most common termination conditions are (a) maximum number of sampled points and (b) maximum time limit exceeded. A discussion of how Sobol' sequences can be generated is beyond the scope of this paper. A good reference is [59], p.311.

The implementation of the SobolOpt algorithm, which is very robust, was carried out by S. Kucherenko and Yu. Sytsko and successively adapted to the *ooOPS* framework. For more details about this algorithm, see [37].

[9] The SobolOpt solver within *ooOPS* shares the same code as the implementation described in Chapter 5.

7.2 Variable Neighbourhood Search

In this section we discuss the implementation of the Variable Neighbourhood Search[10] algorithm for GO presented in Section 2.2. The search space is defined as the hypercube given by the set of variable ranges $x^L \le x \le x^U$. At first we pick a random point $\tilde{x}$ in the search space, we start a local search and we store the local optimum x^*. Then, until k does not exceed a pre-set $k_{\max}$, we iteratively select new starting points $\tilde{x}$ in a neighbourhood $N_k(x^*)$ and start new local searches from $\tilde{x}$ leading to local optima x'. As soon as we find a local minimum x' better than x^*, we update $x^* = x'$, re-set $k = 1$ and repeat. Otherwise the algorithm terminates.

For each $k \le k_{\max}$ consider hyper-rectangles $R_k(x^*)$ similar to $x^L \le x \le x^U$, centered at x^*, whose sides have been scaled by $\frac{k}{k_{\max}}$. More formally, let $R_k(x^*)$ be the hyper-rectangle $y^L \le x \le y^U$ where, for all $i \le n$:

$$\begin{aligned} y_i^L &= x_i^* - \frac{k}{k_{\max}}(x_i^* - x_i^L) \\ y_i^U &= x_i^* + \frac{k}{k_{\max}}(x_i^U - x_i^*). \end{aligned}$$

This construction forms a set of hyper-rectangular "shells" centered at x^*. For $k > 0$, we define the neighbourhoods $N_k(x^*)$ as $R_k(x^*) \backslash R_{k-1}(x^*)$ (observe that $R_0(x^*) = \emptyset$). The neighbourhoods are disjoint, which gives the VNS algorithm a higher probability not to fall in local optima that have already been located. Furthermore, the union of all the neighbourhoods $N_k(x^*)$ is the whole space $x^L \le x \le x^U$. This is a rather unusual features in VNS implementation, specially when VNS is applied to combinatorial problems. Here it is justified by the (partial) continuity of the search space. The neighbourhoods are obviously just used for sampling; the local search itself is performed in the whole space.

Sampling in the neighbourhoods $N_k(x^*)$ is a non-trivial task. Since sampling in hyper-rectangles is easy, one possible solution would be to sample a point in $R_k(x^*)$ and reject it if it is in $R_{k-1}(x^*)$, but this would be highly inefficient. A different strategy was preferred in this implementation.

1. Choose an index $j \le n$ randomly.
2. Sample a random value x'_j in the one-dimensional interval given by projection of $R_k(x^*)$ on the j-th coordinate.
3. Let the projection of $R_{k-1}(x^*)$ on the j-th coordinate be the interval $\rho = [y_j^L, y_j^U]$. If $x'_j \in \rho$, then: if $|x'_j - y_j^L| \le |x'_j - y_j^U|$ let $x'_j = y_j^L$, else let $x'_j = y_j^U$.
4. For all $i \le n$ such that $i \ne j$, sample a random value x'_i in the projection of $R_k(x^*)$ on the i-th coordinate.

[10] The implementation of the VNS solver in *ooOPS* is fundamentally different from the GLOB implementation described in Chapter 6.

The above procedure generates a point $x' = (x'_1, \ldots, x'_n)$ which is guaranteed to be in $N_k(x^*)$. Evidently, $x' \in R_k(x^*)$. Now suppose, to get a contradiction, that $x' \in R_{k-1}(x^*)$. But then for each $i \leq n$ we have $y_i^L \leq x'_i \leq y_i^U$. Since we specifically set one of the x'_i to be outside this interval in steps 1-3 of the algorithm, $x' \notin R_{k-1}(x^*)$ as claimed. Therefore $x' \in N_k(x^*)$. The only problem with this method is that the sampling is not uniformly distributed anymore, as there are areas of $N_k(x^*)$ where there is zero probability of sampling x'. This, unfortunately, affects the convergence proof of the algorithm, since there are unexplored areas. A simpler, more robust strategy is to define each neighbourhood $N_k(x^*)$ as the hyper-rectangle $R_k(x^*)$. This is wasteful (a point might be sampled in $N_{k-1}(x^*)$, which had already been explored at the previous iteration), but the convergence proof holds. Both approaches have been coded in the solver, and selection occurs by modifying an appropriate parameter.

The other main solver parameters control: the minimum k to start the VNS from, the number of sampling points and local searches started in each neighbourhood, an ε tolerance to allow moving to a new x^* only when the improvement was sufficiently high, and the maximum CPU time allowed for the search.

7.3 Spatial Branch-and-Bound

The overall sBB algorithm was discussed in Section 2.3. Below, we consider some of the key steps of the algorithm in more detail.

Bounds tightening

These procedures appear in steps 1 and 2 of the algorithm structure outlined in Section 2.3. They are optional in the sense that the algorithm will, in principle, converge even without them. Depending on how computationally expensive and how effective these procedures are, in some cases convergence might actually be faster if these optional steps are not performed. In the great majority of cases, however, the bounds tightening steps are essential to achieve fast convergence. Two major bounds tightening schemes have been proposed in the literature: optimization-based and feasibility-based.

The optimization-based bounds tightening procedure identifies the smallest range of each variables subject to the convex relaxation of the problem to remain feasible. This ensures that the sBB algorithm will not have to explore hyper-rectangles which do not actually contain any feasible point. Unfortunately, this is a computationally expensive procedure which involves solving at least $2n$ convex NLPs (or LPs if a linear convex relaxation is employed) where n is the number of problem variables. Let $\alpha \leq \bar{g}(x) \leq \beta$ be the set of constraints in the relaxed (convex) problem (α, β are the constraint limits). The following procedure will construct sequences $x^{L,k}, x^{U,k}$ of lower and upper variable bounds which converge to new variable bounds that are at least as tight as, and possibly tighter than x^L, x^U.

1. Set $x^{L,0} \leftarrow x^L$, $x^{U,0} \leftarrow x^U$, $k \leftarrow 0$.
2. Repeat

$$x_i^{L,k} \leftarrow \min\{x_i \mid \alpha \leq \bar{g}(x) \leq \beta \wedge x^{L,k-1} \leq x \leq x^{U,k-1}\} \quad \forall i \leq n;$$
$$x_i^{U,k} \leftarrow \max\{x_i \mid \alpha \leq \bar{g}(x) \leq \beta \wedge x^{L,k-1} \leq x \leq x^{U,k-1}\} \quad \forall i \leq n;$$
$$k \leftarrow k+1.$$

until $x^{L,k} = x^{L,k-1}$ and $x^{U,k} = x^{U,k-1}$.

Because of the associated cost, this type of tightening is normally performed only once, at the first step of the algorithm.

Feasibility-based bounds tightening is computationally cheaper than the one described above, and as such it can be applied at each and every region considered by the algorithm. Variable bounds are tightened by using the problem constraints to calculate extremal values attainable by the variables. This is done by isolating a variable on the left hand side of a constraint and evaluating the right hand side extremal values by means of interval arithmetic.

Feasibility-based bounds tightening is trivially easy for the case of linear constraints. Given linear constraints in the form $l \leq Ax \leq u$ where $A = (a_{ij})$, it can be shown that, for all $1 \leq j \leq n$:

$$x_j \in \left[\max\left(x_j^L, \min_i\left(\frac{1}{a_{ij}}\left(l_i - \sum_{k\neq j}\max(a_{ik}x_k^L, a_{ik}x_k^U)\right)\right)\right),\right.$$
$$\left.\min\left(x_j^U, \max_i\left(\frac{1}{a_{ij}}\left(u_i - \sum_{k\neq j}\min(a_{ik}x_k^L, a_{ik}x_k^U)\right)\right)\right)\right] \quad \text{if } a_{ij} > 0$$

$$x_j \in \left[\max\left(x_j^L, \min_i\left(\frac{1}{a_{ij}}\left(l_i - \sum_{k\neq j}\min(a_{ik}x_k^L, a_{ik}x_k^U)\right)\right)\right),\right.$$
$$\left.\min\left(x_j^U, \max_i\left(\frac{1}{a_{ij}}\left(u_i - \sum_{k\neq j}\max(a_{ik}x_k^L, a_{ik}x_k^U)\right)\right)\right)\right] \quad \text{if } a_{ij} < 0.$$

As pointed out in [71] p.202, feasibility-based bounds tightening can also be carried out for certain types of nonlinear constraints.

Choice of region

The region selection at step 2 follows the simple policy of choosing the region in the list with the lowest lower objective function bound as the one which is most promising for further consideration (recall that the lower bound l calculated in each region is associated to the subregions after branching — see step 7 of the sBB algorithm).

Local solution of the original problem

The most computationally expensive step in the sBB algorithm is typically the call to the local NLP solver to find the upper bound to the objective function value relative to the current region. The two methods described below should at least halve the number of upper bounding problems that are solved during the sBB algorithm. Note that a distinction is made between the variables that are present in the original NLP ("original variables") and those added by the standardization procedure ("added variables" — see Section 2.3).

1. *Branching on added variables.* Suppose that in the sBB algorithm an added variable w is chosen as the branch variable. The current region is then partitioned into two sub-regions along the w axis, the convex relaxations are modified to take the new variable ranges into account, and lower bounds are found for each sub-region. The upper bounds, however, are found by solving the original problem which is not dependent on the added variables. Thus the same exact original problem is solved at least three times in the course of the algorithm (i.e. once for the original region and once for each of its two sub-regions). The obvious solution is for the algorithm to record the objective function upper bounds in each region. Whenever the branch variable is an added variable, avoid solving the original (upper bounding) problem and use the stored values instead.
2. *Branching on original variables.* Even when the branching occurs on an original problem variable, there are some considerations that help avoid solving local optimization problems unnecessarily. Suppose that the original variable x is selected for branching in a certain region. Then its range $[x^L, x^U]$ is partitioned into $[x^L, x']$ and $[x', x^U]$. If the solution of the upper bounding problem in $[x^L, x^U]$ is x^*, and $x^* \in [x^L, x']$, then it is unnecessary to solve the upper bounding problem again in the sub-region $[x^L, x']$ as an upper bound is already available at x^*. Of course, the upper bounding problem still needs to be solved for the other subregion $[x', x^U]$ (see Fig. 9).

Branching

There are many branching strategies [17] available for use in spatial Branch-and-Bound algorithms. Generally, branching involves two steps, namely determining the point (i.e. set of variable values) on which to branch, and finding the variable whose domain is to be sub-divided by the branching operation. Here, we use the solution $\tilde{x}$ of the upper bounding problem (step 4) as the branching point, if such a solution is found; otherwise the solution of the lower bounding problem $\bar{x}$ (step 3) is used. We then use the standard form to identify the nonlinear term with the largest error with respect to its convex relaxation. By definition of the standard form (see Section 2.3), this is equivalent to evaluating the defining constraints at $\bar{x}$ and choosing the one

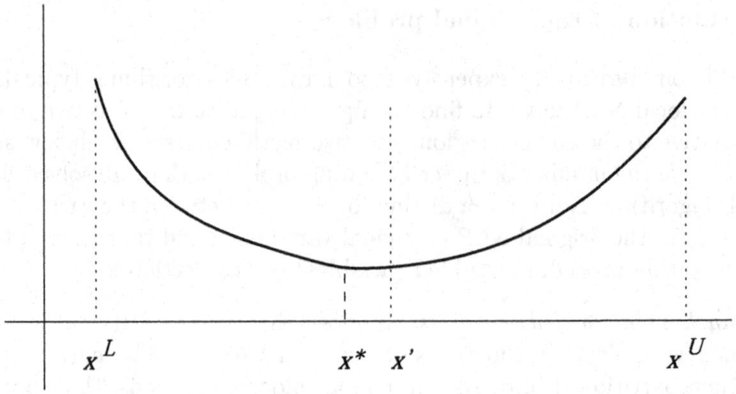

Fig. 9. If the locally optimal solution in $[x^L, x^U]$ has already been determined to be at x^*, solving in $[x^L, x']$ is unnecessary.

giving rise to the largest error in absolute value. In case the chosen defining constraint represents a unary operator, the only variable operand is chosen as the branch variable; if it represents a binary operator, the branch variable is chosen as the one whose value at the branching point is nearest to the mid-point of its range (see [71], p. 205-207). Supposing variable $x_i \in [x_i^L, x_i^U]$ was chosen as a branching variable with branching point $t \in (x_i^L, x_i^U)$, should x_i be an integer variable (i.e. $i \in Z$), the domain $[x_i^L, x_i^U]$ should be split as follows: $[x_i^L, \lfloor t \rfloor], [\lceil t \rceil, x_i^U]$. If x_i is a binary variable, one branching will evidently be sufficient to fix the variable in subsequent subproblems.

Generation of convex relaxation

As has been explained in Section 2.3, the automatic generation of the convex relaxation entails putting the problem in standard form and then replacing each nonlinear defining constraint with a linear convexification thereof.

The problem in standard form consists of:

- the objective function, consisting of one linearizing variable only (a defining constraint for this equation also exist in the problem, obviously);
- the linear constraints, represented by a matrix;
- the nonlinear constraints, represented by triplets of variable indices and an operator label;
- the constraint bounds;
- the variable ranges;
- the variable values.

Since the nonlinear constraints are isolated in the form $x_i = x_j \otimes x_k$ where $\otimes$ is an operator, they can be efficiently represented by the triplet (i, j, k) and an operator label, Λ, which indicates what operator acts on x_j, x_k to produce

x_i. If Λ is a unary operator, the index k is set to a dummy value representing "not a variable". If one of the operands is a constant, the respective variable index is set to "not a variable" and the constant value should be stored in a special purpose data field. If both operands are constants, the triplet can be evaluated and discarded. The result of the evaluation replaces all instances of the variable with index i; this implies the elimination of the problem variable with index i.

The procedure that transforms a problem in standard form is based on the following steps:

1. copy original variable values and bounds, original constraint bounds and linear constraint coefficients from the original problem to initialize the respective standard form data structures;
2. add a constraint $x_i = objectivefunction$ to the problem (separating linear and nonlinear parts) and set i as the variable index representing the objective function;
3. cycle over the original nonlinear parts of the constraints:
   a) recursively split the nonlinear part of the current constraint into triplets;
   b) check if the current triplet already exists. If not, store it, otherwise use the existing triplet and discard it;
4. store all linear triplets (i.e. all triplets where the operand is linear) as linear constraints into the linear constraint matrix and then discard them;
5. eliminate the one-variable constraints (i.e. constraints where the algebraic expression only consists of a one-variable term) and use the bounds to update the respective variable range;
6. eliminate constraints of the form $0 \leq x_i - x_j \leq 0$ and substitute the variable x_j by the variable x_i throughout.

Note that the variable elimination schemes are applied to added variables only. The original variables remain unchanged. This is an important issue as it makes it easy to map original problem variables to relaxed problem variables throughout the sBB execution.

Note also that when linear triplets are stored in the linear constraint matrix, particular care should be taken that the minimum amount of new linear constraints is added to the problem. This is best explained with an example. Let $w_1 = w_2 + w_3$, $w_4 = -w_5$ and $w_2 = w_6 - w_7$; the first triplet is reformulated as the linear constraint $0 \leq w_1 - w_2 - w_3 \leq 0$; the second triplet is reformulated as a new linear constraint $0 \leq w_4 + w_5 \leq 0$ because it is independent of the first one. But the third triplet should *not* be reformulated as a new constraint because w_2 already appears in the first constraint. Instead, w_2 in the first constraint is replaced by $w_6 - w_7$ and the variable w_2 (if it does not appear anywhere else in the problem) is eliminated.

The convexified problem can be described by the same data structure used to represent the original problem. In *ooOPS*, because the implementation uses linear relaxations only, the structure can be simplified by removing all references to nonlinear expressions; also note that since the convexified problem is actually of a different type to that of the original problem, it is stored in a separate object instance. The convexification algorithm is as follows:

1. copy variable bounds, constraint bounds and linear constraint matrix from the problem in standard form;
2. cycle on the standard form triplets:
   a) analyse the current triplet and add its convex envelope constraints to the convexified problem.

Although the above process explained in this section involves quite a lot of data copying between data structures, it is not in fact data replication as the procedures operating on the structures may need to change some of the copied values. The purpose of having different sets of copied values is to allow for full code re-entrancy.

Region storage scheme

In abstract terms, a region in the sBB algorithm is a hypercube in Euclidean space; thus, it is characterized by a list of n variable ranges. However, this characterization means that, to store n variable ranges explicitly one must allocate and manage memory of size $2n$. This means that to create a new region, we have to repeatedly copy $2n$ memory units from the old region to the new one. Because the partitioning always acts on just one branch variable, all of the other variable ranges would be copied unchanged. Furthermore, because the partitioning always produces two subregions, the waste would be doubled.

In view of the above, we only store the new range for each child subregion together with a pointer to the parent region in order to retrieve the other ranges. This gives rise to a tree of regions where each node contains:

- a pointer to the parent region;
- the branch variable when the region was created;
- the branch variable range of this region;
- the branch point of the parent variable range (one of the endpoints of the branch variable range, it indicates whether this is the "upper" region or the "lower" region);
- the objective function lower and upper bounds;
- a flag that signals whether an upper bound is already available for the region prior to calculation (see Section 7.3).

Starting from any particular region, we can derive the complete set of variable bounds for it by ascending the tree via its parent. In order also to allow traversing the tree in a downwards direction (see below), we add another piece of information:

- pointers to the children subregions.

Also note that some of the regions in the list may be discarded from further consideration at some point in the algorithm. However, we cannot just delete the discarded regions from the tree because they hold information about the variable ranges, so we need a discarding scheme which involves no actual deletion. This is very easy to accomplish with a boolean flag that indicates whether a region is active or inactive (discarded):

- a flag that indicates whether the region is active or not.

Control flow in the sBB solver

This section refers to the control flow of the sBB solver in *ooOPS* (also see Sections 3.4, 3.4). At the outset, we assume that the user code has created an `opssolvermanager` for the sBB solver and an `opssystem` binding the sBB solver and the problem. The list below starts after the `Solve()` method has been called.

1. Creation of the `opssolvermanager` for the local solver that will be used to solve the upper bounding problem.
2. Creation of an `opssystem` using the `opssolvermanager` just created and the original problem.
3. Creation of the `convexifiermanager` acting on the original problem.
4. Generation of the convex (linear) relaxation. This is held in a modified `ops` class object which only includes data structures for storing linear objective function and coefficients.
5. Creation of the `opssolvermanager` for the local solver used to solve the lower bounding problem.
6. Creation of the `opssystem` using the `opssolvermanager` just created and the convex relaxation.
7. Iterative process:
   a) Follow the sBB algorithm repeatedly calling the `opssystem` acting on upper and lower bounding problems.
   b) On changing the variable ranges update the lower bounding problem on-the-fly (via the `UpdateConvexVarBounds()` method in the `convexifiermanager`).
8. Deallocation of objects created by the global solver code.

8 Conclusion

In this paper we discussed various aspects of writing general-purpose global optimization software. We first performed a literature review of existing global optimization algorithms and existing global optimization software packages targeted at solving problems in general form (1). The most important issue is that of a sound software architecture and design, which makes it possible to implement very high-level algorithms (including those that call whole sub-algorithms as black box procedures, and those based on symbolic manipulation of mathematical expressions) fairly easily. We suggested a possible standard file format for describing MINLP problems in flat form. Various symbolic computation algorithms have been described. Lastly, we discussed local solvers generally, and performed an in-depth analysis of the three global solvers implemented within *ooOPS*.

Acknowledgments

I would like to express the deepest thanks to Dr. Maria Elena Bruni (author of Chapter 4) for valuable suggestions, and to Dr. Sonia Cafieri for detailed information about the SIF format file. Prof. M. Drazić (co-author of Chapter 6) helped with some parts of the implementation of the VNS solver within *ooOPS*. I also wish to thank the following people, who also helped reviewing parts of this paper: Dr. P. Belotti, Dr. T. Davidović, Dr. S. Kucherenko, Prof. F. Malucelli, Prof. F. Schoen.

References

1. C.S. Adjiman, I.P. Androulakis, C.D. Maranas, and C.A. Floudas. A global optimization method, αBB, for process design. *Computers & Chemical Engineering*, 20:S419–S424, 1996.
2. C.S. Adjiman and C.A. Floudas. Rigorous convex underestimators for general twice-differentiable problems. *Journal of Global Optimization*, 9(1):23–40, July 1996.
3. C.S. Adjiman, I.P. Androulakis, and C.A. Floudas. Global optimization of MINLP problems in process synthesis and design. *Computers & Chemical Engineering*, 21:S445–S450, 1997.
4. C.S. Adjiman, S. Dallwig, C.A. Floudas, and A. Neumaier. A global optimization method, αBB, for general twice-differentiable constrained NLPs: I. theoretical advances. *Computers & Chemical Engineering*, 22(9):1137–1158, 1998.
5. C. S. Adjiman, I. P. Androulakis, and C. A. Floudas. A global optimization method, αBB, for general twice-differentiable constrained NLPs: II. Implementation and computational results. *Computers & Chemical Engineering*, 22(9):1159–1179, 1998.
6. C.S. Adjiman. *Global Optimization Techniques for Process Systems Engineering*. PhD thesis, Princeton University, June 1998.
7. F.A. Al-Khayyal and J.E. Falk. Jointly constrained biconvex programming. *Mathematics of Operations Research*, 8(2):273–286, 1983.
8. I. P. Androulakis, C. D. Maranas, and C. A. Floudas. αBB: A global optimization method for general constrained nonconvex problems. *Journal of Global Optimization*, 7(4):337–363, December 1995.
9. M. Berkelaar. *LP_SOLVE: Linear Programming Code*. `http://www.cs.sunysb.edu/ algorith/implement/lpsolve/implement.shtml`, 2004.

10. I.D.L. Bogle and C.C. Pantelides. Sparse nonlinear systems in chemical process simulation. In A.J. Osiadacz, editor, *Simulation and Optimization of Large Systems*, Oxford, 1988. Clarendon Press.
11. A. Brook, D. Kendrick, and A. Meeraus. Gams, a user's guide. *ACM SIGNUM Newsletter*, 23(3-4):10–11, 1988.
12. J. Brimberg and N. Mladenović. A variable neighbourhood algorithm for solving the continuous location-allocation problem. *Studies in Location Analysis*, 10:1–12, 1996.
13. J.S. Cohen. *Computer Algebra and Symbolic Computation: Mathematical Methods.* AK Peters, Natick, Massachussetts, 2000.
14. J.S. Cohen. *Computer Algebra and Symbolic Computation: Elementary Algorithms.* AK Peters, Natick, Massachussetts, 2002.
15. Process Systems Enterprise. *gPROMS v2.2 Introductory User Guide.* Process Systems Enterprise, Ltd., London, UK, 2003.
16. T.G.W. Epperly and E.N. Pistikopoulos. A reduced space branch and bound algorithm for global optimization. *Journal of Global Optimization*, 11:287:311, 1997.
17. T.G.W. Epperly. *Global Optimization of Nonconvex Nonlinear Programs using Parallel Branch and Bound.* PhD thesis, University of Winsconsin - Madison, 1995.
18. R. Fourer and D. Gay. *The AMPL Book.*
19. C.A. Floudas. *Deterministic Global Optimization.* Kluwer, Dordrecht, 2000.
20. R. Fourer. Personal communication. 2004.
21. J.E. Falk and R.M. Soland. An algorithm for separable nonconvex programming problems. *Management Science*, 15:550–569, 1969.
22. S. Galli. Parsing ampl internal format for linear and non-linear expressions, 2004. Didactical project, DEI, Politecnico di Milano, Italy.
23. P.E. Gill. *User's Guide for SNOPT 5.3.* Systems Optimization Laboratory, Department of EESOR, Stanford University, California, February 1999.
24. A. Griewank. On automatic differentiation. In Iri and Tanabe [30], pages 83–108.
25. K. Hägglöf, P.O. Lindberg, and L. Svensson. Computing global minima to polynomial optimization problems using gröbner bases. *Journal of Global Optimization*, 7(2):115:125, 1995.
26. P. Hansen and N. Mladenović. Variable neighbourhood search: Principles and applications. *European Journal of Operations Research*, 130:449–467, 2001.
27. P. Hansen and N. Mladenović. Variable neighbourhood search. In P. Pardalos and M. Resende, editors, *Handbook of Applied Optimization*, Oxford, 2002. Oxford University Press.
28. P. Hansen and N. Mladenović. Variable neighbourhood search. In F.W. Glover and G.A. Kochenberger, editors, *Handbook of Metaheuristics*, Dordrecht, 2003. Kluwer.
29. ILOG. *ILOG CPLEX 8.0 User's Manual.* ILOG S.A., Gentilly, France, 2002.
30. M. Iri and K. Tanabe, editors. *Mathematical Programming: Recent Developments and Applications.* Kluwer, Dordrecht, 1989.
31. R. B. Kearfott. *GlobSol User Guide.* `http://interval.louisiana.edu/GLOBSOL/what_is.html`, 1999.
32. B.R. Keeping and C.C. Pantelides. Novel methods for the efficient evaluation of stored mathematical expressions on scalar and vector computers. *AIChE Annual Meeting*, Paper #204b, nov 1997.

33. B. Keeping, C.C. Pantelides, James Barber, and Panagiotis Tsiakis. Mixed integer linear programming interface specification draft. *Global Cape-Open Deliverable WP2.3-03*, October 2000.
34. P. Kesavan and P.I. Barton. Generalized branch-and-cut framework for mixed-integer nonlinear optimization problems. *Computers & Chemical Engineering*, 24:1361–1366, 2000.
35. D.E. Knuth. *The Art of Computer Programming, Part II: Seminumerical Algorithms.* Addison-Wesley, Reading, MA, 1981.
36. V. Kovačević-Vujčić, M. Čangalović, M. Ašić, L. Ivanović, and M. Dražić. Tabu search methodology in global optimization. *Computers and Mathematics with Applications*, 37:125–133, 1999.
37. S. Kucherenko and Yu. Sytsko. Application of deterministic low-discrepancy sequences to nonlinear global optimization problems. *Computational Optimization and Applications*, 30(3):297-318, 2004.
38. R. Levine, T. Mason, and D. Brown. *Lex and Yacc.* O'Reilly, Cambridge, second edition, 1995.
39. L. Liberti, P. Tsiakis, B. Keeping, and C.C. Pantelides. *ooOPS*. Centre for Process Systems Engineering, Chemical Engineering Department, Imperial College, London, UK, 1.24 edition, jan 2001.
40. L. Liberti and C.C. Pantelides. Convex envelopes of monomials of odd degree. *Journal of Global Optimization*, 25:157–168, 2003.
41. L. Liberti. *Reformulation and Convex Relaxation Techniques for Global Optimization.* PhD thesis, Imperial College London, UK, March 2004.
42. L. Liberti. Performance comparison of function evaluation methods. In *Progress in Computer Science Research*, Nova Publisher (accepted for publication 2004).
43. L. Liberti, N. Maculan, and S. Kucherenko. The kissing number problem: a new result from global optimization. *Electronic Notes in Discrete Mathematics*, 17:203-207, 2004.
44. L. Liberti and S. Kucherenko. Comparison of deterministic and stochastic approaches to global optimization. *International Transactions in Operations Research*, 12(3), 2005.
45. M. Locatelli and F. Schoen. Simple linkage: Analysis of a threshold-accepting global optimization method. *Journal of Global Optimization*, 9:95–111, 1996.
46. M. Locatelli and F. Schoen. Random linkage: a family of acceptance/rejection algorithms for global optimization. *Mathematical Programming*, 85(2):379–396, 1999.
47. M. Locatelli. Simulated annealing algorithms for global optimization. In Pardalos and Romeijn [56], pages 179–229.
48. A. Makhorin. *GNU Linear Programming Kit.* Free Software Foundation, `http://www.gnu.org/software/glpk/`, 2003.
49. G.P. McCormick. Computability of global solutions to factorable nonconvex programs: Part I — Convex underestimating problems. *Mathematical Programming*, 10:146–175, 1976.
50. M. Mathur, S.B. Karale, S. Priye, V.K. Jayaraman, and B.D. Kulkarni. Ant colony approach to continuous function optimization. *Industrial and Engineering Chemistry Research*, 39:3814–3822, 2000.
51. N. Mladenović, J. Petrović, V. Kovačević-Vujčić, and M. Čangalović. Solving a spread-spectrum radar polyphase code design problem by tabu search and variable neighbourhood search. *European Journal of Operations Research*, 151:389–399, 2003.

52. Numerical Algorithms Group. *NAG Fortran Library Manual Mark 11.* 1984.
53. C.C. Pantelides. *Symbolic and Numerical Techniques for the Solution of Large Systems of Nonlinear Algebraic Equations.* PhD thesis, Imperial College of Science, Technology and Medicine, University of London, May 1988.
54. C.C. Pantelides, L. Liberti, P. Tsiakis, and T. Crombie. Mixed integer linear/nonlinear programming interface specification. *Global Cape-Open Deliverable WP2.3-04*, February 2002.
55. C.C. Pantelides, L. Liberti, P. Tsiakis, and T. Crombie. MINLP interface specification. *CAPE-OPEN Update*, 2:10–13, March 2002.
56. P.M. Pardalos and H.E. Romeijn, editors. *Handbook of Global Optimization*, volume 2. Kluwer, Dordrecht, 2002.
57. J.D. Pintér. *LGO: a Model Development System for Continuous Global Optimization. User's Guide.* Pintér Consulting Services, Halifax, NS, Canada, 2005.
58. J.D. Pintér. Global optimization: software, test problems, and applications In Pardalos and Romeijn [56], pages 515-569.
59. W.H. Press, S.A. Teukolsky, W.T. Vetterling, and B.P. Flannery. *Numerical Recipes in C, Second Edition.* Cambridge University Press, Cambridge, 1992, reprinted 1997.
60. A.H.G. Rinnooy-Kan and G.T. Timmer. Stochastic global optimization methods; part I: Clustering methods. *Mathematical Programming*, 39:27–56, 1987.
61. A.H.G. Rinnooy-Kan and G.T. Timmer. Stochastic global optimization methods; part II: Multilevel methods. *Mathematical Programming*, 39:57–78, 1987.
62. H.S. Ryoo and N.V. Sahinidis. Global optimization of nonconvex NLPs and MINLPs with applications in process design. *Computers & Chemical Engineering*, 19(5):551–566, 1995.
63. H. S. Ryoo and N. V. Sahinidis. A branch-and-reduce approach to global optimization. *Journal of Global Optimization*, 8(2):107–138, March 1996.
64. H. Ryoo and N. Sahinidis. Global optimization of multiplicative programs. *Journal of Global Optimization*, 26(4):387–418, 2003.
65. P. RoyChowdury, Y.P. Singh, and R.A. Chansarkar. Hybridization of gradient descent algorithms with dynamic tunneling methods for global optimization. *IEEE Transactions on Systems, Man and Cybernetics—Part A: Systems and Humans*, 30(3):384–390, 2000.
66. N.V. Sahinidis. Baron: Branch and reduce optimization navigator, user's manual, version 4.0. *http://archimedes.scs.uiuc.edu/baron/manuse.pdf*, 1999.
67. F. Schoen. Random and quasi-random linkage methods in global optimization. *Journal of Global Optimization*, 13:445–454, 1998.
68. F. Schoen. Global optimization methods for high-dimensional problems. *European Journal of Operations Research*, 119:345–352, 1999.
69. F. Schoen. Two-phase methods for global optimization. In Pardalos and Romeijn [56], pages 151–177.
70. H. Schichl. *The Coconut API: Reference Manual.* Dept. of Maths, Universität Wien, October 2003.
71. E.M.B. Smith. *On the Optimal Design of Continuous Processes.* PhD thesis, Imperial College of Science, Technology and Medicine, University of London, October 1996.
72. J.E. Smith. Genetic algorithms. In Pardalos and Romeijn [56], pages 275–362.
73. E.M.B. Smith and C.C. Pantelides. Global optimisation of nonconvex minlps. *Computers & Chemical Engineering*, 21:S791–S796, 1997.

74. R. Storn and K. Price. Differential evolution — a simple and efficient heuristic for global optimization over continuous spaces. *Journal of Global Optimization*, 11:341–359, 1997.
75. E.M.B. Smith and C.C. Pantelides. A symbolic reformulation/spatial branch-and-bound algorithm for the global optimisation of nonconvex MINLPs. *Computers & Chemical Engineering*, 23:457–478, 1999.
76. T.K. Shi, W.H. Steeb, and Y. Hardy. *An Introduction to Computer Algebra Using Object-Oriented Programming*. Springer-Verlag, Berlin, second edition, 2000.
77. G. Schrimpf, J. Schneider, H. Stamm-Wilbrandt, and G. Dueck. Record breaking optimization results using the ruin and recreate principle. *Journal of Computational Physics*, 159:139–171, 2000.
78. B. Stroustrup. *The C++ Programming Language*. Addison-Wesley, Reading, MA, third edition, 1999.
79. Lindo Systems. *Lindo API: User's Manual*. Lindo Systems, Inc., Chicago, 2004.
80. M. Tawarmalani and N.V. Sahinidis. Semidefinite relaxations of fractional programming via novel techniques for constructing convex envelopes of nonlinear functions. *Journal of Global Optimization*, 20(2):137–158, 2001.
81. M. Tawarmalani and N. Sahinidis. Convex extensions and envelopes of semi-continuous functions. *Mathematical Programming*, 93(2):247–263, 2002.
82. M. Tawarmalani and N.V. Sahinidis. Exact algorithms for global optimization of mixed-integer nonlinear programs. In Pardalos and Romeijn [56], pages 1–63.
83. M. Tawarmalani and N.V. Sahinidis. Global optimization of mixed integer nonlinear programs: A theoretical and computational study. *Mathematical Programming*, 99:563–591, 2004.
84. P.L. Toint A.R. Conn, N.I.M. Gould. The SIF reference report. `http://www.numerical.rl.ac.uk/lancelot/sif/`.
85. H. Tuy. *Convex Analysis and Global Optimization*. Kluwer, Dodrecht, 1998.
86. B.W. Wah and T. Wang. Efficient and adaptive Lagrange-multiplier methods for nonlinear continuous global optimization. *Journal of Global Optimization*, 14:1:25, 1999.
87. J. M. Zamora and I. E. Grossmann. A branch and contract algorithm for problems with concave univariate, bilinear and linear fractional terms. *Journal of Global Optimization*, 14:217:249, 1999.

MathOptimizer Professional: Key Features and Illustrative Applications

János D. Pintér[1] and Frank J. Kampas[2]

[1] Pintér Consulting Services, Inc., Halifax, Nova Scotia, Canada
jdpinter@hfx.eastlink.ca, http://www.pinterconsulting.com
[2] WAM Systems, Inc., Plymouth Meeting, PA, USA fkampas@wamsystems.com,
http://www.wamsystems.com

Summary. Integrated scientific-technical computing (ISTC) environments play an increasing role in advanced systems modeling and optimization. MathOptimizer Professional (MOP) has been recently developed to solve nonlinear optimization problems formulated in the ISTC system *Mathematica*. We introduce this software package, and review its key functionality and options. MOP is then used to solve illustrative circle packing problems, including both well-frequented models and a new (more difficult) model-class.

Key words: Integrated computing systems, *Mathematica*, LGO solver suite, MathOptimizer Professional, circle packings, illustrative results.

1 Introduction

Operations Research (O.R.) provides a consistent quantitative framework and techniques, to assist analysts and decision-makers in finding "good" (feasible) or "best" (optimal) solutions in a large variety of contexts. For an overview of prominent O.R. application areas, consult e.g. the 50th anniversary issue of the journal *Operations Research* (2002).

A formal procedure aimed at finding optimized decisions consists of the following key steps.

- Conceptual description of the decision problem at a suitable level of abstraction that retains all essential attributes, but omits secondary details and circumstances.
- Development of a q uantitative model that captures the key elements of the decision problem, in terms of decision variables and functional relationships among them.
- Development and/or adaptation of an algorithmic solution procedure, in order to explore the set of feasible solutions, and to select the best decision.

- Numerical solution of the model and its verification; interpretation and summary of results.
- Posterior analysis and implementation of the decision(s) selected.

The problems tackled by O.R. are often so complex that the correct model and solution procedure may not be clear at the beginning. Therefore, decision makers often must carry out the steps outlined above in an iterative fashion. The analyst repeatedly modifies and refines the model formulation and solution procedure until the model captures the essence of the problem, is computationally tractable, and its numerical solution is applicable in the context of the problem studied.

These considerations make a strong case for using high-level, integrated software tools that can effectively assist in performing all related tasks in a unified framework. This point is particularly valid in modeling nonlinear systems, since their analysis may involve the evaluation of computationally intensive functions, visualization, animation, and so on.

Maple (Maplesoft, 2004a), *Mathematica* (Wolfram Research, 2004), and Matlab (MathWorks, 2004) are prominent, fully integrated scientific-technical computing systems. The capabilities and range of applications of these software products and related application packages are documented in software manuals, hundreds of books, and many thousands of articles and presentations. The current user base of ISTC systems is several million people worldwide.

A concise list of the most significant features and capabilities of ISTC environments includes the following (note that each feature listed below is currently supported by at least one – and sometimes by all – of the three ISTC systems mentioned):

- A broad range of simple and advanced computations with high – or even with arbitrarily high, adjustable – precision
- Support for symbolic calculations
- Extensive set of readily available functions, from programming language standards to special functions and general-purpose, complete numerical procedures (examples of the latter are integration routines, differential equation solvers, and numerical optimization routines)
- Context-specific "point and click" (essentially syntax-free) operations via GUI elements that help to execute various tasks
- Support for concise, transparent code development and maintenance
- Support for several programming styles (procedural, functional, and rule-based paradigms)
- Full programmability (i.e., extendibility by adding new functionality)
- Posterior analysis and implementation of the decision(s) selected
- Advanced technical documentation, desktop publishing, and presentation features
- Interactive and multimedia tools (in-situ evaluation, visualization, animation, sound)

- Built-in, fully integrated help system that includes portable application examples
- Automatic code generation from a given ISTC language to more "traditional, lower-level" programming languages (such as Basic, C, Fortran, Java, and so on)
- Automatic conversions of ISTC system documents to tex, html, xml, ps, pdf (and possibly other) file formats
- Direct links to external application packages, to other software products, and to the Internet
- Portability across a broad range of hardware platforms and operating systems (such as e.g., Windows, Macintosh, Linux, and Unix versions).

Many of these features can be effectively used during the various stages of developing O.R. applications. In particular, data analysis, model formulation, solution strategy and algorithm development, numerical solution, and project documentation can all be put together from the beginning – even in a single unified work document, if the developer wishes to do so. Hence, ISTC environments can increasingly provide a "one-stop" solution, to meet a broad range of needs of researchers, educators and students. This tendency has been receiving growing attention also in the O.R. community: for example, *Mathematica* has been recently reviewed in *ORMS Today* (Sodhi, 2003).

We emphasize here that, although all modeling and computational environments – from Excel spreadsheets, through optimization-specific algebraic modeling languages (such as AIMMS, AMPL, GAMS, LINGO, MPL, and others) to the more general-purpose computing (ISTC) systems – may have a lower program execution speed when compared to a compiled "pure number crunching" system, the overall application development time can often be massively reduced by using higher-level systems, especially when development can be started from scratch. It is instructive to recall in this context the debate that surrounded the early development of programming languages – such as Algol, Basic, C, Fortran, Pascal, etc. – as opposed to machine-level assembly programming.

Within the broad category of modeling and optimization problems, we see particularly strong application potentials for ISTC systems in studying nonlinear systems. For discussions of nonlinear system models and a broad range of their applications, consult e.g. Aris (1999), Bazaraa, Sherali, and Shetty (1993), Beltrami (1993), Bertsekas (1999), Bracken and McCormick (1968), Chong and Zak (2001), Diwekar (2003), Edgar, Himmelblau and Lasdon (2001), Gershenfeld (1999), Grossmann (1996), Hansen and Jørgensen (1991), Hillier and Lieberman (2005), Kampas and Pintér (2005), Murray (1983), Papalambros and Wilde (2000), Pardalos, Shalloway, and Xue (1996), Parlar (2000), Pearson (1986), Pintér (1996, 2005), Rich (1973), Schittkowski (2002), Tawarmalani and Sahinidis (2002), Wilson, Turcotte and Halpern (2003), Zabinsky (2003), and Zwillinger (1989).

2 Global Optimization

As the books listed above well illustrate, nonlinearity is literally ubiquitous in the development (and modeling) of objects, formations and processes, including also living organisms of all scales, as well as various man-made systems. Decision-making (optimization) models that incorporate such a nonlinear system description frequently lead to complex problems that may or provably do have multiple - local and global - optima. The objective of global optimization (GO) is to find the "absolutely best" solution of nonlinear optimization models under such circumstances.

In order to formalize the general global optimization paradigm considered here, we shall use the following notation:

- x decision vector, an element of the real Euclidean n-space $\mathbb{R}^n$
- $f(x)$ objective function, $f : \mathbb{R}^n \to \mathbb{R}$
- D non-empty set of admissible decisions.

The set D is defined by

- xl, xu : explicit, finite bounds of x (an embedding "box" in $\mathbb{R}^n$)
- $g(x)$: m-vector of constraint functions, $g : \mathbb{R}^n \to \mathbb{R}^m$

Applying the notation given above, the GO model can be stated as

$$\min f(x) \qquad x \in D = \{x \mid xl \leq x \leq xu, g(x) \leq 0\}. \tag{1}$$

Note that in (1) all inequalities are interpreted component-wise. Under fairly general analytical conditions, the GO model (1) has a global solution (set). (For example, if D is non-empty, and f, g are continuous functions, then the model has a non-empty set of global solutions X^*). Note that although we know that X^* exists, if we use "traditional" local scope optimization strategies, then – depending on the starting point of the search – we will find only the corresponding locally optimal solution. In order to find (i.e., to properly approximate) the "true" solution, a genuine global scope search effort is needed.

Global optimization is of great theoretical and practical importance, with significant existing and prospective applications. As of today (2004), a few hundred books, thousands of articles and dozens of web sites are devoted to the subject. For detailed discussions of the most prominent GO model types and solution approaches, consult, for example, Horst and Pardalos (1995), and Pardalos and Romeijn (2002), visit also Neumaier (2004). Among the earlier cited books on nonlinear systems models and optimization, e.g. Grossmann (1996), Kampas and Pintér (2005), Pardalos, Shalloway and Xue (1996), Pintér (1996, 2005), Tawarmalani and Sahinidis (2002), Zabinsky (2003) – as well as many others – discuss nonlinear models and real-world applications which require a global solution approach.

3 LGO Solver Suite

The LGO (Lipschitz Global Optimizer) software package serves to find – i.e., to numerically approximate – global and local solutions to nonlinear optimization models. The current LGO implementation incorporates the following solver modules:

- Branch-and-bound global search method (BB)
- Global adaptive random search (single-start) (GARS)
- Multi-start based global random search (MS)
- Constrained local search (LS) by the reduced gradient method.

All three global search methods will generate one (BB, GARS) or several (MS) approximations of the global optimizer point(s), before LGO is automatically switched to local search. The LS option can also be used in stand-alone mode, when started from a user-supplied initial point.

For theoretical details of the underlying global search methodology, consult Pintér (1996). The LS approach used is discussed in numerous textbooks: see for instance, Edgar et al. (2001). The LGO software system itself has been discussed in books and articles: consult e.g., Pintér, (2001, 2002, 2004), and the peer review by Benson and Sun (2000). Therefore here we provide only a brief summary of the solver components. (Since LGO is a commercial software product, many of the implementation details will be omitted.)

The BB module is based on a theoretically established (rigorous) global optimization approach. BB combines set partition steps with deterministic and randomized sampling: this combination also enables a statistical bounding procedure. Note, however, the program runtimes can be expected to grow fast(er) for higher-dimensional and more difficult models, if we want to find a close approximation of the global solution solely by BB. (A similar comment applies also to all other theoretically rigorous global search methods.)

Pure random search is a very simple, "folklore" approach to global optimization that converges to the global solution (set) with probability 1. GARS is an improvement over that passive search approach in the sense that it adaptively attempts to focus the global search effort in the region which – on the basis of the actual sample results – is estimated to contain the global solution point (or, in general, one of these points).

Multi-start (MS) based global search applies a similar search strategy to GARS; however, the total sampling effort is distributed among several searches. Each of these leads to a "promising" starting point for subsequent local search. Typically, this approach takes the most computational effort (due to its multiple local searches); however – especially in more difficult models – it often finds the best numerical solution (Pintér, 2003).

In all global search modes an exact penalty (merit) function serves to aggregate the objective and constraint functions. Obviously, this assumes that the model functions are "acceptably" scaled. (A constraint penalty parameter can be adjusted via an LGO solver options file, to assist scaling.)

Ideally – and also in many actual model-instances – all three global search methods (with the added LS) will give the same answer, except perhaps small numerical differences. In practice, especially when solving more difficult problems, the LGO user may wish to try all three global search options (in conjunction with the subsequent local search), to see which method gives the best results.

The local search (LS) component is based on the generalized reduced gradient (GRG) algorithm. In general, this search strategy is "only" locally convergent: therefore its use is recommended following one of the global searches, except in local optimization contexts. The application of LS, as a rule, results in a solution that is feasible and satisfies the Karush-Kuhn-Tucker local optimality conditions.

The solver suite approach – based on three different global solvers – supports the robust and efficient numerical solution of nonlinear models. The entirely derivative-free methods implemented in LGO enable the handling of merely computable model functions: this is of particular relevance with respect to applications, in which higher order (gradient, Hessian, etc.) information is impossible, difficult, or too costly to obtain. LGO can be used even to handle "black box" models provided (only) as object files, dynamic link libraries, or executable programs.

The LGO solver suite is currently available for C and Fortran compiler platforms, with customized links to Excel, GAMS, MPL Maple, *Mathematica* and Matlab (via TOMLAB). For specific descriptions of the versions not discussed here, see e.g. Frontline Systems and Pintér Consulting Services (2001), GAMS Development Corporation and Pintér Consulting Services (2003), Maplesoft (2004b), TOMLAB Optimization Inc. and Pintér Consulting Services (2004). MPL/LGO is also offered – in a demo (size-limited) version – with the latest edition of the classical O.R. textbook by Hillier and Lieberman (2005). LGO, in its various implementations, has been used in commercial applications, as well as in a variety of research and educational environments for more than a decade.

4 MathOptimizer Professional

The MathOptimizer Professional software product is based on an external LGO solver implementation that is seamlessly linked to the *Mathematica* platform. In other words, MathOptimizer Professional offers a combination of *Mathematica*'s sophisticated application development tools with core LGO solver functionality. This leads to a numerical performance that – in terms of both solution quality and solver speed – is comparable to other (compiler-based or optimization modeling language-related) LGO implementations, especially when models are more difficult and/or computationally intensive. In this section, we review the key features of MOP. Further details and an extensive list of practically motivated examples are discussed in the user manual

(Pintér and Kampas, 2003), as well as in our forthcoming book (Kampas and Pintér, 2005).

The functionality of MOP can be summarized by the following steps:

- Optimization model formulation, in a *Mathematica* document (notebook)
- Automatic export and translation of the model into C or Fortran code
- Compilation of the generated code into a dynamic link library (DLL)
- Call to the external LGO engine: the latter is a "ready-made" executable program that is now linked to the model-dependent DLL
- Automatic model solution and report generation by LGO
- Import and display of results into the calling *Mathematica* notebook.

We refer to the approximately 150-page (printed) manual for further details.

It should be noted that the approach outlined supports automatically "only" the solution of models defined by *Mathematica* functions that can be directly converted into (C or Fortran) program code. This, however, still allows the handling of a fairly broad range of continuous nonlinear optimization models (including, of course, all models that could be directly written in C or Fortran). Other implementations with extended functionality are also possible.

One "side benefit" of using MOP is that models built in *Mathematica* can be directly used to generate corresponding C or Fortran test models. This is particularly advantageous in case of larger model-instances.

MathOptimizer Professional (MOP) – and its solver function callLGO – is launched by the *Mathematica* statement

```
Needs["MathOptimizerPro`callLGO`"];
```

The basic functionality of **callLGO** can be queried by the following *Mathematica* statement: see the auto-generated reply immediately below. (The format of this reply is slightly edited for the present purposes.)

```
?callLGO

callLGO[obj_, cons_List, varswithbounds_List, opts___]:
obj is the objective function,
cons is a list of the constraints,
varswithbounds are the variables and their bounds in the format
{{variable, lower bound, initial value for local search, upper
bound...} or {{variable, lower bound, upper bound}...}.

Function return is the value of the objective function, a list of
rules giving the solution, and the maximum constraint violation.

See Options[callLGO] for the options and also see the usage statements
of the various options for their possible values. For example, enter
?Method for the possible settings of the Method option.
```

Table 1 summarizes the current callLGO option list, with added notes. All options can be changed by users, following MOP specifications.

Option name and default setting	Additional notes
ShowSummary→False	Display (or not) LGO report file
Method→MSLS	Alternatives: BBLS, GARSLS, LS
MaxEvals → ProblemDetermined	Global search effort, set by default to $1000(n+m)$ (global search phase stopping criterion)
MaxNoImprov→ProblemDetermined	Global search effort without sufficient improvement, set by default to $200(n+m)$ (global stopping criterion)
PenaltyMul→1	Penalty multiplier
ModelName→LGO Model	Model-dependent name (can be chosen by user)
DllCompiler→BC	Currently supported compilers: Borland C, Lahey Fortran, Microsoft C, Salford Fortran
ShowLGOInputs→False	Display (or not) LGO input files
LGORandomSeed→0	Set internally (can be reset by user)
TimeLimit→300	Seconds (global search phase stopping criterion)
TOBJFGL→-1000000	Target objective function value in global search phase (global search phase stopping criterion)
TOBJFL→-1000000	Target objective function value in local search phase (local search phase stopping criterion)
MFPI→ 10^{-6}	Merit function precision improvement tolerance (local search phase stopping criterion)
CVT→ 10^{-6}	Accepted constraint violation tolerance (local search phase stopping criterion)
KTT→ 10^{-6}	Kuhn-Tucker condition tolerance (local search phase stopping criterion)

Table 1. MathOptimizer Professional: callLGO options.

As indicated above, `callLGO` is activated by a statement of the form
`callLGO[f,{g},{x,xl,xn,xu}, options]`

Here the notations `f`, `g`, `x`, `xl` and `xu` directly correspond to the symbols defined in (1). In addition, `xn` (`xl`≤`xn`≤`xu`) is a user-supplied nominal solution — or a default setting, if `xn` is absent — that is used by LGO in its initial local search mode; finally, `options` denotes the calling parameters of the function `callLGO`.

The following simple example serves to illustrate the basic MOP functionality, as it appears to the user in default mode. Consider the model

$$\min x^2 + 2y^2; \quad x + y \geq 1; \quad -2 \leq x \leq 2; \quad -2 \leq y \leq 2.$$

This optimization problem is solved by the next *Mathematica* statement that leads to the answer shown immediately below:

```
callLGO[x^2+2*y^2,{x+y>=1},{{x,-2,0,2},{y,-2,0,2}}]
```

The answer received is a *Mathematica* list (as shown by the curly braces and comma separators):

```
{0.6666666666666666, {x --> 0.6666666667,
y --> 0.3333333333}, 5.551115123125783*^-17}
```

Here the first list element is the optimal objective function value found, followed by the list of corresponding variable assignments, and the maximal constraint violation at the solution point. (More details are shown automatically in the generated LGO report that can also be displayed in the notebook.) An extensive set of interesting GO challenges and practically motivated numerical examples are discussed also in the MOP User Guide.

In the numerical examples discussed here *Mathematica* versions 5.01 or 5.1 are used in conjunction with the Microsoft Visual C/C++ (MSVC, version 6.0) or the Salford FORTRAN 95 (FTN95) compiler. Furthermore, in all cases Pentium 4 processor based machines running under Microsoft Windows XP Professional version are used. Let us also note here that the RAM requirements of MOP *per se* are rather modest, at least for small or mid-size models; e.g., a personal computer with 256 MB RAM is certainly adequate to handle MOP models with up to (at least) 1000 variables and 1000 constraints. In principle, arbitrarily large models can be handled using virtual memory, given sufficient hard drive space and time.

5 Illustrative applications: solving sphere packing models

Object packing models are aimed at finding the best non-overlapping arrangement of a set of objects in a container object. This general modeling paradigm can be specified in many ways, leading to interesting – and typically quite difficult – models. In addition to a more theoretical interest directed towards specific, analytically tractable problem-instances, there is also an obvious practical motivation to solve packing models.

In our recent and ongoing studies, we have found that this general model-class can be used to test and benchmark global optimization software. Several special cases and model-instances will be discussed below, to illustrate the potentials of numerical global optimization in solving object packing problems.

5.1 Packing Identical Size Circles in the Unit Square

The problem can be stated as follows: given the unit square and a positive integer k, find the maximal radius r of k identical circles that can be placed into the square, without overlaps (evidently, $r = r(k)$).

This problem, as well as some other related circle packing models, has become a fascinating subject for both professional researchers and amateurs, at least in recent decades. There exists a significant body of literature (books, articles, dissertations, and web sites) discussing uniform circle packings. The information available includes proofs (only) for a small number of special cases (namely, for k=2,...,9, 14, 16, 25, and 36 circles), or computer aided optimality proofs, with guaranteed bounds (for up to k=30). In many other cases, only "best known" constructions are known.

To illustrate some rigorous bounding results, Csendes and his colleagues have applied interval arithmetic based GO methodology to prove bounds for best circle packings. The websites of Csendes (2004) and Markót (2004) list their related publications.

Referring to the general case, Graham noted in an interview (Albers, 1996) that he does not expect to know the true (i.e., proven optimal) solution for placing 1000 equal size circles in the unit square. The reason for his "learned skepticism" is that there is no unifying theory, and that intuition may fail: for example, in the k=49-circle case the seemingly obvious "seven-by-seven" configuration is not optimal, consult e.g. Specht (2004).

Without going into further details regarding this area of research, let us mention the thesis of Melissen (1997): he provides a detailed review of uniform circle packing model statements and key analytical results, with more than 350 topical references. The website of Specht (2004) is another rich source of information related to uniform size circle packings (in the unit square, the unit circle, and in rectangles): the site also includes references. We will compare our numerical results to those listed at this website as best (proved or postulated).

We also wish to emphasize that our sole purpose here is to illustrate the applicability of numerical global optimization (specifically, of LGO and thereby of MathOptimizer Professional) to difficult packing models, even without specifying or postulating any prior structure. More extensive studies should be based on more detailed and sophisticated modeling than what we are presenting here. We will cite only some of our existing numerical results: more details are available upon request, and will appear elsewhere.

Let us denote by c_i=(x_i, y_i) the center of circle $i = 1, \ldots, k$: the coordinate pairs (x_i, y_i) and the radius r of these circles are the decision variables. The constraint that the circles i and j do not overlap means that the sum of their radii is not greater than the distance between the centers:

$$||c_i - c_j|| \geq 2r \qquad (2)$$

here $||c_i - c_j|| = \sqrt{(x_i - x_j)^2 + (y_i - y_j)^2}$ for each $1 \leq i < j \leq k$.

Note that each instance of the inequality in (2) is a non-convex constraint (since the norm function is convex). For k-circle configurations, we have $k(k-1)/2$ such constraints.

We shall consider the unit square that is centered at the origin. Additional constraints postulate that the circles are inside the enclosing square. These are derived from simple geometrical considerations; a possible formulation is:

$$|x_i| + r \leq 0.5 \quad \text{and} \quad |y_i| + r \leq 0.5 \quad \text{for } 1 \leq i \leq k. \qquad (3)$$

This way, in addition to the non-convex constraints we have $2k$ non-linear convex constraints. (Alternative formulations are also possible, e.g. by replacing the latter constraints by linear ones. However, we have been using the constraints (3) in our numerical experiments.). Under the conditions (2)-(3), our objective is to

$$\text{maximize} r. \tag{4}$$

We have developed a *Mathematica* model (a function) that is directly based on relations (2)-(4). This function is parameterized by the number of circles: therefore inserting a positive integer value k in the function provides the corresponding k-circle model instance. As noted earlier, MathOptimizer Professional automatically translates this *Mathematica* model to C or Fortran form, and then the external LGO solver is invoked to solve it. For illustration, the 20-circle solution found is shown in Fig. 1.

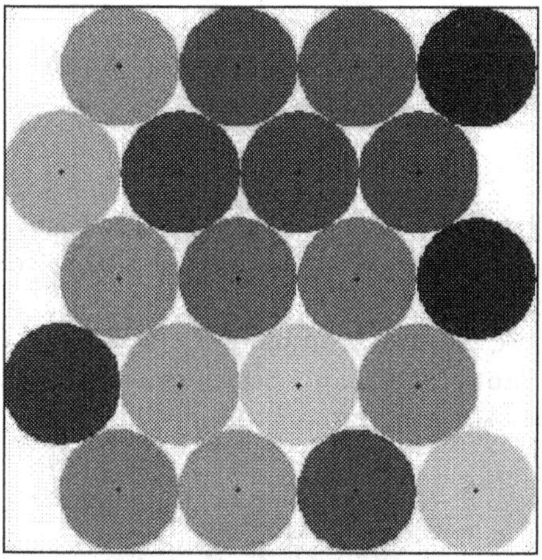

Fig. 1. Packing 20 uniform size circles in the unit square.

Note that this figure is directly imported from the *Mathematica* notebook document where the model formulation and all calculations have also been done. The optimized value of the circle radius $r = r(20)$ found by MOP is $r \sim 0.1113823476$.

The radius found agrees well (to about 10^{-10} absolute precision) with the value 0.111382347512 posted at `http://www.packomania.com`. Note that such — arguably minor — imprecision can be due to several factors: one of the significant factors is that LGO applies central finite-difference based gradient approximation in its local search phase. This adds some error to that of standard floating point precision calculations.

The computer used to solve this example has a 3.2 GHz Pentium 4 processor and is running Windows XP; we used the Salford Fortran 95 compiler

(Salford Software, 2004). The corresponding runtime is about 19 seconds. Note that runtimes may change slightly even when repeating the same run, due to hardware and OS status changes: however, the timing cited gives an impression of the MOP solver speed. (LGO per se runs faster, of course.) Recall that the 20-circle model instance of (2)-(4) has 41 decision variables, and 230 nonlinear constraints of which 190 are non-convex. Further detailed numerical results will appear e.g. in our forthcoming book (Kampas and Pintér, 2005).

5.2 Packing Identical Size Circles in the Unit Circle

This problem can be stated as follows: given the unit circle and a positive integer k, find the maximal radius $r = r(k)$ of k identical circles that can be placed into the circle, without overlaps.

Applying straightforward modifications of model (2)-(4), the adapted model can be written as

$$\text{maximize } r \tag{5}$$

$$||c_i - c_j|| \geq 2r \qquad 1 \leq i < j \leq k \tag{6}$$

$$r + ||c_i|| \leq 1 \qquad 1 \leq i \leq k. \tag{7}$$

As above, $||c_i - c_j|| = \sqrt{(x_i - x_j)^2 + (y_i - y_j)^2}$ and $||c_i|| = \sqrt{x_i^2 + y_i^2}$. This model has $2k+1$ decision variables, k convex nonlinear constraints, and $k(k-1)/2$ non-convex constraints.

For illustration, we cite the solution of the 20-circle model instance (that has 41 decision variables and 210 constraints of which again 190 are non-convex). Fig. 2 shows the configuration obtained.

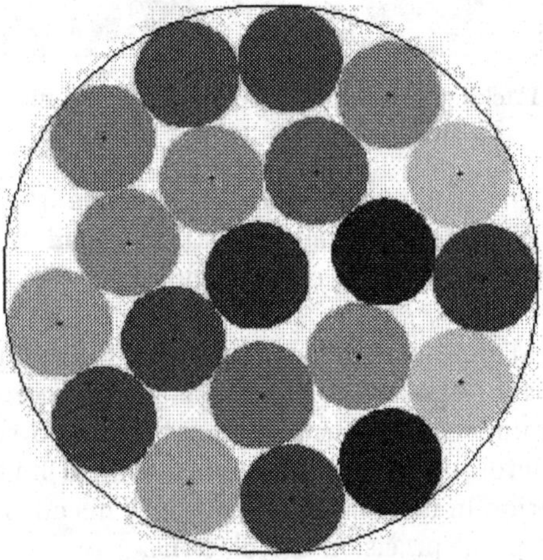

Fig. 2. Packing 20 uniform size circles in the unit circle.

The radius $r = r(20)$ found by MOP in this example equals 0.1952240114. This value agrees to at least 10^{-9} absolute precision with the best known result cited at www.packomania.com (0.1952240110...) The corresponding runtime is approximately 43 seconds, on the same computer as mentioned above.

5.3 Packing Non-Uniform Size Circles in an Optimized Circle

In this section, we introduce a new class of object packing models. Our objective is to find the minimal size circle that contains a given non-overlapping circle configuration that is made up by (in principle, arbitrary size) circles. To our best knowledge, this model has not been studied before by others in a GO setting: we also think that such models can be significantly more difficult than the more specific cases discussed in the preceding two sections.

We shall denote by r_i the radius of circle i for i=1,..., k. With a straightforward generalization of (5)-(7), we obtain the following model:

$$\text{minimize } r \tag{8}$$

$$||c_i - c_j|| \geq r_i + r_j \qquad 1 \leq i < j \leq k \tag{9}$$

$$r_i + ||c_i|| \leq r \qquad 1 \leq i \leq k. \tag{10}$$

Notice that now r is the unknown radius of the circumscribing circle that is minimized: its value depends on the set of circle radii $\{r_i\}$. Similarly to (5)-(7), model (8)-(10) has $2k$+1 decision variables, k convex nonlinear constraints, and $k(k-1)/2$ non-convex constraints.

To illustrate this model, in the last numerical example presented here we will pack circles of radius $r_i = \sqrt{\frac{1}{i}}$ for $i = 1, \ldots, k$ into a circumscribing circle. Notice that in this case the total area of the embedded circles is slowly divergent as k goes to infinity: therefore the optimized radius also will be unbounded as a function of k (packings with bounded total area may also be of interest, of course). Fig. 3 shows the optimized circle arrangement found for k=20.

The radius of the circumscribing circle $r = r(20)$ in this case approximately equals 2.12545. The corresponding runtime is about 47 seconds, on the machine mentioned before. Comparing this runtime with the previous one (that was 43 seconds for packing 20 identical size circles) one can see that MOP (i.e., LGO) handles the more general model with fairly little extra computational effort.

Although obviously all numerical test results depend also on certain solver parameterizations, we think that the examples presented indicate the capabilities and potentials of MOP. (The same default solver settings were used in all examples reviewed here, without any "tweaking".)

Let us remark finally that we have attempted to solve a large variety of circle packing model instances using also other "general purpose" commercial optimization software products (and applying all solver options with

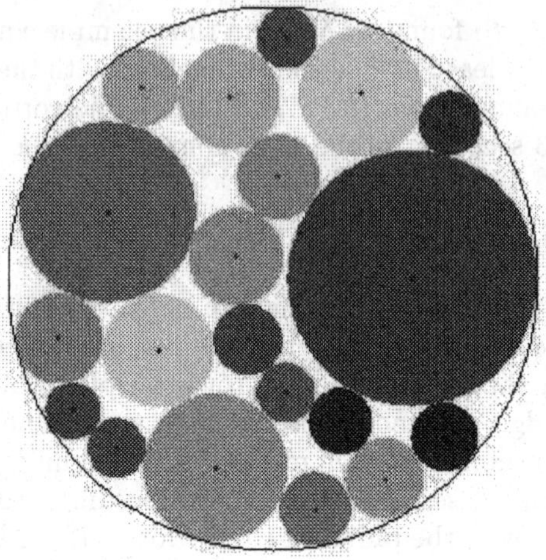

Fig. 3. Packing 20 uniform size circles in the unit circle.

default settings, the same way MOP was used). The solvers tested specifically included *Mathematica*'s built-in constrained optimization function (NMinimize), and several third party packages. Our comparative results consistently have demonstrated the relative strength and efficiency of MOP, both in terms of solution quality and runtime. These results will appear in a forthcoming paper, as well as in Kampas and Pintér (2005).

6 Conclusions

In addition to perhaps more "traditional" development environments – such as compiler platforms, spreadsheets, and algebraic modeling languages – integrated scientific-technical computing systems will play an increasing role in advanced systems modeling and optimization.

In order to meet related user demands, MathOptimizer Professional has been recently developed to handle nonlinear optimization problems formulated in *Mathematica*. MOP operations are based on an easy-to-use *Mathematica* interface to the LGO solver suite. Following a brief introduction to the key features of MathOptimizer Professional, we illustrate its usage by solving relatively small, yet non-trivial circle packing problems. More detailed numerical results and comparative assessments will appear elsewhere.

For over a decade, the core LGO solver suite has been applied in a large variety of research and professional contexts: consult, e.g., Pintér (1996, 2001, 2002, 2003, 2004, 2005), with numerous further references to such applications. In recent years, LGO has become a solver engine option available for use with an increasing number of modeling environments. Currently these include essentially "all" C and Fortran compilers, Excel spreadsheets, the GAMS and MPL modeling language, and the integrated scientific-technical computing systems Maple, *Mathematica* and MATLAB. (Further development is in progress.) The current LGO implementations have been used to solve models in up to a few thousand variables and constraints. We expect that MathOptimizer Professional will enable the solution of sizable, sophisticated *Mathematica* models with efficiency comparable to that of compiler platform based nonlinear solvers.

Acknowledgments

The research of JDP related to *Mathematica* software development – including MathOptimizer Professional – has been supported by Wolfram Research, by providing quality software and professional documentation. We also wish to thank Mark Sofroniou of Wolfram Research for his permission to use (and to modify in our MOP development work) his `Format.m` package.

References

1. Albers, D.J. (1996) A nice genius. *Math Horizons*, November 1996 issue, 18-23.
2. Aris, R. (1999) *Mathematical Modeling: A Chemical Engineer's Perspective.* Academic Press, San Diego, CA.
3. Bazaraa, M.S., Sherali, H.D. and Shetty, C.M. (1993) *Nonlinear Programming: Theory and Algorithms.* Wiley, New York.
4. Beltrami, E. (1993) *Mathematical Models in the Social and Biological Sciences.* Jones and Bartlett, Boston.
5. Benson, H.P. and Sun, E. (2000) LGO – versatile tool for global optimization. *ORMS Today 27* (5), 52-55. See also `http://www.lionhrtpub.com/orms/orms-10-00/swr.html`.
6. Bertsekas, D.P. (1999) *Nonlinear Programming.* (2nd Edition.). Athena Scientific, Cambridge, MA.
7. Bracken, J. and McCormick, G.P. (1968) *Selected Applications of Nonlinear Programming.* Wiley, New York.
8. Chong, E.K.P. and Zak, S.H. (2001) *An Introduction to Optimization.* (2nd Edition.) Wiley, New York.
9. Csendes, T. (2004) `http://www.inf.u-szeged.hu/~csendes/publ.html`.
10. Diwekar, U. (2003) *Introduction to Applied Optimization.* Kluwer Academic Publishers, Dordrecht.
11. Edgar, T.F., Himmelblau, D.M., and Lasdon, L.S. (2001) *Optimization of Chemical Processes.* (2nd Edition.) McGraw-Hill, NY.
12. Frontline Systems and Pintér Consulting Services (2001) *LGO Global Solver Engine for Excel – Premium Solver Platform.* Frontline Systems, Inc., Incline Village, NV. http://www.solver.com/xlslgoeng.htm.
13. GAMS Development Corporation and Pintér Consulting Services (2003) *GAMS/LGO User Guide.* GAMS Development Corporation, Washington, DC. `http://www.gams.com/solvers/lgo.pdf`.

14. Gershenfeld, N.A. (1999) *The Nature of Mathematical Modeling.* Cambridge University Press, Cambridge, UK.
15. Grossmann, I., Ed. (1996) *Global Optimization in Engineering Design.* Kluwer Academic Publishers, Dordrecht.
16. Hansen, P.E. and Jørgensen, S.E., Eds. (1991) *Introduction to Environmental Management.* Elsevier, Amsterdam.
17. Horst, R. and Pardalos, P.M., Eds. (1995) *Handbook of Global Optimization, Vol. 1.* Kluwer Academic Publishers, Dordrecht.
18. Hillier, F. and Lieberman, G.J. (2005) *Introduction to Operations Research.* (8th Edition.) McGraw-Hill, New York.
19. Kampas, F.J. and Pintér, J.D. (2005) *Advanced Optimization – Scientific, Engineering, and Economic Applications with Mathematica Examples.* Elsevier, Amsterdam. (To appear.)
20. Maplesoft (2004a) *Maple.* Maplesoft, Inc., Waterloo, ON. `http://www.maplesoft.com`.
21. Maplesoft (2004b) *Global Optimization Toolbox.* Maplesoft, Inc. Waterloo, ON. `http://www.maplesoft.com/products/toolboxes/globaloptimization/index.aspx`.
22. MathWorks (2004) *MATLAB.* The MathWorks, Inc., Natick, MA.
23. Markót, M.Cs. (2004) `http://www.inf.u-szeged.hu/~markot/publ.html`.
24. Melissen, J.B.M. (1997) *Packing and Covering with Circles.* Ph.D. Dissertation, University of Utrecht.
25. Murray, J.D. (1983) *Mathematical Biology.* Springer-Verlag, Berlin.
26. Neumaier, A. (2004) Global Optimization website. `http://www.mat.univie.ac.at/~neum/glopt.html`.
27. *Operations Research: 50th Anniversary Issue* (2002) INFORMS, Linthicum, MD.
28. Papalambros, P.Y. and Wilde, D.J. (2000) *Principles of Optimal Design.* Cambridge University Press, Cambridge, UK.
29. Pardalos, P.M. and Resende, M.G.C., Eds. (2002) *Handbook of Applied Optimization.* Oxford University Press, Oxford.
30. Pardalos, P.M. and Romeijn, H.E., Eds. (2002) *Handbook of Global Optimization, Vol. 2.* Kluwer Academic Publishers, Dordrecht.
31. Pardalos, P.M., Shalloway, D. and Xue, G. (1996) *Global Minimization of Nonconvex Energy Functions: Molecular Conformation and Protein Folding.* DIMACS Series, Vol. 23, American Mathematical Society, Providence, RI.
32. Parlar, M. (2000) *Interactive Operations Research with Maple: Models and Methods.* Birkhäuser, Boston.
33. Pearson, C.E. (1986) *Numerical Methods in Engineering and Science.* Van Nostrand Reinhold, New York.
34. Pintér, J.D. (1996) *Global Optimization in Action.* Kluwer Academic Publishers, Dordrecht.
35. Pintér, J.D. (2001) *Computational Global Optimization in Nonlinear Systems: An Interactive Tutorial.* Lionheart Publishing, Atlanta, GA.
36. Pintér, J.D. (2002) Global optimization: software, test problems, and applications, Chapter 15 (pp. 515-569) in: Pardalos and Romeijn, Eds. *Handbook of Global Optimization. Vol. 2.* Kluwer Academic Publishers, Dordrecht.
37. Pintér, J.D. (2003) GAMS/LGO nonlinear solver suite: key features, usage, and numerical performance. (Submitted for publication.)

38. See also at `http://www.gams.com/solvers/solvers.htm#LGO`.
39. Pintér, J.D. (2004) *LGO – An Integrated Model Development and Solver Environment for Continuous Global Optimization. User Guide*. Pintér Consulting Services, Inc., Halifax, NS, Canada.
40. Pintér, J.D. (2005) *Applied Nonlinear Optimization in Modeling Environments.* CRC Press, Boca Raton, FL. (To appear.)
41. Pintér, J.D. and Kampas, F.J. (2003) *MathOptimizer Professional. User Guide.* Pintér Consulting Services, Inc., Halifax, NS, Canada.
42. Rich, L.G. (1973) *Environmental Systems Engineering.* McGraw-Hill, Tokyo.
43. Salford Software (2004) *Salford Fortran 95.* `http://www.salfordsoftware.co.uk/`.
44. Schittkowski, K. (2002) *Numerical Data Fitting in Dynamical Systems.* Kluwer Academic Publishers, Boston / Dordrecht / London.
45. Specht, E. (2004) `http://www.packomania.com`.
46. Tawarmalani, M. and Sahinidis, N.V. (2002) *Convexification and Global Optimization in Continuous and Mixed-integer Nonlinear Programming.* Kluwer Academic Publishers, Dordrecht.
47. TOMLAB Optimization Inc. and Pintér Consulting Services (2004) *TOMLAB /LGO User Guide.* `http://tomlab.biz/docs/TOMLAB_LGO.pdf`.
48. Wilson, H.B., Turcotte, L.H., and Halpern, D. (2003) *Advanced Mathematics and Mechanics Applications Using MATLAB.* Chapman & Hall / CRC Press, Boca Raton, FL.
49. Wolfram Research (2004) *Mathematica.* Wolfram Research, Inc., Champaign, IL. `http://www.wolfram.com`.
50. Zabinsky, Z.B. (2003) *Stochastic Adaptive Search for Global Optimization.* Kluwer Academic Publishers, Dordrecht.
51. Zwillinger, D. (1989) *Handbook of Differential Equations.* (3rd Edn.) Academic Press, New York.

Variable Neighborhood Search for Extremal Graphs 14: The AutoGraphiX 2 System

M. Aouchiche[1], J.M. Bonnefoy[2], A. Fidahoussen[2], G. Caporossi[3], P. Hansen[3], L. Hiesse[4], J. Lacheré[5], and A. Monhait[6]

[1] Ecole Polytechnique de Montréal, Canada mustapha.aouchiche@gerad.ca
[2] ISIMA, Clermont-Ferrand, France
[3] GERAD and HEC Montréal, Canada
{gilles.caporossi,pierre.hansen}@gerad.ca
[4] Institut d'Informatique d'Entreprise, Clamart, France
[5] Ecole Polytechnique, Univ. de Nantes, France
[6] CUST, Univ. Blaise Pascal, Clermont-Ferrand, France

Summary. The AutoGraphiX (AGX) system for computer assisted or, for some of its functions, fully automated graph theory was developed at GERAD, Montreal since 1997. We report here on a new version (AGX 2) of that system. It contains many enhancements, as well as a new function for automated proof of simple propositions. Among other results, AGX 2 led to several hundred new conjectures, ranking from easy ones, proved automatically, to others requiring longer unassisted or partially assisted proofs, to open ones. Many examples are given, illustrating AGX 2's functions and the results obtained.

Key words: Graph theory, automated system, computer-assisted, AGX, automated proof, conjecture, refutation.

1 Introduction

Computers have been extensively used in graph theory and its applications to various fields since the fifties of the last century. The main use was computation of the values of graph invariants, *i.e.*, quantities such as the independence and chromatic numbers, the radius or the diameter of a graph, which do not depend on the labelling of its vertices or edges. In addition to such tasks of intelligent number-crunching (which imply the design of exact algorithms or heuristics as well as their efficient implementation with well-chosen data-structures [34, 35, 40, 41]), computers can also be used for graph drawing [17, 18] and for advancing the theory itself, *i.e.*, finding in a computer-assisted or sometimes fully automated way conjectures, proofs and refutations. See [27] for a survey and discussion of systems designed for that purpose, focussed on

the operational systems Graph [14, 15], Graffiti [22, 23] and AutoGraphiX [10, 11], as well as the forthcoming book *Graphs and Discovery* [24], in particular chapters 4, 5, 6, 9 and 10 which discuss those systems. The systems VEGA [38], Graph Theorist [19, 20, 21] and HR [12] are also relevant (although the last one is more often applied to algebra and number theory).
The AutoGraphiX (AGX) system was developed at GERAD, Montreal since 1997. It addresses the following problems:

(*i*) Find a graph G satisfying given constraints;
(*ii*) Find a graph G maximizing or minimizing a given invariant, possibly subject to constraints;
(*iii*) Find a conjecture, which may be algebraic, *i.e.*, a relation between graph invariants, or structural, *i.e.*, a characterization of extremal graphs for some invariant;
(*iv*) Corroborate, refute and/or strengthen or repair a conjecture;
(*v*) Suggest ideas of proof.

The AGX system was first described in [10]; three ways it uses to fully automate conjecture making are presented in [11]. Applications to graph theory are given in [1, 6, 16, 29]; applications to chemical graph theory in [7, 8, 25, 30, 31], and developments of these results in [9, 26].
AGX was enriched in various ways over the years. It appeared however that a new version, in C++, comprising a series of enhancements and some new features should be written. A main goal was to follow closely the way a graph theorist proceeds with his work, both in the discovery of new conjectures and in the proof or refutation of them. As explained below, this makes various tasks much more efficient and points the way towards obtaining first successes in other ones such as automated proof in graph theory.
The main ideas behind AGX are that
(*i*) all problems listed above can be expressed as parametric constrained optimization ones on an infinite family of graphs, and
(*ii*) a *generic* heuristic can be used for solving all of them. More precisely, letting $i(G)$ denote an invariant of G, or a formula involving several invariants which is itself an invariant, $\mathcal{G}_n$ the set of all graphs with n vertices, $\mathcal{G}_{n,m}$ the set of all graphs with n vertices and m edges, one solves heuristically the problem

$$Min/Max\{i(G), G \in \mathcal{G}_n\} \quad \text{or} \quad Min/Max\{i(G), G \in \mathcal{G}_{n,m}\}.$$

In practice only moderate values of n and m will be considered. In this paper we report on AGX 2, the new version of AGX which results from work done in the three last years. The paper is organized as follows: interactive features of AGX 2 are described and illustrated in a series of figures in the next section. The algebraic structure of AGX 2 is presented in Section 3, together with lists of currently available invariants and operators. A new optimization routine is explained in Section 4. Ways to solve the five problems listed above

are detailed in Section 5. A new feature, *i.e.*, a routine for automated proof of simple conjecture is presented in Section 6. Examples illustrating various types of results obtained are given in section 7. Brief conclusions are given in Section 8.
In order to test extensively AGX 2, we selected a form of problem of reasonable difficulty, *i.e.*, we consider problems which lead to some very easy conjectures, more complex ones needing a few pages of proof, as well as some open conjectures which we are, up to now, unable to prove. We adopted the following AGX form:
For all connected graphs $G = (V, E)$ with $n = |V|$, find conjectures of the form

$$l(n) \leq i_1(G) \oplus i_2(G) \leq u(n) \tag{1}$$

where $i_1(G)$ and $i_2(G)$ are invariants; $\oplus$ is one of the four operations $+, -, \times, /$; $l(n)$ and $u(n)$ are lower and upper bounding functions of the order n of G which are *best possible*, *i.e.*, such that for each value of n (except possibly very small ones where border effects appear) there is a graph G for which the bound is tight.
Note that this form extends that of the well-known Nordhaus-Gaddum [37] relations in that

- $i_1(G)$ and $i_2(G)$ are two different invariants instead of $i_2(G)$ being equal to $i_1(\bar{G})$, where $\bar{G}$ is the complementary graph of G (in which an edge joins vertices u and v if and only if no edge does so in G);
- operations $-$ and $/$ are considered in addition to $+$ and $\times$;
- it is required that the relations be best possible (instead of this being only desirable).

Over 700 conjectures or observations (*i.e.*, results proved by the system) have been obtained by AGX 2 in a fully automated way and it is not hard to obtain many more. Several examples given below have the form (1).

2 AGX 2 Interactive functions

This section describes the interactive functions of AGX 2, from the syntax of the problem to the analysis of results.

Control center

When launching AGX 2, using the command "AGX 2", a dialogue box appears (Figure 1). It contains a list of problems (defined by the user) and a set of buttons, each representing a function (Optimize single, Optimize batch, View results...).

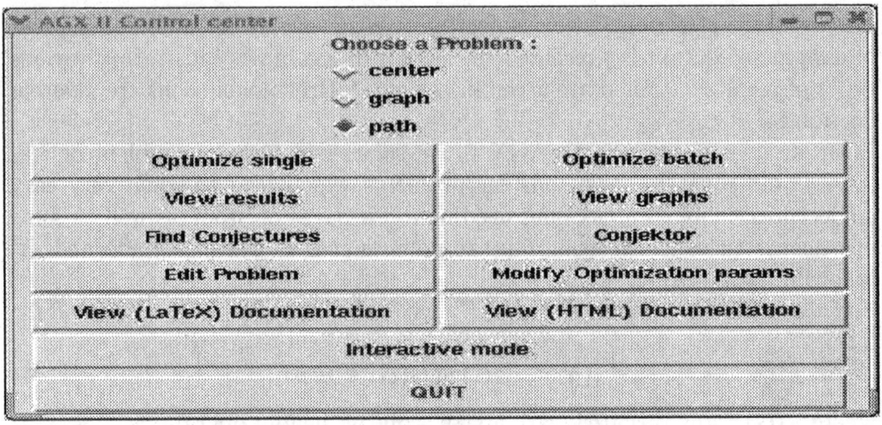

Fig. 1. AGX 2 Control Center

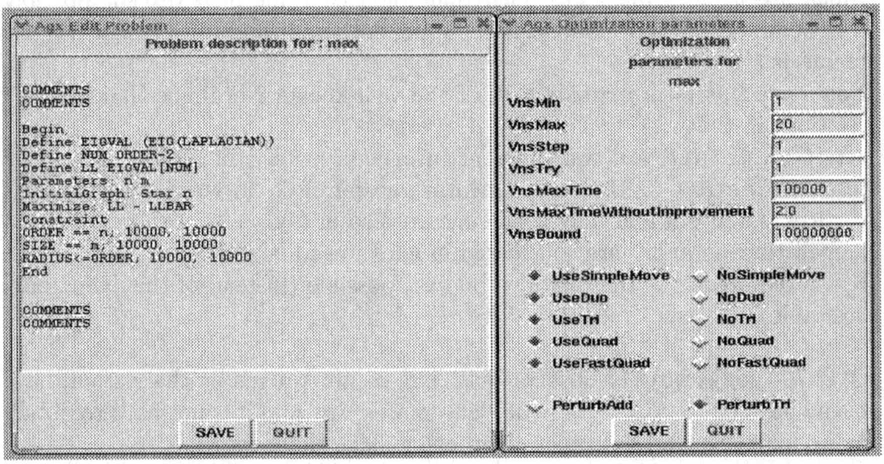

Fig. 2. Problem structure (left) and optimization parameters (right)

Editing a problem

An AGX 2 problem file (see Figure 2) is a ".def" file, edited first using a text editor such as Nedit or Emacs. Instructions begin with the keyword *Begin* and end with *End.* We may also add comments before *Begin* or after *End.* The components of the problem file are:

• *Define*: An optional part that enable us to define a new invariant using existing ones and predefined operations (we can get the list of all existing invariants and all possible operations using the *View Documentation* button). The keyword *Define* is followed by the name of the invariant and then the expression of its definition (in Figure 2 LL is the second smallest eigenvalue of the Laplacian matrix). In order to avoid runtime errors, names must not: *(i)* be any reserved word such as a function name (SQRT or LOG for example) or an invariant name, *(ii)* end by the sequence "BAR" as it is reserved to the computation of an invariant for the complementary graph, *(iii)* contain any non-alphanumeric character as this may induce errors in the latex report generation or in the syntax analysis of expressions.

- *Parameters*: Names of the problem parameters.
- *InitialGraph*: Any optimization process needs an initial solution; AGX 2 needs an initial graph. *InitialGraph* is followed by the keyword corresponding to the initial graph (we can get the list of all possible graphs using the *View Documentation* button). In Figure 2 (left), the initial graph is a star on n (a parameter) vertices.
- *Minimize* or *Maximize* followed by the expression of the objective function. In Figure 2, the objective is to maximize $LL - LLBAR$ (the algebraic connectivity, second smallest eigenvalue of the Laplacian matrix, of a graph G minus the algebraic connectivity of the complement of G).
- *Constraint* (optional) followed by a list of constraints (with a constant and a linear penalty for violation) ending with ";". In the present example the constraints are the values of n and m and $RADIUS <= ORDER$ which imposes the connectedness of G.

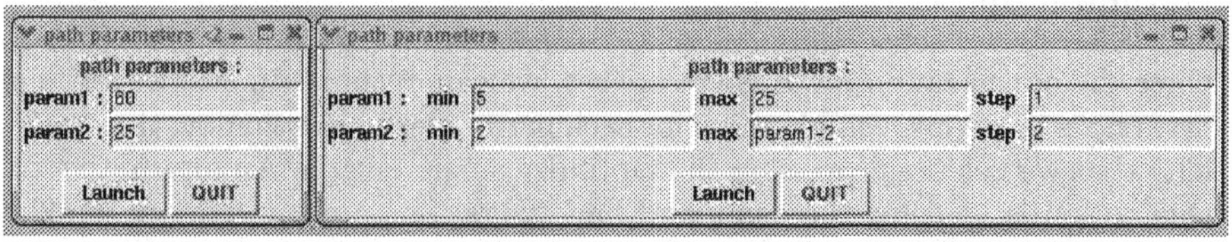

Fig. 3. Optimization single (left) and optimization batch (right)

Optimization parameters

Before starting the optimization process (which will be described in details in Section 4), we can modify parameters of the Variable Neighborhood Search optimization heuristic by pressing the *Modify Optimization params* button. When the window on the right in Figure 2 appears, we can modify the minimum (VnsMin) or the maximum (VnsMax) size of a perturbation, the perturbation increment (VnsStep), the number of trials before incrementation (VnsTry), the maximum time allowed for the optimization (VnsMaxTime), the time without improvement of the solution (VnsMaxTimeWithoutImprovement) or VnsBound, that is used when refuting conjectures (the program stops if a counter-example is generated; if you do not want to use this feature, simply set VnsBound to 100000000).

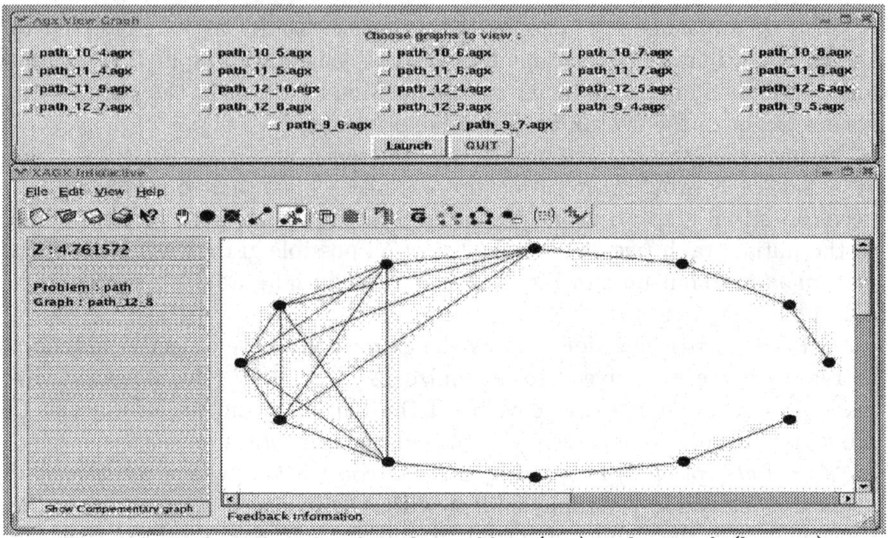

Fig. 4. List of graphs related to the problem (top) and a graph (bottom)

Optimization

Once the optimization parameters are set, we can optimize in two ways. The first one is by giving a single value for each parameter in the window (Figure 3, left) that we get when pressing the Optimize single button. An alternative to this option is to enter the command line "xagx -opt prob param1 param2", where "prob" is the problem name and "param1 param2" are the current values of the problem parameters (here two). The second option is to specify a range of values for each parameter in the window that appears when pressing the Optimize batch button (Figure 3, right).

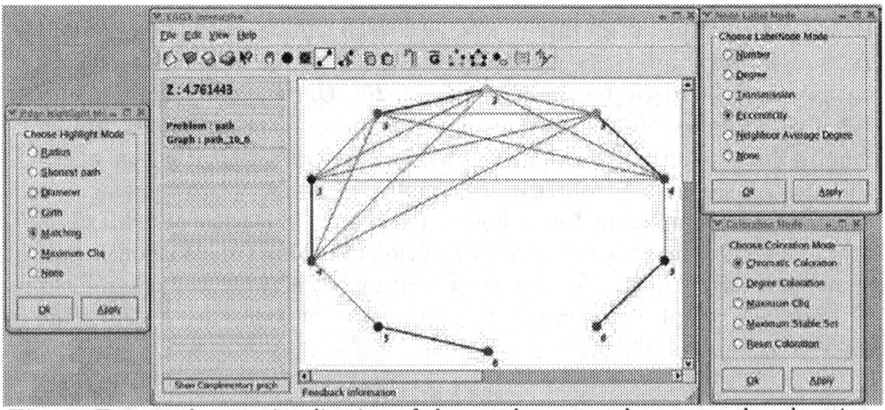

Fig. 5. To get a better visualization of the graph we can choose a node coloration (right bottom), edge highlighting (left) and vertex labelling (right top)

View graphs

When we press the *View graphs* button, the list (Figure 4, top) of all available graphs corresponding to the selected problem is displayed. We just have to select the desired graph and press *Launch* to display it (Figure 4, bottom), and then modify it interactively if needed (possible modifications are described below).

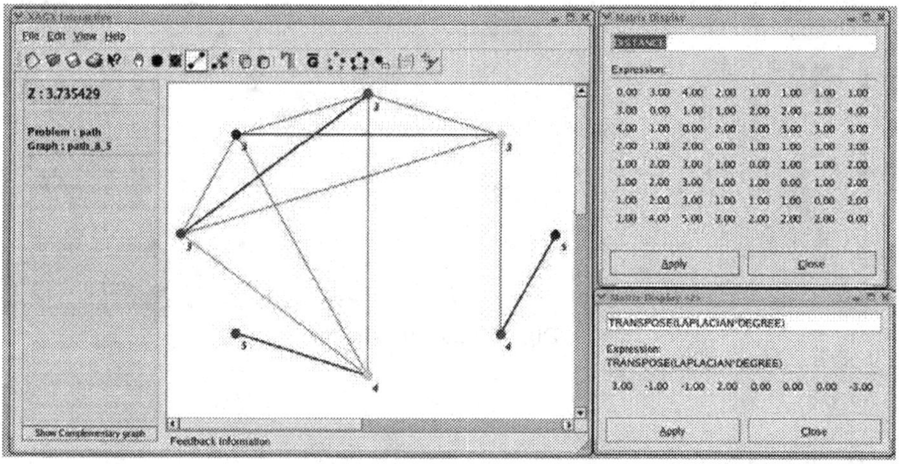

Fig. 6. Display matrices (right, top) and vectors (right, bottom)

Interactive mode

In addition to loading an existing graph (that is a result of the optimization process), we can generate or draw a new one. After starting AGX 2 *Interactive* by pressing the *Interactive mode button* in the control center (or using the line command xagx -i") we can built a graph"as we want it" using different possible operations: add, move or remove a vertex, add or delete an edge, swap between the graph and its complement. Each of these operations is represented by a button in the top of the window. Moreover, it is possible to display informations (Figure 5) about the current graph in three ways. (*i*) On the graph itself such as chromatic or degree coloration, edge highlighting (for shortest paths, radius, diameter, girth,...), labelling of the vertices by simple numbering, their degrees, transmissions (*i.e.*, sum of distances to all other vertices) or distances from a selected vertex. (*ii*) Display selected invariants, from a list that appears in a window by pressing the corresponding button, and their values for the current graph and/or its complement. The invariants and their values appear in the left margin of the window. (*iii*) Open a new window on which we can display (Figure 6) vectors or matrices (one at a

time) related to the current graph such as degree, eigenvalue, eccentricity or transmission vector and adjacency, distance or Laplacian matrix.
All the informations related to the current graph discussed above are automatically updated when modifying the graph.

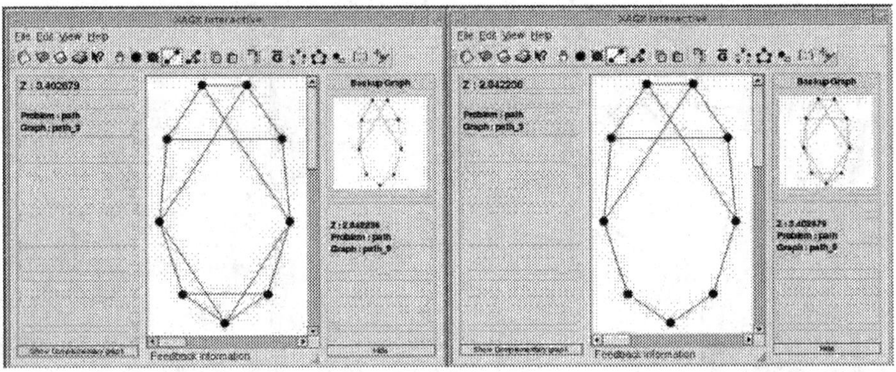

Fig. 7. Backup window

Backup window

The main aim of changing interactively a graph is to get a better one (with respect to our objective), so we need to keep the current best one to make comparisons. It is the task of the backup window (see Figure 7). The backup window in the right margin contains the current "referential" graph, the values of the selected invariants and of the objective function. Once the "referential" graph is kept, we can modify as desired or as needed the current graph and make comparisons between both graphs. It is possible to swap between the two graphs or replace any one by the other.

View results

Pressing the *View results* button in the control center dialogue box gives us the tools to analyze the results. First, a small dialogue box asks us for choosing one (to get a 2D curve) or two (3D curve) parameters among those of the problem, and a filter (optional, when no filter is selected all available graphs are taken) to select among the resulting graphs those satisfying some conditions (connected, complements connected, feasible with respect to the problem constraints, ranges for the parameters). Then (Figure 8) the objective appears as a (2D or 3D) curve or a set of points in 2 or 3 dimensional space. The curve or the point set (it is possible to swap between the two representations)

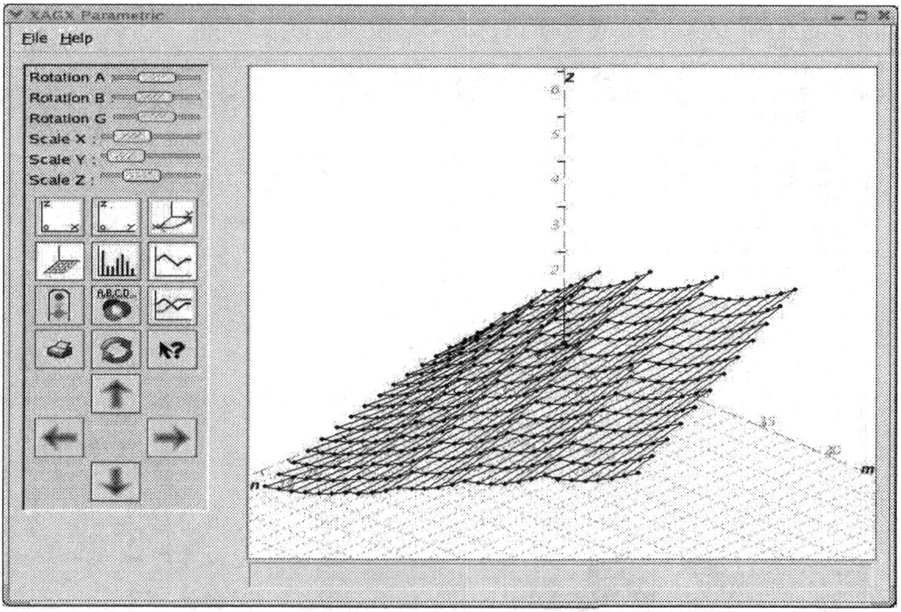

Fig. 8. Viewing results in a 3D representation

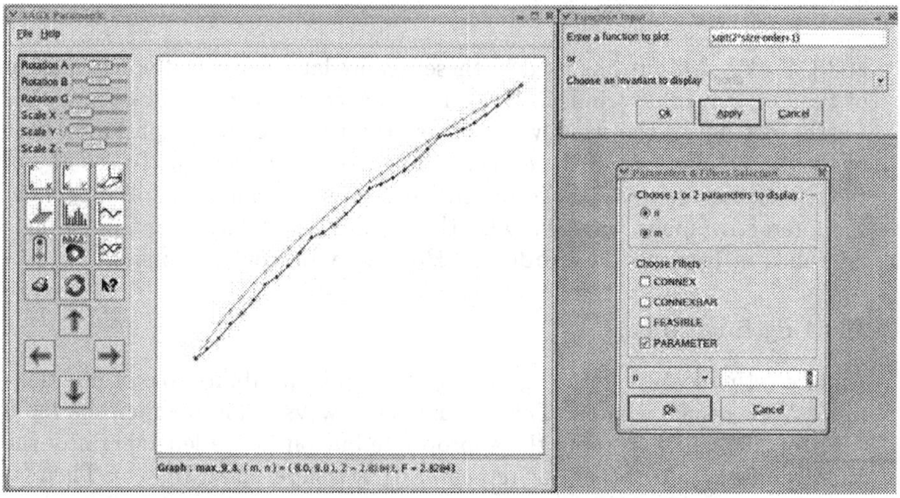

Fig. 9. A comparison with the known Hong bound

gives the objective function values for the chosen parameters ranges and represents the graphs that satisfy the chosen conditions only. We can perform many geometric manipulations that help us to view and analyze the results: rescaling the axis (Scale X, Scale Y, Scale Z), rotation with respect to any axis (Rotation A, Rotation B, Rotation C), move the axis to the right, left, top or bottom, projecting on the XOZ or the YOZ plane, swapping between OX

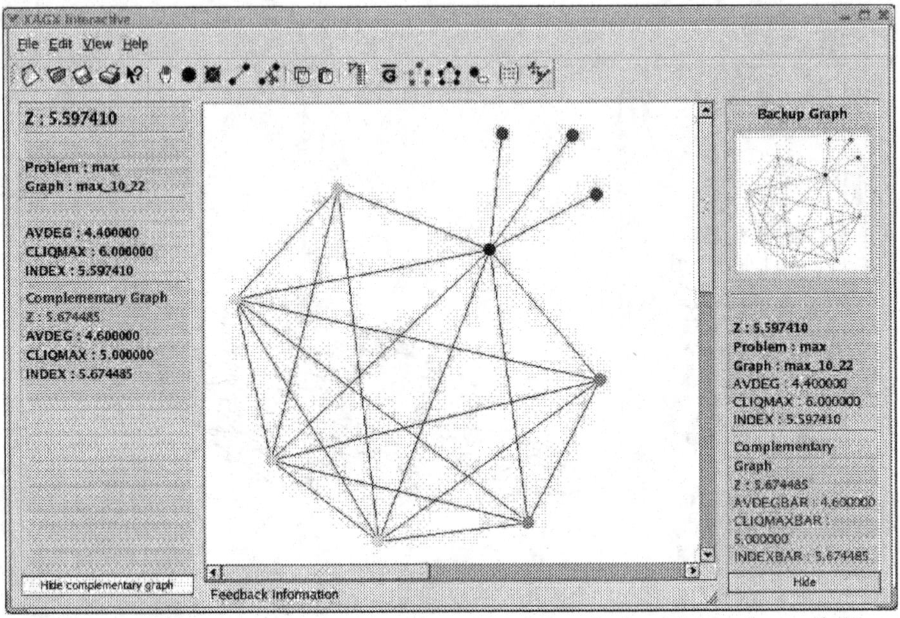

Fig. 10. An extremal graph for the index and values of selected invariants of the graph and its complement

and OY axis. All buttons used in these manipulations are in the left margin of the (View results) window. If the mouse gets over a point, the status bar (in the bottom of the window) indicates the name of the graph and values of the parameters and of the objective function, in this graph. To see any of the represented graphs, we click on the corresponding point, and it appears in the interactive mode window. Once the graph is available, all manipulations described in **Interactive mode** and **Backup window** are possible.

Plotting functions

An interesting feature in the interactive AGX 2, is the ability to plot functions in the View results window. This is done in two ways. First, we open the *Input function* dialogue box using the appropriate button in the left margin of the View results window (that one representing two superposed curves). Then (*i*) we display the value curve of an invariant chosen from a list of available ones or (*ii*) we enter a function that is expressed using the syntax rules of the objective function described in **Editing a problem**, and the corresponding curve is displayed. Plotting functions is a useful tool for curve fitting and deriving conjectures interactively. The curves in Figure 9 represent the Hong bound (top) on the index, *i.e.* the largest eigenvalue of the adjacency matrix, of a graph ([33]: $\lambda_1 \leq \sqrt{2m - n + 1}$, where m and n are the size and the order of a graph respectively), and (bottom) the bound given by AGX. An extremal graph corresponding to the AGX bound (Figure 9) is displayed in Figure 10.

Fig. 11. Selecting invariants to look for conjectures

Find conjectures

Once the optimization process is achieved, in addition to the View graphs and results options we try to find conjectures starting from the control center. First, we press the *Find Conjectures* button, then we are asked to choose invariants from the list of all of them (existing in the system or defined by the user in the editing problem phase). Once the invariants are chosen and the launch button pressed, AGX 2 will try to find affine relations between the selected invariants, based on the available graphs, found during the optimization or those modified interactively (if saved). If such relations exist, AGX 2 gives a basis of all affine relations satisfied by all available graphs. Note that, because of the numerical stability of the method (used to find the affine relations), the number of invariants to select must be at most half the number of available graphs.

3 Algebraic syntax used in AutoGraphiX

Each term used in AGX is assigned a name and an expression, the first to allow its use in any expression, and the second to make its computation possible. Expressions are stored in Reverse Polish Notation (RPN) which is computed by the parser from the standard algebraic expression provided by the user. Any expression may involve invariants or operators using graphs, matrices, vectors or numbers. Once a term is defined, it may be used as any hardcoded invariant. Some special expressions are generated by AGX and have reserved words to be accessed from; these are "OBJ", representing the objective function, "LHSn" or"RHSn" representing respectively the left or the right hand side of the nth constraint. Terms are recorded by the user using the keyword "define" in the problem definition file. By the use of flags in the data structure, AGX never computes twice the same term for the same graph; the definition of custom term is thus a way to speed up the optimization process. There is no restriction on the nature of a term as long as it remains an object that could be handled by AGX; thus a term may represent a graph, a matrix a vector or a number. For example, the *distance-k domination*, minimum number of vertices to select such that each vertex of the graph G is either selected or at a distance of at most k from a selected vertex, may be defined by the user as the classical domination number, which corresponds to the case $k = 1$, on the graph G' obtained from G by adding edges to connect vertices that are at a distance smaller than or equal to k in the original graph. Constructing this graph G' is called **KTRANSITIVE** in AGX and it uses the number k as argument. The following line may be used in AGX to define $k - domination$:

$$define\ DISTKDOMIN\ DOMINATION(KTRANSITIVE(k))$$

where k may be a parameter or be replaced by the desired value. To improve the performance of the optimization in AGX, a growing number of invariants are hardcoded, some of which could also be defined by the user as an expression using the others. A list of the main invariants, vectors, matrices or graphs presently available in AGX is given in Table 1, and operators in Table 2.

AVDEG : Average degree of the graph.
AVDIST : Average distance between pairs of vertices.
CHEMENERGY : $\sum_{i=1}^{n/2} 2eig[i] + eig[\frac{n+1}{2}]$ if n odd where $eig[i]$ is the ith eigenvalue of the adjacency matrix.
CHROMATIC : Chromatic number, minimum number of colors needed so that each vertex is assigned a color and two adjacent vertices do not share the same color.
CLIQMAX : Size of the maximum clique.
DEGMAX : Maximum degree.
DEGMIN : Minimum degree.
DOMINATION : Minimum number of vertices to select so that each vertex of G either is selected or is adjacent to a selected vertex.
DIAMETER : Maximum distance between 2 vertices; maximum eccentricity of a vertex.
ECONN : Edge connectivity number: minimum number of edges that must be removed to disconnect the graph.
ENERGY : $\sum_{i=1}^{n} \|eig[i]\|$ where $eig[i]$ is the ith eigenvalue of the adjacency matrix.
GIRTH : Size of the smallest elementary cycle
INDEX : Largest eigenvalue of the adjacency matrix.
K4 : Number of K_4 (cliques on 4 vertices).
MATCHING : Maximum cardinality matching

NCONN : Node connectivity number: minimum number of vertices that must be removed to disconnect the graph.
ORDER : Number of vertices.
SIZE : Number of edges.
STABLEMAX : Size of the maximum stable set.
SEPARATOR : Difference between the first and the second eigenvalue of the adjacency matrix.
SUMDEG2 : $\sum_{i=1}^{n} d_i^2$ where d_i is the degree of vertex i.
RADIUS : Radius of a graph; minimum eccentricity of a vertex.
RANDIC : Randić index: $\sum_{ij \in E} \frac{1}{\sqrt{d_i d_j}}$.
TRIANGLES : Number of triangles in the graph.
VARDEG : Degree variance
WIENER : $\sum_{i=1}^{n} \sum_{j=i+1}^{n} d_{ij}$.
DEGREE : (Vector) Degrees vector.
TRANS : (Vector) Transmissions vector.
NAVDEG : (Vector) Average Degree of neighbors vector.
ADJACENCY : (Matrix) Adjacency matrix
DISTANCE : (Matrix) Distances matrix
LAPLACIAN : (Matrix) Laplacian matrix
KTRANSITIVE : (Graph) Graph G' obtained from G by adding edges connecting vertices that are at distance less than k in G.

Table 1: Invariants available in AGX

Operators using 2 arguments:
* : matrix multiplication; output = matrix .* : term by term matrix multiplication; output = matrix ./ : term by term matrix division; output = matrix + : addition; output = matrix - : subtraction; output = matrix **ONE** : construction of a matrix of ones; output = matrix **FREQ** : number of times a value occurs in the matrix; output = number **FREQVECTOR** : vector of frequencies of the values of the argument matrix; output=vector
Operators using 1 argument:
FLOOR : term by term floor; output = matrix **CEIL** : term by term ceiling; output = matrix **TRANSPOSE** : transpose of the matrix; output = matrix **MINROW** : minimum row by row; output = column vector **MAXROW** : maximum row by row; output = column vector **MINCOL** : minimum column by column; output = row vector **MAXCOL** : maximum column by column; output = row vector **AVGCOL** : average column by column; output = column vector **AVGROW** : average row by row; output = column vector

VARCOL : variance column by column; output = row vector
VARROW : variance row by row; output = row vector
SUMROW : sum row by row; output = column vector
SUMROW : sum row by row; output = column vector
ABS : term by term absolute value; output = matrix
SUM : sum of elements of matrix; output = number
MODE : mode of the matrix; output = number
in case of ties, the first encountered value is considered
MAX or SUP : maximum of the matrix; output = number
MIN or INF : minimum of the matrix; output = number
ID : construction of an identity matrix; output = matrix
EIG : computation of the eigenvalues of a square matrix; output vector
example: (EIG(ADJACENCY))[2] gives the third eigenvalue of the adjacency matrix (don't forget that indices start from 0)
MEAN : computation of the average of a matrix; output = number
VARIANCE : computation of the variance of a matrix; output = number
STDEV : computation of the standard deviation of a matrix output = number
MINPEIG : minimum positive eigenvalue of a matrix; output = number
MAXNEIG : maximum negative eigenvalue of a matrix; output = number
RANK : computation of the rank of a square matrix; output = number
SQRT : term by term square root of a matrix; output = matrix

Table 2: Operators available in AGX

4 Optimization using Variable Neighborhood Search

Optimization in AGX 1 is performed by specializing the Variable Neighborhood Search (VNS) [32, 36] metaheuristic (general framework to build heuristics) to the problem of finding extremal graphs. VNS extends local search methods in order to get out of local optima in a systematic way, avoiding cycling through some random moves.

Initialization:
Select the set of neighborhoods structures N_k, $k = 1, \dots k_{max}$ that will be used in the search and a stopping condition; find an initial solution x.
Repeat until condition is met:

- Set $k \leftarrow 1$;
- **Until** $k = k_{max}$**,do:**
  (a) Generate a point x' at random from the kth neighborhood of x ($x' \in N(x)$);
  (b) Apply *Variable Neighborhood Local Search (VNLS)* with x'

as initial solution; denote with x'' the obtained local optimum;
(c) If the solution thus obtained is better than the incumbent, move there ($x \leftarrow x''$) and continue the search with $N_1(x)$ (k=1); otherwise, set $k \leftarrow k+1$.
done

Table 3: Rules of Variable Neighborhood Search.

Local search methods proceed by performing a sequence of local transformations in an initial solution which improve each time the objective function value until a local optimum is found. That is, at each iteration, an improved solution x' in the neighborhood $N(x)$ of the current solution x is obtained. VNS generalizes local search in two ways: several neighborhoods are considered instead of one, and they are used first in an exhaustive search of the smaller ones, and then in a random search of the larger ones. Contrary to other metaheuristics such as Simulated Annealing or Tabu search, VNS does not follow a trajectory but explores increasingly distant neighborhoods of the current solution, and jumps from there to a new one if and only if an improvement has been made. In this way, often favorable characteristics of the current incumbent solution, *e.g.* that most variables are already at their optimal value, are kept and used to obtain promising neighborhood solutions, from which a further local search is performed.
Rules of VNS are recalled in Table 3.

The stopping condition may be: maximum number of iterations, maximum CPU time allowed or maximum number of iterations between improvements. Rules of the routine VNLS used in VNS are given in Table 4.

Initialization:
Select the set of neighborhood structures N'_k, $k = 1, \dots k'_{max}$ that will be used. Consider an initial solution x;
Main Step:
Set $k \leftarrow 1$ and *improved* $\leftarrow$ *false* ;
Until $k = k'_{max}$, **do :**
(a) Find the best neighbor x' of x in $N'_k(x)$.
(b) If the solution x' so obtained is better than x,
set $x \leftarrow x'$ and *improved* $\leftarrow$ *true*.
(c) Set $k \leftarrow k+1$;
(d) if $k = k'_{max}$ and *improved* $=$ *true*
set $k \leftarrow 1$.
done

Table 4: Rules of Variable Neighborhood Local Search (Variable Neighborhood Descent).

Usually, the neighborhoods $N_k(x)$ in VNS will be nested, while the neighborhoods $N'_k(x)$ in VNLS need not be so. In the VNS applied to finding extremal graphs these neighborhoods correspond to removing or adding $1, 2, \cdots k$ edges. In VNLS they correspond to a series of elementary local changes: rotation of an edge (keeping one vertex fixed and changing the other), removal or addition of an edge, displacement of an edge, detour (replacing an edge by 2-path), 2-Opt and several others. While it is possible to change this last set within VNS (*e.g.* using different neighborhoods at the beginning and the end of the search), it is most often kept constant.

4.1 VNS optimization in AGX 2

The optimization algorithm implemented in AGX 2 is designed in the same way as in AGX 1 except that the VNLS routine uses the new Adaptive Local Search (ALS) described below as one of its descent methods. However, as ALS only deals with transformations of moderate size (moves involving at most 4 vertices), for some special problems, the user may still need to use specific moves not covered by ALS in the descent. In this case, the special move needs to be implemented and used in the VNLS routine as it was the case in AGX 1. Up to now, problems handled with AGX 2 did not need this feature and choosing only potentially useful moves is no more needed as the algorithm is designed to achieve this task in most cases. In order to do so, the program uses a list of moves that have proved to be improving at least for one graph. Using a special data structure, the program learns during the optimization process and constructs a list of improving moves to use. Even if a small number of moves is used during the optimization, the algorithm is designed to ensure that no other move (involving 4 vertices or less) could improve the local optimum found.
The description of the adaptive neighborhood search is done in three parts; in the first one the data structure is described, then a basic local search using this structure is explained and finally, based upon this local search algorithm, the adaptive local search is described.

Data structure and encoding

Let G=(V,E) be a labelled graph on n vertices $(v_1, v_2, \ldots v_n)$ and G' be an induced subgraph of G on n' vertices. To each subgraph G' is associated a pattern that is defined by the following procedure:
- relabel the vertices of G' from v'_1 to $v'_{n'}$ in the way that preserves the order of the labels of these vertices in G;
- from the lexicographical ordering of the possible edges of G', compute the corresponding binary vector by writing a 1 if the edge is present and a 0 otherwise.
Example 1:

Edge	binary vector	pattern
(1,2)	000001	1
(1,3)	000010	2
(1,4)	000100	4
(2,3)	001000	8
(2,4)	010000	16
(3,4)	100000	32

If $n' = 4$, the possible edges are the following:

Each subgraph G' is characterized by a unique label (called pattern) from 0 to $2^{\frac{n'(n'-1)}{2}}$ as shown on the example below:
pattern 0: empty subgraph
pattern 1: $E = \{(1,2)\}$
pattern 2: $E = \{(1,3)\}$
pattern 3: $E = \{(1,2),(1,3)\}$
:
pattern 13 (1+4+8): $E = \{(1,2),(1,4),(2,3)\}$
:
pattern 18 (2:16): $E = \{(1,3),(2,4)\}$
:
pattern 33 (1+32): $E = \{(1:2),(3,4)\}$
:
pattern 63: complete subgraph on 4 vertices.

General local search algorithm

Using this scheme, one may define a local search by the following algorithm.

Initialization:
Set *improved* ← *true*;
While *improved* = *true* **do:**
  set *improved* ← *false*
  For each subgraph G' **on** n' **vertices do:**
    - Compute the pattern number corresponding to G'.
    - Consider replacing G' in G by the subgraph corresponding to each possible alternative pattern.
    - If this change improves the objective function:
      *(a)*apply this change
      *(b)*set *improved* ← *true*
  done
done

A local optimum is found; it is not possible to improve the current graph by any transformation that involves at most n' vertices

Table 5: General local search algorithm

Adaptive local search

Unfortunately, as soon as n' is not very small, the local search is time consuming and is almost impossible to use for large graphs.
Furthermore, most of the alternative patterns are useless depending on the initial pattern, the objective function and the constraints.
To improve the optimization algorithm in AGX 2, we aim at trying only the alternative patterns that may improve the graph. In order to perform a fast search, AGX uses a knowledge base indicating only the potentially improving moves. This information is stored as a binary matrix T indicating whether changing the pattern (i) to (j) needs to be considered. The t_{ij} entry is set to 1 if replacing the pattern i by j was found interesting at least once. Note that most of the local moves involved in AGX 1 were special cases of moves for AGX 2 (for example, changing the pattern 33 to pattern 18 is a 2-opt move). The adaptive neighborhood search for n' given may be defined as follows.

Step 1:
If it is the first time the algorithm is run on this problem:
- initialize $T = \{t_{ij} = 0\}$.
If the program was already run:
- load the last version of the matrix T for the problem under study.

Step 2:
Set $f \leftarrow false$ (this flag indicates that no pattern was added to the list at this iteration).
For each subgraph of the current graph with n' vertices do:
Let p_i be the corresponding pattern
for each alternative pattern p_j do:
if replacing the subgraph p_i by the pattern p_j would improve the current solution:
update the matrix T by setting $t_{ij} \leftarrow true$
set $f \leftarrow true$.
done
done

Step 3:
If $f = true$: Update the matrix T to take symmetry into account:
for each $t_{ij} = true$ do:

for each pattern i' obtained from i by relabelling the vertices do:
let $p_{j'}$ be the pattern obtained from p_j by the same relabelling
set $t_{i'j'} \leftarrow true$.
done
done
If $f = false$:
Stop; a local optimum is found.
Save the matrix T for the next time AGX will be running on the same problem.

Step 4:
set $improved \leftarrow true$
while $improved = true$ **do:**
set $improved \leftarrow false$
For each subgraph G' of G on n' vertices do:
let p_i be the corresponding pattern
For each alternative pattern p_j:
if $t_{ij} \leftarrow true$ **do:**
if replacing p_i by p_j in the current graph improves the solution:
apply the change
$improved \leftarrow true$
done
done
done
Go to Step 2

Table 6: Adaptive local search.

5 AutoGraphiX Tasks

In this section we discuss the ways in which AGX 1 and AGX 2 address the five problems listed in the introduction.

Find a graph satisfying given constraints

Let $i_1(G), i_2(G), \cdots i_k(G)$ denote k invariants of a given graph G and $b_1, b_2, \cdots b_k$ real numbers. AGX makes a search for a graph satisfying given constraints of the form

$$i_l(G) \begin{cases} \leq b_l & \text{if } l \in I^- \\ = b_l & \text{if } l \in I^= \\ \geq b_l & \text{if } l \in I^+ \end{cases} \qquad (2)$$

where $I^- \cup I^= \cup I^+ = \{1, 2, ...k\}$.
For this aim, the constraints in 2 are transformed into

$$h_l(G) = \begin{cases} max\{i_l(G) - b_l, 0\} & \text{if } l \in I^- \\ |i_l(G) - b_l| & \text{if } l \in I^= \\ max\{b_l - i_l(G), 0\} & \text{if } l \in I^+ \end{cases} \tag{3}$$

Thus the problem of finding a graph G satisfying (2) is equivalent to the following optimization problem on a family of graphs $\mathcal{G}_n$ (or $\mathcal{G}_{n,m}$),

$$\min_{G \in \mathcal{G}_n} f(G) = \sum_{l=1}^{k} h_l(G)$$

Any graph G with $f(G) = 0$ satisfies constraints (2).
The constraint set (2) can be generalized by replacing one or more invariants by an expression of several invariants.

Find graphs with an optimal or near-optimal value for an invariant subject to constraints

Here the problem is to find a graph G among those satisfying given constraints, as chosen, that minimizes (or maximizes) the value of a given invariant $i(G)$. The problem can be stated as follows

$$\min_{G \in \mathcal{G}_n} f(G) = i(G) + M \sum_{l=1}^{k} h_l(G)$$

where M is a constant large enough to ensure that any graph G' not satisfying one or more constraints has a larger objective function value than any graph satisfying all of them.

Refute a conjecture

Let us consider a conjecture of the form $h(G) \leq g(G)$ where $h(G)$ and $g(G)$ are functions of one or several invariants of a graph G. *AutoGraphiX* then solves heuristically the following problem

$$\min_{G \in \mathcal{G}_n} f(G) = g(G) - h(G).$$

If a negative value of the function $f(G)$ is found, *i.e.*, a graph G' such that $f(G') < 0$, then the conjecture is disproved and G' is a counter-example.

Suggest or reinforce a conjecture

In general, we use the *AutoGraphiX* system to solve an optimization problem of the form $min_{G \in \mathcal{G}_n} f(G)$ where n is a parameter (such as the order and we can use several parameters), f a function of one or more invariants of G and $\mathcal{G}_n$ the set of all graphs of parameter n. The analysis of the graphs obtained (under the assumption that they are extremal) suggest us relations, so conjectures, between the invariants considered in the function f. There are several ways to generate conjectures [11].

Give an idea of proof

It is the most difficult among all *AutoGraphiX* tasks and those of any other system used for assisted or automated proofs. The observation of the progression process followed by AGX to the extremal graph can help to prove the result. In particular if best results are obtained with a single transformation it suffices to prove that this transformation can be applied and improves the objective function value for any non optimal graph in the family considered [29].

6 Automated proofs

Computer-assisted proofs are currently quite frequent in graph theory and have led to many successes, the most prominent of which is the four-color theorem [2, 3, 4, 5, 39]. In contrast, fully automated proofs are rare, although a couple of approaches have been explored. The system *Graph* of Cvetković and his collaborators [14, 15] has pioneered the computer-assisted approach to graph theory. It also comprises a component, *THEOR*, for automated proof of graph theoretical properties. *THEOR* uses a special first-order predicate calculus, called "arithmetic graph theory", together with a resolution-based prover. It appears however that full formalization of graph theory is a difficult task, and even simple properties require long proofs. To illustrate, it takes 10 lines to prove that

"If the graph is connected, then the graph is trivial or there is no point x such that x is isolated".

A different approach is explored in the *Graph Theorist* system of Epstein [19, 20, 21]. This knowledge intensive, specific domain learning system uses algorithmic description of classes of graphs such as connected, acyclic or bipartite. It mainly uses theory-driven discovery of concepts, conjectures and theorems based upon search heuristics.

Neither approach was powerful enough to prove new properties. However, as these systems were developed in the 80th, progress both in automated theorem proving and in computers, suggest further work along such lines might be fruitful.

In the *AutoGraphiX 2* system a different approach has been followed: instead of formalizing each graph theory concept, one works at a more global level, *i.e.*, on the one hand recognizing classes of extremal graphs from values of invariants and exploiting their properties and, on the other hand, combining relations on graph invariants. Both approaches appear to be closer than the previous ones to the usual way a graph theorist works, and have proved to be successful: several hundred new conjectures could thus be proved. They are usually easy, but may be useful, *e.g.* to enrich the database of relations for further proofs.
The first approach implemented in AGX 2 is illustrated in the automated report presented in Table 7 below. Two invariants are selected for comparison: the *average degree* $\bar{d}$ of a graph G and the *edge connectivity* κ_e, *i.e.*, the smallest number of edges to remove in order to disconnect G. Then, the lower and upper bounds for each invariant are introduced as an initial knowledge, together with a characterization of the corresponding families of extremal graphs. Note that while these results are easy to obtain, they could also be derived in an assisted way with AGX 2.
Then the two invariants are compared using the four operations $-, +, /$ and $\times$, and a rule-based approach is applied to check whether the intersection of the relevant sets of families of extremal graphs is empty or not in each case. For the first conjecture, on $\bar{d} - \kappa_e$, this intersection is empty both for the lower bound (trees and complete graphs) and for the upper bound (complete graphs and graphs with a cut edge). However, the relation $\kappa_e \leq \bar{d}$ is well-known and recorded in the system. The extremal graphs are complete ones, among others. For the upper bound, the numerical method of AGX [11] obtains automatically the algebraic expression $\frac{4.0+n^2-(4.0)n}{n}$. The extremal graphs are cliques on $n-1$ vertices with an appended edge. AGX 2 does not mention this fact in the automated report, as a routine for recognizing this class of graphs has not yet been included (it is an elementary task: checking if if the minimum degree $\delta = 1$ and the size $m = \frac{(n-1)(n-2)}{2} + 1$). But these extremal graphs can be displayed with AGX 2's interactive function and recognized at a glance. This provides the idea of an easy proof of the upper bound of *Conjecture 1*.

Proof:
For given κ_e,

$$m \leq \frac{(n-1)(n-2)}{2} + \kappa_e,$$

$$\bar{d} = \frac{2m}{n} \leq \frac{n^2 - 3n + 2 + 2\kappa_e}{n},$$

$$\bar{d} - \kappa_e = \frac{n^2 - 3n + 2 - (n-2)\kappa_e}{n} \leq \frac{n^2 - 4n + 4}{n}$$

and the bound is attained if and only if G is a clique on $n-1$ vertices with an appended edge. □

Invariants

This report contains conjectures obtained by AGX between the two following graphical invariants : $\overline{d}$ (average degree) and κ_e (edge connectivity). Suppose G is a connected graph with $n \geq 3$ vertices, then selected invariants have the following bounds :

$$2 - 2/n \leq \overline{d} \leq n - 1$$

The lower bound (resp. upper bound) is reached for the trees (resp. the complete graphs).

$$1 \leq \kappa_e \leq n - 1$$

The lower bound (resp. upper bound) is reached for graphs with a cut edge (resp. the complete graphs).

Conjectures

Suppose G is a connected graph with average degree $\overline{d}$, and edge connectivity κ_e. Let n denote the number of vertices of the graph.
Then AGX obtains the following results :

Conjecture 1
For all connected graphs G having at least 3 vertices, we have :

$$0 \leq \overline{d} - \kappa_e \leq \frac{4.0 + n^2 - (4.0)n}{n}$$

Proof :
No proof: the lower bound has been obtained automatically by AGX.

Proposition 2
For all connected graphs G having at least 3 vertices, we have :

$$3 - 2\frac{1}{n} \leq \overline{d} + \kappa_e \leq -2 + 2n$$

The lower bound (resp. upper bound) is reached for the trees (resp. the complete graphs).

Proof :

For connected graphs having at least 3 vertices, we have $2 - 2/n \leq \overline{d}$ and the bound is reached for the trees. For connected graphs having at least 3 vertices, on a $1 \leq \kappa_e$ and the bound is reached for graphs with a cut edge. The intersection of both families of graphs are the trees. By summing, we obtain the lower bound.
For connected graphs having at least 3 vertices, we have $\overline{d} \leq n - 1$ and the bound is reached for the complete graphs. For connected graphs having at least 3 vertices, we have $\kappa_e \leq n - 1$ and the bound is reached for the complete graphs. Both families of graphs are the same, so extreme graphs are the complete graphs. By summing, we obtain the upper bound.

Conjecture 3
For all connected graphs G having at least 3 vertices, we have :

$$1 \leq \overline{d}/\kappa_e \leq \frac{4.0 + n^2 - (3.0)n}{n}$$

Proof :
No proof, the lower bound has been obtained automatically by AGX.

Proposition 4
For all connected graphs G having at least 3 vertices, we have :

$$2 - 2\frac{1}{n} \leq \overline{d} \cdot \kappa_e \leq (-1 + n)^2$$

The lower bound (resp. upper bound) is reached for the trees (resp. the complete graphs).
Proof :
The proof is the same than for Proposition 2, by multiplying and not summing.

Table 7: An *AutoGraphiX* automated report.

The next case, *i.e.*, $\overline{d} + \kappa_e$ is more favorable: AGX 2 finds that the sets of families of extremal graphs have non empty intersection (trees for the lower bound and complete graphs for the upper bound). Thus summing the expressions of $\overline{d}$ and κ_e for these classes of graphs automatically gets and proves the two algebraic expressions for the bounds. So *Proposition 2* gives all requested informations in a fully automated way.
The cases of *Conjecture 3* and *Proposition 4* are similar to those of *Conjecture 1* (with the same extremal graphs) and *Proposition 2*.
A slightly more complicated argument, called the *second value based approach* leeds to automated profs in some further cases. Let us illustrate it by a simple case, *i.e.*, the lower bound $D + \kappa_e \geq 3$ where D denotes the *diameter* of the

graph G , *i.e.*, the largest distance between a pair of vertices of G. As $D \geq 1$ and $\kappa_e \geq 1$, one only gets $D+\kappa_e \geq 2$ as a direct consequence. However $D=1$ is only true for the complete graph K_n, for which $\kappa_e = n-1$. Considering the next value of D, *i.e.*, $D=2$ one can have $\kappa_e = 1$ *e.g.* for a star. So $D+\kappa_e \geq 3$. The second approach implemented in AGX 2 consist in combining inequalities. These are (i) lower and upper bounds on invariants; (ii) relations implied by the definitions *e.g.* $\delta \leq \bar{d} \leq \Delta$ where δ and $\bar{d}$ have been defined above and Δ denotes the maximum degree of G or $D \leq 2r$ where D has been defined and r denotes the *radius* of G, *i.e.*, the minimum among all vertices of G of its *eccentricity*; (iii) bounds derived from the previous ones by simple rules e.g. if $0 < l_1 \leq i_1 \leq u_1$ and $0 < l_2 \leq i_2 \leq u_2$ then

$$l_1 + l_2 \leq i_1 + i_2 \leq u_1 + u_2 \ , \ l_1 \times l_2 \leq i_1 \times i_2 \leq u_1 \times u_2,$$
$$l_1 - l_2 \leq i_1 - i_2 \leq u_1 - u_2 \ , \ l_1/l_2 \leq i_1/i_2 \leq u_1/u_2.$$

When applied to a set of 259 conjectures provided by the system HR [13], AGX 2 could prove over 85% of them.

7 Some examples

In this section, we illustrate by examples various uses of AGX 2 related to finding conjectures.

Example 2: At the *Computer and Discovery* workshop, held in Montreal, June 02-05 2004, M. Klin asked if the system AGX 2 could find the graph with order $n = 14$, girth $g = 6$ and as many edges as possible. This was tried immediately and AGX 2 provided a first graph with $m = 18$ edges, then a second one with $m = 19$ and within one minute of computing time a third graph with $m = 21$. This graph is represented in Figure 12 and is known to be optimal and unique for the problem posed.

Example 3: This is a straightforward one. *Let G be a connected graph with at least $n = 3$ vertices, domination number β and maximum degree Δ. Then*

$$\beta + \Delta \leq n$$

and the bound is attained for complete graphs (among others).
The proof is easy: let a vertex v of maximum degree belong to a minimum domination set. Then $1 + \Delta$ vertices, *i.e.*, those of v and its neighbors are covered. Adding the remaining vertices to the dominant set gives $\beta \leq 1+n-\Delta-1 = n-\Delta$ and the bound follows. For complete graph K_n, $\beta = 1$ and $\Delta = n-1$ so $\beta + \Delta = n$. □

Recall that the *eccentricity* of a vertex v in a graph G is the maximum distance from that vertex to another one. The *radius* r of G is its minimum eccentricity.

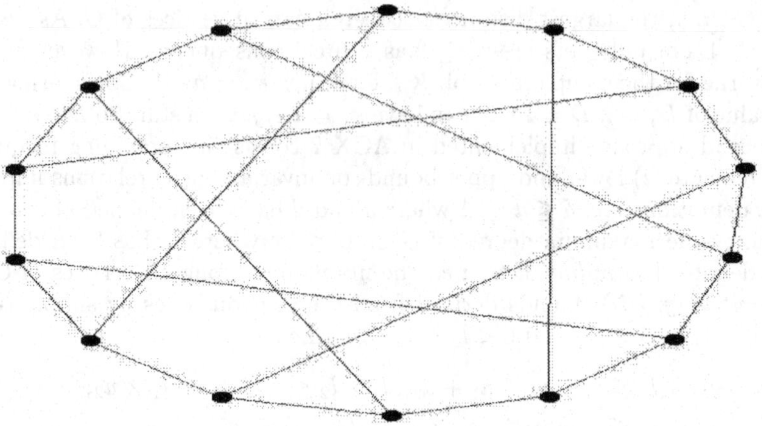

Fig. 12. Graph with $n = 14, g = 6$ and $m = 21$, the maximum number of edges

Example 4: *Let G be a connected graph of order $n \geq 3$ with domination number β and radius r. Then*

$$\beta + r \leq \lfloor \frac{5n+4}{6} \rfloor$$

and the bound is attained by a tree of diameter $n-2$ if $n = 3 \, mod \, [6]$ and by a path otherwise
This conjecture is open. The structural part is obtained automatically and the algebraic expression for the upper bound follows easily.

The *clique number* ω of G is the cardinality of its largest complete subgraph.

Example 5: *Let G be a connected graph of order $n \geq 3$ with domination number β and clique number ω. Then*

$$\beta + \omega \leq n + 1$$

and the bound is tight for complete graphs. Moreover

$$\beta \cdot \omega \leq \lceil \frac{n}{2} \rceil \cdot \lfloor \frac{n}{2} \rfloor$$

and the bound is tight for graphs composed of a clique with $\lceil \frac{n}{2} \rceil$ vertices and $\lfloor \frac{n}{2} \rfloor$ pending edges each incident with a different vertex of the clique.
The first relation was obtained automatically, and the second one from the structural result. Both conjectures are easily proved:
For the upper bound on the sum, if G is complete then $\beta + \omega = n + 1$. If not, due to connectedness of the graph, there exist a vertex v in a maximum clique S of G that has at least a neighbor that does not belong to S. So the set of vertices composed of v and all vertices that are neither in S nor neighbors of

v is a dominant set of cardinality at most $n - \omega$, *i.e.*, $\beta + \omega \leq n$.
Thus $\beta + \omega \leq n + 1$ with equality if and only if the graph is complete.
For the upper bound on the product, if G is complete then $\beta \cdot \omega = n$. If not, as it is shown above, $\beta + \omega \leq n$ and then we can solve the problem as the integer program of maximize $\beta \cdot \omega$ subject to $\beta + \omega \leq n$. Thus the bound follows and it is easy to see that the bound is reached for the graph described in the statement. □

Some conjectures are of medium difficulty. Recall that Randić index of a graph $G = (V, E)$ is defined by $Ra = \sum_{ij \in E} \frac{1}{\sqrt{d_i d_j}}$, and the independence number α of G is the cardinality of the largest set of vertices pairwise non adjacent.

Example 6: *Let G be a connected graph of order $n \geq 5$ with independence number α and Randić index Ra. Then*

$$Ra \cdot \alpha \leq \lceil \frac{3n-2}{4} \rceil \sqrt{\lceil \frac{3n-2}{4} \rceil \lfloor \frac{n+2}{4} \rfloor}.$$

The bound is attained for complete bipartite graphs K_{pq} with $p = \alpha = \lceil \frac{3n-2}{4} \rceil$ and $q = \lfloor \frac{n+2}{4} \rfloor$.

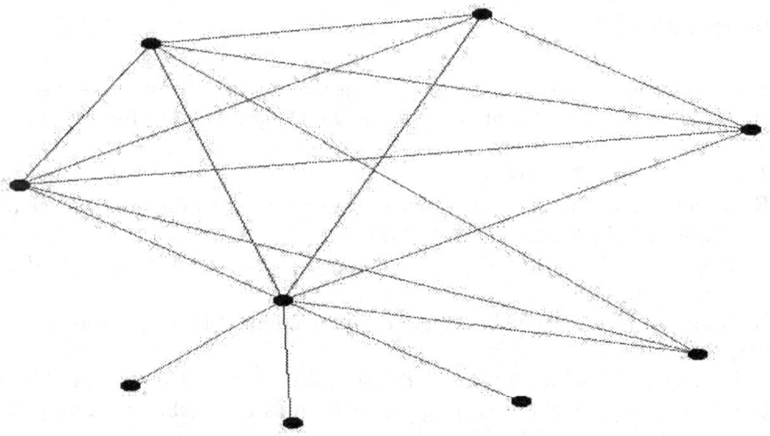

Fig. 13. A fanned pineapple with $n = 9$ and $m = 16$

In some cases only a structural conjecture could be found.
Let us define a *pineapple* as a clique together with one or several pending edges all incident with the same vertex of the clique (see Figure 10). A pineapple is *fanned* if one or several edges are added to it between a pending vertex and other vertices of the clique (see Figure 13).

Example 7: *For all connected graphs G of order $n \geq 5$ and size m, the index λ_1 is maximum for a pineapple or a fanned pineapple (depending on the value of m).*
This structural conjecture is open.

8 Conclusion

The system AGX 2 is a new and enhanced version of AGX with an improved optimization routine, better ways to find and prove conjectures and several new interactive functions. It mimics the work of the graph theorist and helps him in various ways: easy interactive manipulation of a graph, rapid computation of invariants and formulas involving several of them, display of results, obtention of relation between invariants in several ways and finally automated proof of easy relations. It is a useful tool, being currently applied to a series of specialized problems, *i.e.,* study of particular invariants and comparison between invariants on various classes of graphs. There are also many avenues for future research and development, in particular examining how the many as yet unexplored forms of conjectures in graph theory listed in [28] could be automated.

References

1. M. Aouchiche, G. Caporossi, and P. Hansen. Variable neighborhood search for extremal graphs 8: Variations on graffiti 105. *Congr. Numer.*, 148:129–144, 2001.
2. K. Appel and W. Haken. Every planar map is four colorable. Part I. Discharging. *Illinois J. Math.*, 21:429–490, 1977.
3. K. Appel and W. Haken. Every planar map is four colorable. Part II. Reducibility. *Illinois J. Math.*, 21:491–567, 1977.
4. K. Appel and W. Haken. The four color proof suffices. *Mathematical Intelligencer*, 8:10–20, 1986.
5. K. Appel and W. Haken. Every planar map is four colorable. *Contemp. Math.*, 98:1–741, 1989.
6. S. Belhaiza, N. M. M. Abreu, P. Hansen, and C. S. Oliveira. Variable neighborhood search for extremal graphs 11: Bounds on algebraic connectivity. *In D. Avis, A. Hertz and O. Marcotte (editors)* Graph Theory and Combinatorial Optimization, *Dordrecht: Kluwer (2005, to appear.*
7. G. Caporossi, D. Cvetković, I. Gutman, and P. Hansen. Variable neighborhood search for extremal graphs 2: Finding graphs with extremal energy. *Journal of Chemical Information and Computer Sciences*, 39:984–996, 1999.
8. G. Caporossi, I. Gutman, and P. Hansen. Variable neighborhood search for extremal graphs 4: Chemical trees with extremal connectivity index. *Computers and Chemistry*, 23:469–477, 1999.
9. G. Caporossi, I. Gutman, P. Hansen, and L. Pavlović. Graphs with maximum connectivity index. *Computational Biology and Chemistry*, 27:85–90, 2003.

10. G. Caporossi and P. Hansen. Variable neighborhood search for extremal graphs 1 The autographix system. *Discrete Math.*, 212, no. 1-2:29–44, 2000.
11. G. Caporossi and P. Hansen. Variable neighborhood search for extremal graphs 5 Three ways to automate finding conjectures. *Discrete Math.*, 276 no. 1-3:81–94, 2004.
12. S. Colton. Automated theory formation in pure mathematics. *Springer: London*, 2002.
13. S. Colton. *Private communication.* 2004.
14. D. Cvetković and I. Pevac. Man-Machine Theorem Proving in Graph Theory. *Artificial Intelligence*, 35:1–23, 1988.
15. D. Cvetković and S. Simić. Graph theoretical results obtained by the support of the expert system "graph" - an extended survey. *in [24].*
16. D. Cvetković, S. Simić, G. Caporossi, and P. Hansen. Variable neighborhood search for extremal graphs 3: On the largest eigenvalue of color-constrained trees. *Linear and Multilinear Algebra*, 49 no. 2:143–160, 2001.
17. G. di Battista, P. Eades, R. Tamassia, and I.G. Tollis. Algorithms for drawing graphs: an annotated bibliography. *Computational Geometry: Theory and Applications 4*, 5:235–282, 1994.
18. G. di Battista, P. Eades, R. Tamassia, and I.G. Tollis. Graph drawing: Algorithms for the visualization of graphs. *Prentice Hall*, 1999.
19. S. L. Epstein. On the discovery of mathematical theorems. *In Proceedings of the Tenth International Joint Conference on Articial Intelligence*, pages 194–197, 1987.
20. S. L. Epstein. Learning and Discovery: One System's Search for Mathematical Knowledge. *Comput. Intell.*, 4:42–53, 1988.
21. S. L. Epstein and N. S. Sridharan. Knowledge Presentation for Mathematical Discovery: Three Experiments in Graph Theory. *J. Applied Intelligence*, 1:7–33, 1991.
22. S. Fajtlowicz. On conjectures of graffiti. *Discrete Math.*, 72 no. 1-3:113–118, 1988.
23. S. Fajtlowicz. Written on the Wall. version 09-2000 (regularily updated file accessible via e-mail from clarson@math.uh.edu), 2000.
24. S. Fajtlowicz, P. Fowler, P. Hansen, M. Janowitz, and F. Roberts (editors). Graphs and discovery. *DIMACS Series in Discrete Math. and Theoretical Computer Science, AMS*, forthcoming, 2005.
25. P. W. Fowler, P. Hansen, G. Caporossi, and A. Soncini. Variable neighborhood search for extremal graphs 7: Polyenes with maximum homo-lumo gap. *Chemical Physics Letters*, 342:105–112, 2001.
26. I. Gutman, O. Miljković, G. Caporossi, and P. Hansen. Alkanes with small and large randić connectivity indices. *Chemical Physics Letters*, 306:366–372, 1999.
27. P. Hansen. How far is, should and could be conjecture-making in graph theory an automated process? *in [24].*
28. P. Hansen, M. Aouchiche, G. Caporossi, H. Mélot, and D. Stevanovic. What forms do interesting conjectures have in graph theory? *In [24].*
29. P. Hansen and H. Mélot. Variable neighborhood search for extremal graphs 9: Bounding the irregularity of a graph. *in [24].*
30. P. Hansen and H. Mélot. Variable neighborhood search for extremal graphs 6: Analyzing bounds for the connectivity index. *J. of Chem. Inf. Comp. Sc.*, 43 (1):1–14, Jan-Feb 2003.

31. P. Hansen, H. Mélot, and I. Gutman. Variable neighborhood search for extremal graphs 12: a note on the variance of bounded degrees in graphs. *Match, to appear*, 2005.
32. P. Hansen and N. Mladenović. Variable neighborhood search: Principles and applications. *European J. Oper. Res.*, 130 no. 3:449–467, 2001.
33. H. Hong. Bounds of eigenvalues of graphs. *Discrete Math.*, 123:65–74, 1993.
34. D. Knuth. The Stanford graphbase: a platform for combinatorial computing. *Addison - Wesley, Reading, Massachusetts*, 1993.
35. K. Mehlhorn and S. Nähger. Leda: A platform for combinatorial and geometric computing. *Communications of the ACM*, 38 (1):96–102, 1995.
36. N. Mladenović and P. Hansen. Variable neighborhood search. *Comput. Oper. Res.*, 24 no. 11:1097–1100, 1997.
37. E. A. Nordhaus and J. W. Gaddum. On complementary graphs. *Amer. Math. Monthly*, 63:175–177, 1956.
38. T. Pisanski and A. Zitnik. Interactive conjecturing with vega. *in [24]*.
39. N. Robertson, D. Sanders, P. Seymour, and R. Thomas. The four-color theorem. *Journal of Combinatorial Theory, Ser. B*, 70:2–44, 1997.
40. S. Skiena. Implementing discrete mathematics: Combinatorics and graph theory with mathematica. *Addison-Wesley*, 1990.
41. Vega. A system for manipulating discrete mathematical structures. *Project Homepage:* `http://vega.ijp.si/`.

From Theory to Implementation: Applying Metaheuristics.

I.J. García del Amo, F. García López, M. García Torres,, B. Melián Batista, J.A. Moreno Pérez, and J.M. Moreno Vega

Departamento de Estadística, I.O. y Computación
Universidad de La Laguna, 38271 La Laguna, Spain
{igdelamo,fcgarcia,mgarciat,mbmelian,jamoreno,jmmoreno}@ull.es

Summary. Metaheuristics are strategies to design heuristic procedures to find high quality solutions to an optimization problem. This chapter focuses on the implementation aspects of heuristic algorithms based on metaheuristics, using an object oriented approach. This programming paradigm takes advantage of the common parts shared by codes that implement different metaheuristic procedures. We give a class hierarchy for metaheuristics that permits quickly generate algorithms from existing metaheuristic codes for specific problems by extending a few classes and adding the problem functionality. It also allows the development of new metaheuristic algorithms without programming from scratch the basis of the procedure. It consists of selecting an appropriate class with the closest functionality, and extending it to add the core of the algorithm. The purpose of this hierarchy is thus to provide an extensible model for a quick implementation of metaheuristics and the problem structures associated with them.

Key words: Metaheuristic, implementation, API, variable neighbourhood search, genetic algorithms.

1 Introduction

Metaheuristics are strategies to design heuristic procedures. Since the first time the word *metaheuristic* appeared in the seminal paper of Tabu Search by Fred Glover in 1986 [8], there have been a lot of papers, reviews and books on Metaheuristics [29, 37, 31, 2, 12]. The classification of metaheuristics is usually based on the kind of procedures for which they are designed. For example, there are constructive metaheuristics like GRASP [30], evolutive metaheuristics like Genetic Algorithms [28] or neighborhood metaheuristics like the classical greedy local search. However, other possible classifications of metaheuristics are given by the computational tool or technique considered fundamental for the procedure, like Neural Networks [27] or Ant Colony Systems [3]. Some of the proposed algorithms are designed following not only one metaheuristic, but several of them. Moreover, some proposed metaheuristics are formed by mixed strategies that combine different kinds of tools, in such a way that they are either hybrid metaheuristics obtained by a combination of them, or they can be considered a simplified version of them obtained by ignoring some of the tools applied. It is also usual for many metaheuristic to have some modifications or adaptations to special circumstances that have been proposed to provide different versions or extensions of the metaheuristics. For these reasons the field of metaheuristics is continuously growing with new proposals that are becoming efficient and effective for an increasing number of difficult optimization problems.

However, the most relevant classification of them is to separate the metaheuristics based in populations of solutions from the metaheuristics based on a single solution. Among the single solution based metaheuristics (or point-based solutions, as we will refer to from now on), we should emphasize the importance of some of them, like the Greedy Search, the Random Search, the Local Search [38], the Guided Local Search [33], the Simulated Annealing [22], the Tabu Search [13] or the Variable Neighborhood Search [21]. On the other hand, among the population based metaheuristics, some of the most important ones are the Ant Colony Systems [3], the Scatter Search [24], the Estimation of Distribution Algorithms [25] or the Genetic Algorithms [28]. Other classifications appear in [32] and [39].

1.1 Variable Neighborhood Search

Variable Neighborhood Search (VNS) [17, 18, 20, 19, 21] is a recent and effective metaheuristic for solving combinatorial and global optimization problems that is capable of escaping from the local optima by systematically changing the neighborhood structures within the search. VNS proceeds using a descent method to reach a local minimum, then explores, either systematically or at random, increasing neighborhoods of this solution. In each iteration, one or several points within the current neighborhood are used as an initial solution for a local descent. The procedure jumps from the current solution to a new one if and only if a better solution has been found.

It has been empirically observed that, for many problems, local minima with respect to one or several neighborhoods are relatively close to each other. Therefore, a local optimum often provides some information about the global one. This may for instance be several variables with the same value, but usually it is unknown which ones are such. An organized study of the neighborhood of this local optimum is applied, until a better one is found.

Variable Neighborhood Descent (VND) is a deterministic version of VNS based on the fact that a local minimum with respect to one neighborhood structure is not necessary so for another. Thus, each time the descent reaches a local optimum for a set of moves, the method changes the set of moves; it changes the neighborhood each time it is trapped by the neighborhood structure. The method thus takes advantage from combining several descent heuristics and it stops at a local minimum with respect to all the neighborhood structure. Since a global minimum is a local minimum with respect to all possible neighborhood structures, the probability of reaching to the global optimum increases.

Another simple application of the VNS principle is the Reduced Variable Neighborhood Search (RVNS). From an initial solution, a point is chosen at random in the first neighborhood. If its value is better than the current one, the search is re-centered there. Otherwise, the search proceeds to the next neighborhood. After all neighborhoods have been considered, it starts again with the first, until a stopping condition is satisfied. Usual stopping criteria are based on a maximum computing time since the last improvement, or a maximum number of iterations.

In the previous two methods, we see how to use variable neighborhoods to descent to a local optimum and to find promising regions for near-optimal solutions. Merging the tools for both tasks leads to the General Variable Neighborhood Search scheme that uses VND to improve each solution sampled by RVNS. However, the basic VNS scheme (Figure 1) is obtained by combining a local search with systematic changes of neighborhoods around the local optimum found.

BVNS method

1. Find an initial solution x; choose a stopping condition;
2. Repeat until the stopping condition is met:
   (1) Set $k \leftarrow 1$;
   (2) Repeat the following steps until $k = k_{max}$:
   a) *Shaking.* Generate a point x' at random from the kth neighborhood of x ($x' \in \mathcal{N}_k(x)$);
   b) *Local search.* Apply some local search method with x' as initial solution; denote with x'' the so obtained local optimum;
   c) *Move or not.* If the local optimum x'' is better than the incumbent x, move there ($x \leftarrow x''$), and continue the search with $\mathcal{N}_1$ ($k \leftarrow 1$); otherwise, set $k \leftarrow k + 1$;

Fig. 1. Basic Variable Neighborhood Search Method

1.2 Scatter Search

Scatter Search (SS) [10, 11, 14, 15, 24] is an evolutionary algorithm that combines good solutions from a reference set ($RefSet$) to construct new ones exploiting the knowledge of the problem at hand. Genetic Algorithms [28] are also evolutionary algorithms in which a population of solutions evolves by using the mutation and crossover operators. These operators have a significant reliance on randomization to create new solutions. Unlike the population in Genetic Algorithms, the *RefSet* of solutions in Scatter Search is relatively small.

The principles of the Scatter Search metaheuristic were first introduced in the 1970s as an extension of formulations for combining decision rules and problem constraints. This initial proposal generates solutions considering characteristics of several parts of the solution space [7]. Scatter Search has an implicit form of memory, which can be considered as an inheritance memory, since it keeps track of the best solutions found during the search, and selects their good features to create new solutions. The Scatter Search Template, proposed by Glover in 1998 [10], summarizes the general description of Scatter Search given in [9].

Scatter Search consists of five components processes: *Diversification Generation Method*, that generates a set of diverse solutions, *Improvement Method*, that improves a solution to reach a better solution, *Reference Set Update Method*, which builds and updates the reference set consisting of *RefSetSize* good solutions, *Subset Generation Method*, to produce subsets of solutions of the reference set, and *Solution Combination Method*, that combines the solutions in the produced subsets. A comprehensive description of the elements of Scatter Search can be found in [10, 11, 14, 15, 24, 24].

The basic Scatter Search procedure (see Figure 2) starts generating a large set of diverse solutions *Pop*, which is obtained using the *Diversification Generation Method.* This procedure creates the initial population (*Pop*),which must be a wide set consisting of diverse and good solutions. Several strategies can be applied to get a population with these properties. The solutions to be included in the population can be created, for instance, by using a random procedure to achieve a certain level of diversity. An *Improvement Method* is applied to each solution obtained by the previous method reaching a better solution, which is added to *Pop.*

A set of good representative solutions of the population is chosen to generate the reference set (*RefSet*). The good solutions are not limited to those with the best objective function values. The considered reference set consists of *RefSetSize*1 solutions with the best objective function values and *RefSetSize*2 diverse solutions. Then $RefSetSize = RefSetSize1 + RefSetSize2$. The reference set is generated by selecting first the *RefSetSize*1 best solutions in the population and secondly adding *RefSetSize*2 times the most diverse solution in the population.

Several subsets of solutions from the *RefSet* are then selected by the *Subset Generation Method.* The *Solution Combination Method* combines the solutions in each subset using their good features. Then, the *Improvement Method* is applied to the result of the combination to get an improved solution. Finally, the *Reference Set Update Method* uses the obtained solution to update the reference set.

```
procedure Scatter Search
begin
    Create Population;
    Generate Reference Set;
    repeat
        repeat
            Subset Generation Method;
            Solution Combination Method;
            Improvement Method;
        until (StoppingCriterion1);
        Reference Set Update Method;
    until (StoppingCriterion2);
end.
```

Fig. 2. Scatter Search Metaheuristic Pseudocode

1.3 From Theory to Practice

After this review of current *state-of-the-art* on metaheuristics, it is time to ask the main question this chapter aims to answer. How can we implement these metaheuristics? Moreover, is there a way of programming them such that we could implement more metaheuristics than those seen here, without having to code them all from the beginning? We believe that the answer is "yes". But before getting into the work, we need to comment first on some necessary concepts, for instance, Object Oriented Programming (OOP).

Although it is not the purpose of this chapter to explain OOP paradigm and all of its features, we shall provide a brief description of it to contextualize the reading. Interested readers should refer to [1] for more information about OOP standards and features.

The OOP paradigm is a relatively new one. Although the first programs following its guidelines were developed in later 1960's, it has not been until 1990's that they had become widely spread. Traditional programming deals with functions (code) and variables (data), emphasizing the difference between them. We could say that traditional programming is a "code-based" approach, as programmers should rely on the efficiency and correctness of the code to resolve the task it was programmed to. OOP, on the other hand, focuses on dealing with objects, which are functionality units containing both data and code to manage it. This provides OOP with desirable properties such as modularity, encapsulation, abstraction, polymorphism or inheritance. We can say that OOP is a "design-based" paradigm, as it relies on the architecture of the program and the objects that live and interact in it, considering the code of the objects virtually irrelevant. The challenge in OOP therefore is of designing a sane object system ([35]).

Considering the problem of designing methods to implement several metaheuristics and making them extensible and easy to code, we can naturally think of OOP as an appropriate candidate for doing it.

- We want to code common parts just one time, so we do not have to program all the metaheuristic again every time we want to use a new one. This can be done by encapsulating common code into superclasses, and creating subclasses which inherit from them for metaheuristic-dependent code. This also has the advantage that if we detect an error in a part of the code of a superclass, we only have to correct it once, and all the subclasses will be updated.
- We want the code to be easily extensible. As we mentioned earlier, what really matters in OOP is the way in which objects relate to each other (i.e., WHAT they do), and not the specific code they use to do it (i.e., HOW they do it). For example, if we have a *problem* object, and a *metaheuristic* object, we want the *metaheuristic* object to get the *problem* object and to produce a *solution* object. We only care about what are the specifications of the *metaheuristic*. If the requirements are met, we should not care about how the *metaheuristic* object gets the solution. So, in theory, it does not matter if the *metaheuristic* object uses the VNS search or the Scatter Search, as long as it produces a valid solution for the problem. Thus, to extend the classes to include a new metaheuristic, we only have to create a subclass of the appropriate class and re-implement the necessary methods to use the new algorithm. We are changing the insides, but for an external viewer, it will still remain as a *metaheuristic* object.

2 Class hierarchy

In this section we will explain in detail each of the classes that form the hierarchy we propose (Figures 3, 4 and 5). We will start talking about the classes that define the structure of a problem and a solution (classes *Problem* and *Solution* respectively). Next, we will explain the general properties and methods we consider every metaheuristic should have (conforming the base class *Metaheuristic*). Then, we will continue with an explanation of why metaheuristics should be separated depending on if they are population-based or point-based. Finally, we will end with the specific details of two examples of metaheuristics already reviewed in the previous section: a point-based one (VNS) and a population-based one (Scatter Search). After this, we will comment the class *StopCriterion* and its relationship with metaheuristic classes. Note that, although classes *MhT_VNS* and *MhT_SS* are represented in Figure 4, they inherit from classes *PointBased* and *PopulationBased* respectively, which are represented in Figure 3.

The explanation of each class will be preceded by an enumeration of its attributes and methods, along with a short, general description of it and some relevant comments that we think are useful to understand the purpose of the class.

In the enumeration section, we will present firstly the attributes of the class, and then, its methods. Every attribute and method is preceded by one of these three symbols:

- "+": expresses that the attribute/method is public.
- "–": expresses that the attribute/method is private of the class.
- "#": expresses that the attribute/method is protected (that is, only the class or some subclass of it can access it).

Apart from these symbols, if a method or class has its name written in *italics*, that means that the method or class is abstract, and therefore it has to be redefined by a subclass (if it is an abstract method) or a subclass needs to be created for an object to be instantiated (if it is an abstract class).

Every attribute/method will end by a colon followed by the type of the attribute/method. For example, " : bool" means that the attribute is of type *bool* or that the method returns a *bool* value.

For convention, we will consider that the accessor methods of every class (i.e., *get* and *set*) return copies of the attribute (if it is a *getter*) or make a copy of the parameter before assigning it to the attribute (if it is a *setter*). This assumption is for preserving data encapsulation and integrity, so that every method can safely work with the object's data without interfering with other methods. Nevertheless, we understand that, in some cases, working with copies can be simply unaffordable (for example, in problems in which a solution is formed by a high number of elements). In these cases the reader is advised to implement carefully these methods to avoid strange behavior (i.e., freeing object's attributes, modifying the current solution in a *local search* procedure, etc).

2.1 Class Problem

Problem
– **isMaxProblem** : bool
+ **isMaxProblem**() : bool + **setIsMaxProblem**(isMax : bool) : void + **areEqual**(solution1 : Solution, solution2 : Solution) : bool + **firstSolutionIsBetter**(solution1 : Solution, solution2 : Solution) : bool + ***evaluate**(solution : Solution) : double*

The *Problem* class is probably the most important of all, but obviously, it is also the most problem-dependent one. Of course, there is no doubt that as class named *Problem* should be strongly problem-dependent. And that is why it is so difficult to generalize: we cannot assume anything about the attributes it contains, because they depend on the type of problem we are talking about. We can only assume that the problem should contain some kind of structure which explicitly enumerates all the elements that can form the solution, or at least, a way to obtain them. But that is all, we cannot know *a priori* if it is a list of objects, a function to obtain them, if we need more attributes to completely describe the problem... So, at most, we can define one attribute and some methods, which are listed below:

Attributes:

- **isMaxProblem**. Attribute that determines if the problem is a maximization or minimization problem.

Methods:

- **isMaxProblem**(). Method to get the value of the attribute *isMaxProblem.*
- **setIsMaxProblem**(isMax : bool). Method to set the value of the attribute *isMaxProblem*
- **areEqual**(solution1 : Solution, solution2 : Solution). Method to compare if two solutions have the same score. If a solution has not been evaluated yet, the method should call *evaluate* (explained below) and save the score obtained into the solution, then compare. The method should return *true* only if both solutions have the same score, and *false* in any other case.
- **firstSolutionIsBetter**(solution1 : Solution, solution2 : Solution). Similar to the method *areEqual* described above, but it checks if the first solution has a higher score than the second one. The method should return *true* only if the first solution has an strictly higher score than the second one.
- ***evaluate****(solution : Solution)*. Abstract method to evaluate (give a score) to a solution. This method should be implemented by the subclass that specifies the problem.

2.2 Class Solution

Solution
– **score** : double
+ **getScore()** : double + **setScore(score : double)** : void

The *Solution* class is as problem-dependent as the *Problem* one, because it has to provide a correct arrangement for some elements of the problem in order to conform a solution to it. And that is exactly why it cannot be generally defined with much detail. We cannot know how this elements should be placed, it could be linearly, in which case we would use a vector or an array, or maybe it could need a more complex structure like a tree or a priority queue.

So, the specific structure to handle the elements must be left to a subclass that knows more about the problem. The only property that we think any solution should have, at least, is a representative value of the fitness of the solution.

Attributes:

- **score**. Attribute that reflects the fitness of the solution. This attribute is intended as a variable to store the value returned by the method *evaluate* of the class *Problem*. If the solution has not been yet evaluated, it should contain a *not a number* (NaN) value.

Methods:

- **getScore**(). Method to get the value of the attribute *score*.
- **setScore**(score : double). Method to set the value of the attribute *score*.

2.3 Class Metaheuristic

Metaheuristic
– **bestSolution** : Solution – **problem** : Problem – **iteration** : int – **iterationOfBestSolution** : int – **elapsedTime** : double – **elapsedTimeOfBestSolution** : double – **stopCriterion** : StopCriterion
+ **getBestSolution()** : Solution + **getProblem()** : Problem + **getIteration()** : int + **getIterationOfBestSolution()** : int + **getElapsedTime()** : double + **getElapsedTimeOfBestSolution()** : double + **getStopCriterion()** : StopCriterion + **setStopCriterion(stopCriterion : StopCriterion)** : void + **setBestSolution(solution : Solution)** : void + **setProblem(problem : Problem)** : void + **resetIteration()** : void + **resetElapsedTime()** : void + *runSearch() : void* # **setIteration(iteration : int)** : void # **increaseIteration()** : void # **setIterationOfBestSolution(iteration : int)** : void # **setElapsedTimeOfBestSolution(elapsedTime : double)** : void

This class is the base class from which every metaheuristic will inherit. It has a few attributes to control the execution of the metaheuristic, and a main method *runSearch* to look for a solution to a problem.

The sequence of use of this class in the general case should be as follows:

1. Inform the metaheuristic class of the problem we are considering by setting the *problem* attribute to the appropriate value.
2. Set a stop criterion for the search.
3. Call the *runSearch* method.
4. When the search is finished, get the best solution found.

This sequence can be altered in special cases, for example, when we already have a solution for the problem and we want the metaheuristic to improve it. In that case, before calling *runSearch*, we would have to set the *bestSolution* attribute to the solution object we have. The metaheuristic should then return a solution which is, at least, as good as the one provided, if not better.

The class also has attributes for posterior statistical analysis, such as the number of iterations run, the iteration in which the best solution was found, or the elapsed time of search until that moment.

Attributes:

- **bestSolution**. Attribute to store the best solution found until the moment. Normally, this attribute will be unset before the beginning of the search, but if a solution is provided, the metaheuristic should try to *continue* the search of the best solution from that point. In any case, the search method should not update the solution unless the new solution found has a higher score than the one provided.
- **problem**. This is an object containing the problem which we are searching for a solution.
- **iteration**. The current iteration of the *runSearch* main loop. The metaheuristic should reset this value to 0 each time the method *runSearch* is called.
- **iterationOfBestSolution**. Iteration in which the best solution was found.
- **elapsedTime** : A time-stamp for several usages. Normally, this attribute would be reset before the beginning of the search and will be updated at the finish of the search, containing the number of time units since the last reset ($search\_stop\_time - reset\_time$).
- **elapsedTimeOfBestSolution**. A time-stamp for the moment in which the best solution was found ($best\_solution\_time - reset\_time$).
- **stopCriterion**. Object to determine if the search should stop at a given moment or should continue searching for a better solution.

Methods:

- **getBestSolution()**. Method to get the best solution found by the metaheuristic.
- **getProblem()**. Method to get the problem object.

- **getIteration**(). Method to get the current iteration of the search.
- **getIterationOfBestSolution**(). Method to get the iteration of the best solution. If a best solution hasn't yet been found, it should return a non-numeric value.
- **getElapsedTime**(). Method to get the elapsed time (in time units) since the last reset.
- **getElapsedTimeOfBestSolution**(). Method to get the elapsed time (in time units) since the las reset until the moment the best solution was found.
- **getStopCriterion**(). Method to get the *StopCriterion* object of the metaheuristic.
- **setStopCriterion**(stopCriterion : StopCriterion). Method to set the *StopCriterion* object of the metaheuristic.
- **setBestSolution**(solution : Solution). Method to set the best solution found until the moment (for example, to continue a search).
- **setProblem**(problem : Problem). Method to set the *Problem* object.
- **resetIteration**(). Method to reset the iterations for the search.
- **resetElapsedTime**(). Method to reset the time from which we will count.
- ***runSearch()***. Abstract method to search for a solution. This method must be implemented by a subclass. The implementation should also reset the iterations and the elapsed time at the beginning of the method.
- **setIteration**(iteration : int). Method to set the current iteration. This method can only be called by an object of class Metaheuristic or subclass of it. An external object should never update this variable.
- **increaseIteration**(). Method to increase the current iterations.
- **setIterationOfBestSolution**(iteration : int). Method to set the iteration in which the best solution was found.
- **setElapsedTimeOfBestSolution**(elapsedTime : double). Method to set the time in which the best solution was found.

2.4 Class PointBased

PointBased
− **currentSolution** : Solution − **newSolution** : Solution
+ **getCurrentSolution()** : Solution + **getNewSolution()** : Solution + **runSearch()** : void # **setCurrentSolution(solution : Solution)** : void # **setNewSolution(solution : Solution)** : void # ***initializeParameters() : void*** # ***generateInitialSolution() : void*** # ***generateNewSolution() : void*** # ***acceptNewSolution() : bool***

 # acceptanceUpdateParameters() : void
 # rejectionUpdateParameters() : void

The *PointBased* class is one of the two subclasses of *Metaheuristic* we will implement. This class implements its methods taking in mind that a point-based metaheuristic will only obtain one solution per iteration, and if it is better than the best it has found, it updates this best one with the new solution obtained.

This schema represents the core of this class, and it is shown mainly in the *runSearch* method, that was abstract in the *Metaheuristic* class, and is now defined in this class to follow the former guidelines. The *runSearch* method calls several internal methods (see fig 6) that are declared abstract, in order to permit a subclass to define them in a way that matches the metaheuristic specific algorithm.

Attributes:

- **currentSolution**. Solution with which the metaheuristic is currently working.
- **newSolution**. Temporary variable in which the newly created solution is stored. If after generating a new solution it is accepted, then the current solution is replaced by the new one.

Methods:

- **getCurrentSolution**(). Method to get the current solution of the metaheuristic.
- **getNewSolution**(). Method to get the newly created solution in each iteration of the metaheuristic.
- **runSearch**(). Implementation of the method to search for a solution, specially adapted to point-based metaheuristics, in which some abstract methods are used.
- **setCurrentSolution**(solution : Solution). Method to set the current solution.
- **setNewSolution**(solution : Solution). Method to set the newly created solution in each iteration of the metaheuristic.
- ***initializeParameters()***. Abstract method to initialize some parameters. The method is declared abstract, as we cannot know *a priori* how the metaheuristic needs to be initialized. A subclass that implements a specific metaheuristic, should implement also this method.
- ***generateInitialSolution()***. Abstract method to generate the initial solution. It is declared abstract, because the way in which an initial solution has to be generated depends not only on the metaheuristic, but also on the problem. Anyway, when this method is implemented, it should store the new solution in the *bestSolution* attribute. And also, if the *bestSolution* attribute is already set (for example, when we want the metaheuristic to

continue a search from a solution), the method should not modify this value, but return immediately, leaving the attribute "as is".

- ***generateNewSolution().*** Abstract method to generate a new solution when a previous solution exists. The method is abstract for the same reason as the previous method *generateInitialSolution*. This method should store the new solution in the attribute *newSolution*, and if it needs the previous existing solution, it can access it through the *currentSolution* attribute.
- ***acceptNewSolution().*** Abstract method to decide if the newly created solution is accepted to become the current solution. This method is provided because some metaheuristics accept every new solution, but others check the new solution for some properties, and do not accept it if it does not conform to them.
- ***acceptanceUpdateParameters().*** Abstract method to call when the new solution is accepted.
- ***rejectionUpdateParameters().*** Abstract method to call when the new solution is rejected.

2.5 Class PopulationBased

PopulationBased
– **initialPopulationSize** : int – **maxPopulationSize** : int – **currentPopulation** : Population – **newPopulation** : Population
+ **getInitialPopulationSize()** : int + **getMaxPopulationSize()** : int + **getCurrentPopulation()** : Population + **getNewPopulation()** : Population + **getBestSolutionInPopulation(population : Population)** : Solution + **setInitialPopulationSize(size : int)** : void + **setMaxPopulationSize(size : int)** : void + **runSearch()** : void # **setCurrentPopulation(population : Population)** : void # **setNewPopulation(population : Population)** : void # ***initializeParameters()*** *: void* # ***generateInitialPopulation()*** *: void* # ***generateNewPopulation()*** *: void* # ***acceptNewPopulation()*** *: bool* # ***acceptanceUpdateParameters()*** *: void* # ***rejectionUpdateParameters()*** *: void*

This class is the counterpart of the *PointBased* class, but for populations of solutions. These classes (*PointBased* and *PopulationBased*) were specifically designed to be as similar as possible, so that a parallelism between them could be naturally established. For example, if the class *PointBased* has a method called *generateNewSolution*, the class *PopulationBased* should have its equivalent called *generateNewPopulation*.

In Figure 7 we can see the code of the method *runSearch* for the *PopulationBased* class, and there is shown clearly the strong similarities that exist between this method and the respective one from *PointBased* class (6).

Just one more comment. In this class (and implicitly in all of its subclasses), there is an assumption for the object/type *Population*. For implementation purposes, we can simply consider it as an array, vector, list, etc. of *Solution* objects. The only requirements are that it can handle a set of solutions, granting the insertion, access and removal of each of them.

Attributes:

- **initialPopulationSize**. Number of elements (solutions) that should be contained in the initial population. The method to generate it, though abstract, should honor this value when implemented.
- **maxPopulationSize**. The maximum number of elements (solutions) that any population should contain. This value has to be observed every time a new population is created.
- **currentPopulation**. The population the metaheuristic is working on in the current iteration.
- **newPopulation**. The new population generated in each iteration of the metaheuristic. As with the *PointBased* class, if this method requires the previous population, it can access it through the *currentPopulation* attribute.

Methods:

- **getInitialPopulationSize**(). Method to get the size of the initial population.
- **getMaxPopulationSize**(). Method to get the maximum size of any population.
- **getCurrentPopulation**(). Method to get the population the metaheuristic is currently working on.
- **getNewPopulation**(). Method to get the new population created in each iteration of the metaheuristic.
- **getBestSolutionInPopulation**(population : Population). Method that looks for the solution with the highest score among all the population, and then returns it.
- **setInitialPopulationSize**(size : int). Method to set the attribute *initialPopulationSize*.
- **setMaxPopulationSize**(size : int). Method to set the attribute *maxPopulationSize*.
- **runSearch**(). Implementation of the method to search for a solution, specially adapted to population-based metaheuristics, in which in each iter-

ation the metaheuristic explores a set of solutions. As in the *PointBased* class, some abstract methods are used.

- **setCurrentPopulation**(population : Population). Method to set the attribute *currentPopulation*.
- **setNewPopulation**(population : Population). Method to set the attribute *newPopulation*.
- ***initializeParameters()***. Abstract method to initialize some metaheuristic-dependent parameters.
- ***generateInitialPopulation()***. Abstract method to generate the initial population. As its *PointBased* counterpart, it is declared abstract, because the way in which an initial population has to be generated depends on the problem. If the *bestSolution* attribute is already set (for example, when we want the metaheuristic to continue a search from a solution), this solution should be included, or at least used to generate, the initial population.
- ***generateNewPopulation()***. Abstract method to generate a new population when a previous population exists. As the method *generateNewSolution* of the *PointBased* class, this method should store the new population in the attribute *newPopulation*, and if it needs the previous existing population, it can access it through the *currentPopulation* attribute.
- ***acceptNewPopulation()*** *: bool*. Method to determine if the new population is accepted to substitute the current population. Usually, populations must have some properties in order to avoid degeneration of the solutions, and if it does not, the population is rejected.
- ***acceptanceUpdateParameters()*** *: void*. Abstract method to call when the new population is accepted.
- ***rejectionUpdateParameters()*** *: void*. Abstract method to call when the new population is rejected.

2.6 Class MhT_VNS

MhT_VNS

− **k** : int
− **kMax** : int

+ **getK()** : int
+ **getKMax()** : int
+ **setKMax(kMax : int)** : void
# **setK(k : int)** : void
# **increaseK()** : void
# **generateNewSolution()** : void
# **acceptNewSolution()** : bool
# **acceptanceUpdateParameters()** : void
# **rejectionUpdateParameters()** : void
# ***initializeParameters()*** *: void*
# ***generateInitialSolution()*** *: void*

 # shake() : void
 # improveSolution(solution : Solution) : void

This class is the first one we will see that represents the implementation of a specific metaheuristic. The *MhT_VNS* class uses attributes and methods that are exclusive of the *VNS* metaheuristic, although it is still declared *abstract* because of problem-dependent issues and the existence of several variants of the *VNS* general algorithm.

This class is intended to provide the general schema of the *VNS* metaheuristic, but at the same time, allow a subclass to customize some aspects of the algorithm, like, for example, the local search (here called *improveSolution*) or the shake procedure (see Figure 1). For example, a subclass of *MhT_VNS* could implement the *improveSolution* as a strictly local search, other could use a global search, and other could even use another metaheuristic.

For implementing this class the key concepts are, as in every other metaheuristic, to identify firstly if it is a *point based* or a *population based* metaheuristic, and secondly, where do the algorithm fit in the methods provided by the super-class (in this case, *PointBased*).

For example, the *generateNewSolution* method could consist in a *shake* and a *local search* (see Figure 8). The method to test if a new solution is accepted is simply a comparison between the new solution and the best solution found until the moment, and if has a higher score, it is accepted (Figure 9). If a solution is accepted, K is reset to 1 (Figure 10), and if it is rejected, K is increased (Figure 11).

Attributes:

- **k**. This attribute controls the current size of the neighborhood of the solution to explore in the *improveSolution* phase. This attribute varies from 1 to $kMax$.
- **kMax**. Maximum value for the k attribute.

Methods:

- **getK**(). Method to get the attribute k.
- **getKMax**(). Method to get the attribute $kMax$.
- **setKMax**(kMax : int). Method to set the attribute $kMax$.
- **setK**(k : int). Method to set the attribute k.
- **increaseK**(). Method to increase the value of the attribute k.
- **generateNewSolution**(). Method to generate the new solution from an existing one. See Figure 8.
- **acceptNewSolution**(). Method to determine if a new solution is accepted. See Figure 9.
- **acceptanceUpdateParameters**(). Method to call in case a new solution is accepted. See Figure 10.
- **rejectionUpdateParameters**(). Method to call in case a new solution is rejected. See Figure 11.

- *initializeParameters().* Abstract method to initialize parameters. This methods has to be implemented by a subclass that knows more details about the problem.
- *generateInitialSolution().* Abstract method to generate an initial solution. This methods has to be implemented by a subclass that knows more details about the solution.
- *shake().* Abstract method to provide the metaheuristic with a way to escape from a local minimum solution. It is declared abstract to allow a subclass to implement different ways of *shaking*, and also, because to be able to shake a solution, the method needs to know more details about the solution.
- *improveSolution(solution : Solution).* Abstract method to improve a solution within a k-sized neighborhood. Like the *shake* method, this method is defined abstract both for allowing several implementations and because there is a need for more information on the problem and the solution. For a possible implementation of a local search procedure, see Figure 13.

2.7 Class MhT_SS

MhT_SS

− **refSetSize** : int
− **refSetSize1** : int
− **refSetSize2** : int
− **subsetSize** : int
− **refSet** : Population
− **newRefSet** : Population
− **stopCriterionRefSet** : StopCriterion
− **stopCriterionPopulation** : StopCriterion

+ **getRefSetSize()** : int
+ **getRefSetSize1()** : int
+ **getRefSetSize2()** : int
+ **getSubsetSize()** : int
+ **getRefSet()** : Population
+ **getNewRefSet()** : Population
+ **getStopCriterionRefSet()** : StopCriterion
+ **getStopCriterionPopulation()** : StopCriterion
+ **setRefSetSize(size : int)** : void
+ **setRefSetSize1(size : int)** : void
+ **setRefSetSize2(size : int)** : void
+ **setSubsetSize(size : int)** : void
+ **setStopCriterionRefSet(stopCriterion : StopCriterion)** : void
+ **setStopCriterionPopulation(stopCriterion : StopCriterion)** : void
# **setRefSet(refSet : Population)** : void
# **setNewRefSet(refSet : Population)** : void
# **generateNewPopulation()** : void

acceptNewPopulation() : bool
acceptanceUpdateParameters() : void
rejectionUpdateParameters() : void
initializeParameters() *: void*
generateInitialPopulation() *: void*
generateRefSet() *: void*
selectSubset() *: void*
combineSolutions(subset : Population) *: Population*
improveSolutions(subset : Population) *: Population*
updateRefSet() *: void*

This class is the other example of an specific metaheuristic we will see, but, as we mentioned earlier, instead of being *point based*, as was *VNS*, this metaheuristic is classified as *population based.* It is also defined abstract because, to be able to create an instantiable class, we need more information about the problem.

To implement this class, the first thing we have to do is find the methods of the *PopulationBased* class in which to insert the *Scatter Search* specific code (see Figure 2 for the pseudocode). We only have to take in mind that the main loop for the *runSearch* method in *PopulationBased* consists in generating a new population in each iteration. Remember that the *Scatter Search* is based in the generation and update of a reference set in each iteration, not the population itself. This means that we have to think a little how to split the code of the algorithm in order to fit in the abstract methods used by *runSearch.*

An implementation of the classical *Scatter Search* will typically let the reference set evolve, and when it is done, return the best solution in it. So, in practice, it only uses one population. This fact has some implications, like, for example, that most of the code of the algorithm has to be embedded in the *generateNewPopulation* method. Another consequence is that other methods and attributes are meaningless, like *acceptNewPopulation* (which shall always return *true*), or the *Metaheuristic* attribute *stopCriterion* (which, as it refers to the evolution of the population, we want it to stop in the first iteration, so in fact, it has to return always *true*). More sophisticated implementations may use other stop criteria that would allow also the evolution of populations. For further reference on *Scatter Search* implementations, see [24].

Attributes:

- **refSetSize**. Attribute that determines the size of the complete reference set (*good* solutions + *diverse* solutions).
- **refSetSize1**. Attribute that determines the number of *good* solutions that will be in the reference set.
- **refSetSize2**. Attribute that determines the number of *diverse* solutions that will be in the reference set.

- **subsetSize**. Attribute to determine the number of solutions that will be taken from the reference set to be combined. A typical value for this attribute is 2.
- **refSet**. Object containing the reference set of solutions.
- **newRefSet**. Object containing the new reference set of solutions generated in each iteration.
- **stopCriterionRefSet**. Object representing the stop criterion for the loop in which the new reference set is generated. When this stop criterion determines that the loop should stop, it will have generated a new reference set.
- **stopCriterionPopulation**. Object representing the stop criterion for the loop in which new reference sets are being generated. When this stop criterion determines that the loop should stop (usually because the new reference sets generated lack of good solutions or diverse solutions), a new reference set will have to be created from the population.

Methods:

- **getRefSetSize**(). Accessor method to get the value of the attribute *refSetSize.*
- **getRefSetSize1**(). Accessor method to get the value of the attribute *refSetSize1.*
- **getRefSetSize2**(). Accessor method to get the value of the attribute *refSetSize2.*
- **getSubsetSize**(). Accessor method to get the value of the attribute *subsetSize.*
- **getRefSet**(). Accessor method to get the value of the attribute *refSet.*
- **getNewRefSet**(). Accessor method to get the value of the attribute *newRefSet.*
- **getStopCriterionRefSet**(). Accessor method to get the value of the attribute *stopCriterionRefSet.*
- **getStopCriterionPopulation**(). Accessor method to get the value of the attribute *stopCriterionPopulation.*
- **setRefSetSize**(size : int). Accessor method to set the attribute *refSetSize.*
- **setRefSet1Size**(size : int). Accessor method to set the attribute *refSet1Size.*
- **setRefSet2Size**(size : int). Accessor method to set the attribute *refSet2Size.*
- **setSubsetSize**(size : int). Accessor method to set the attribute *subsetSize.*
- **setStopCriterionRefSet**(stopCriterion : StopCriterion). Accessor method to set the attribute *stopCriterionRefSet.*
- **setStopCriterionPopulation**(stopCriterion : StopCriterion). Accessor method to set the attribute *stopCriterionPopulation.*
- **setRefSet**(refSet : Population). Accessor method to set the attribute *refSet.*

- **setNewRefSet**(refSet : Population). Accessor method to set the attribute *newRefSet.*
- **generateNewPopulation**(). Method to generate a new population for the search. This method contains most of the code of the algorithm, because *Scatter Search* is based more in the evolution of the reference set than the evolution of the population. So, this method also uses other methods that will be explained bellow, like *combineSolutions*, or *updateRefSet.* See Figure 12 for the code of this method.
- **acceptNewPopulation**(). This method decides if a new population is accepted or not. In the classical implementation of *Scatter Search*, the process of finding a solution is based on the evolution of the reference set, not the population. So, in the practice, this method does not really has to check if the new population is better than the older, it should always accept the new one.
- **acceptanceUpdateParameters**(). Method to perform the needed operations when a population is accepted. As no new population is intended to be created, this method should contain no code.
- **rejectionUpdateParameters**(). This method is like the *acceptanceUpdateParameters* above, so it should contain no code.
- ***initializeParameters().*** Abstract method to initialize different variables that may be needed by a subclass, depending on the specific problem.
- ***generateInitialPopulation().*** Abstract method to generate the initial population that has to be defined by a subclass, as it may need problem-specific data.
- ***generateRefSet().*** Abstract method to generate a reference set from a population of solutions. This method should generate a set of size *refSetSize*, composed of *refSetSize1* good solutions and *refSetSize2* diverse solutions. A way to measure the diversity of a solution will also have to be implemented by a subclass.
- ***selectSubset().*** Abstract method to get the next subset of solutions from the reference set to be combined and improved. The subset selected should contain *subsetSize* solutions.
- ***combineSolutions****(subset : Population).* Abstract method to combine the subset of solutions selected by the previous method. This method should produce a new solution from the subset.
- ***improveSolutions****(subset : Population).* Abstract method to improve the new subset of solutions obtained by the method *combineSolutions.*
- ***updateRefSet().*** Abstract method that has to decide which of the solutions of the new reference set created should replace some of the solutions of the old reference set.

2.8 Class StopCriterion

StopCriterion
+ *stop(mh : Metaheuristic)* *: bool*

The *StopCriterion* class is one of the most remarkable features of this implementation. When talking about a stop criterion from the traditional point of view, we should expect a method of a subclass of *Metaheuristic* to handle this stop condition, having knowledge of the specific metaheuristic and problem used.

In the context of OOP, we have used a class approach instead. The explanation is simple: think for example of a test in which we wanted to compare the effectiveness of different metaheuristics, and we wanted to stop their search after 1 second. Why do we have to implement a subclass for every metaheuristic evaluated, repeating the same code for the stop criterion in each of them? This rises the risk of introducing errors, and at the same time, we are duplicating code, breaking the principles of encapsulation and modularity. All of this can be avoided by using a unique object *stop criterion* that will return *true* if the metaheuristic has been running for a second. This has several advantages:

- The *StopCriterion* class reduces drastically the number of subclasses that need to be created. If it were not for the *StopCriterion* class, we would have to create a subclass of each metaheuristic only to produce a different stopping condition of each of the loops in its search method. Think, for example, of the *MhT_SS* class, which uses at least, three different stop criteria. If we wanted to test just two stop criteria (for example, one based in the elapsed time, and another based in the number of iterations), that would imply, from the traditional approach, eight different subclasses of that metaheuristic to combine all the possibilities for the stop criteria. With the use of a *StopCriterion* class, there is no need for subclasses of the metaheuristic, this could be done simply by instantiating an object of each of the stop criteria and combining them in all the possible ways.
- The use of this class also increases the versatility of a metaheuristic, because with a sole implementation of this metaheuristic, its functionality can be fine-tuned in execution time, simply by changing its *stopCriterion* object. With the example of the *MhT_SS* class, is obvious that not only eliminates the necessity for subclasses of the metaheuristic, but also allows the *MhT_SS* class to exhibit different behaviors or "flavors" in execution time.

Of course, the use of the *StopCriterion* class is not exempt of disadvantages. For example, as the *StopCriterion* is an external class to *Metaheuristic* (i.e., it does not inherits from Metaheuristic), all the accessor methods to get the value of an attribute (the *getters*) need to be declared public in order to allow *StopCriterion* to check the values of the attributes. Moreover, we need to enable public accessors to variables that in other cases would normally not even exist, because the *StopCriterion* class may need access to some internal variables in order to determine if the stop condition is met. Anyway, despite this disadvantages, we recommend the use of this class for its benefits.

The *StopCriterion* class is defined abstract, with only one method, *stop()*, that has to be implemented by an specific subclass.

Methods:

- ***stop****(mh : Metaheuristic).* Method to call when is needed to know if a metaheuristic should stop searching. This method receives a parameter, a *Metaheuristic* object, to which the *StopCriterion* will ask for some attributes in order to know if the stop condition is met.

2.9 Class GeneralStopCriterion

GeneralStopCriterion
– **maxTime** : double – **maxIterations** : int
+ **stop(mh : Metaheuristic)** : bool + **getMaxTime()** :double + **getMaxIterations()** : int + **setMaxTime(maxTime : double)** : void + **setMaxIterations(maxIterations : int)** : void

This class inherits from *StopCriterion*, and is aimed to determine a metaheuristic's stopping condition without depending on the problem nor the specific metaheuristic considered.

Before we continue, it is convenient to explain in more detail the differences among possible stopping criteria:

- **Problem-dependent stop criteria**. The stop criterion uses information that is specific of the problem, usually concerning the quality of the best solution obtained by the metaheuristic. For example, if we are dealing with a minimization problem, and we know which is the theoretical minimum value a solution can reach, we could stop searching when a solution is within a certain range above that minimum.
- **Metaheuristic-dependent stop criteria**. In this case, the stop criterion uses information about the metaheuristic itself. For example, in the *Scatter Search*, we need to stop some loops if the metaheuristic has reviewed all the specified combinations of solutions of the reference set, or when the quality of the solutions of the reference set lowers from a certain point.

- **Problem-Metaheuristic-dependent stop criteria**. This case is a mixture of the two previous cases, when the stop criterion uses both problem-dependent and metaheuristic-dependent information. An example of this could be to stop the search of a *VNS* metaheuristic either when the quality of the solution is above a certain level, or when the value of K has reached the top ($KMax$) and has not been reseted to 1 for a long time (which could indicate that the solution is good, as we are always moving the maximum distance between neighborhoods without finding a better solution).
- **Independent stop criteria**. This is a generic case in which the stop criterion accesses information that do not depend on the problem nor the metaheuristic used. The typical information used here is the number of iterations of the metaheuristic or the elapsed time since the beginning of the search.

This class is an implementation of an *independent stop criterion*, that can be configured to use the iterations, the elapsed time, or both.

Attributes:

- **maxTime**. Attribute that determines the maximum time a metaheuristic is allowed to run. If this attribute is not going to be used, it should contain a NaN value.
- **maxIterations**. Similar to the attribute above, but it determines instead the maximum number of iterations a metaheuristic is allowed to run.

Methods:

- **stop**(mh : Metaheuristic). Method that returns *true* if the metaheuristic has been running for more than *maxTime* or has looped through more than *maxIterations* iterations.
- **getMaxTime**(). Method to get the attribute *maxTime*.
- **getMaxIterations**(). Method to get the attribute *maxIterations*.
- **setMaxTime**(maxTime : double). Method to set the attribute *maxTime*.
- **setMaxIterations**(maxIterations : int). Method to set the attribute *maxIterations*.

3 Implementation: The *p*-Median Problem

The p-selection problems constitute a wide class of hard combinatorial optimization problems whose solutions consist of p items from a universe U. The standard moves for this class of problems are the interchange moves. An interchange move consists of replacing an item in the solution by another one out of the solution. A very representative p-selection problem is the p-median location problem whose standard version is explained below. The p-median problem is a well known location problem (see [26] or [4]) that have been proved $\mathcal{NP}$-hard [23]. This problem has often been used to test metaheuristics, among them parallel VNS [5] and Scatter Search [6].

Let $U = \{u_1, u_2, ..., u_n\}$ be the set of the locations of a finite set of users that are also potential locations for p facilities. Let D be the $n \times n$ matrix whose entries contain the distances $d_{ij} = D(u_i, u_j)$ between the points u_i and u_j, for $i, j = 1, ..., n$. The distance between a set of points $X \subset U$ and a point $u_i \in U$ is stated as follows:

$$D(X, u_i) = \min_{u_j \in X} D(u_j, u_i).$$

The cost function for a set of points X is the sum of the distances to all the points in U; i.e.,

$$f(X) = \sum_{u_i \in U} \min_{u_j \in X} D(u_j, u_i) = \sum_{u_i \in U} D(X, u_i).$$

The p medians selected from U constitute the set that minimizes this cost function. The optimization problem is then stated as follows:

$$\text{minimize} \sum_{u_i \in U} \min_{u_j \in X} D(u_i, u_j)$$

where $X \subseteq U$ and $|X| = p$.

Using a solution coding that provides an efficient way of implementing the moves and evaluating the solutions is essential for the success of any search method. A solution X can be represented by an array $x = [u_i : i = 1, ..., n]$ where u_i is the i-th element of the solution for $i = 1, 2, ..., p$, and the $(i-p)$-th element outside the solution for $i = p + 1, ..., n$. Let X_{ij} denote the solution obtained from X by interchanging u_i and u_j, for $i = 1, ..., p$ and $j = p+1, ..., n$.

For the p-selection problems, as the p-median problem, the local search procedure is based on the explained interchange moves. The usual greedy local search is implemented by choosing iteratively the best possible move among all interchange moves. The code of a local search that can be used in the *improveSolution* method of the *MhT_VNS* class is given in Figure 13. Here, the function *improved* tests if the new solution improves the current one or not. The *exchange* method is defined in Figure 14

In order to use the class hierarchy explained above to solve the p-median problem, we define the problem and solution objects for this problem. These classes inherit from their respective superclasses *Problem* and *Solution*, defined previously.

We call the problem class *PMedian_Problem* that is declared as follows:

PMedian_Problem

- **locations** : List
- **n** : int

− **p** : int

+ **getLocations()** : List
+ **getN()** : int
+ **getP()** : int
+ **setLocations(List : locations)** : void
+ **setN(n : int)** : void
+ **setP(p : int)** : void
+ **evaluate**(solution : Solution) : double

---

where *locations* is a list (vector, array, etc) of points, n is the size of that list, i.e., the number of possible locations, and p is the number of facilities we want to allocate. The methods are simply accessors to the attributes, except for the *evaluate* method, which calculates the sum of the distances of every point in the *locations* list to its nearest facility.

The solution class will be called *PMedian_Solution*, and will be defined as follows:

PMedian_Solution

− **facilities** : List

+ **getFacilities()** : List
+ **setFacilities(List : facilities)** : void

---

where the list *facilities* denotes the points where the facilities will be allocated. In this case, where the facility points are only allocated in the given location points, it is easier for all the functions to deal with the *PMedian_Solution* class if the *facilities* attribute contains all the possible locations, ordered in the way we mentioned above. That is, the first p elements of the list are the points selected for the facilities, and the final $n - p$ points are the discarded. The methods are simply accessors to the attributes.

3.1 VNS for the *p*-median

PMedian_VNS

---

initializeParameters() : void
generateInitialSolution() : void
shake() : void
improveSolution() : void

The basic VNS pseudocode 1 can be applied to any problem by providing the implementation of the initialization procedure, the shaking method, the local search and the function to test whether the solution is improved or not.

The *shake* procedure consists of, given the size k for the shake, choosing k times two points, u_i in the solution and u_j outside the solution at random, and performing the corresponding interchange move (see Figure 15).

The *improveSolution* method is implemented using the *Local Search* defined previously using the basic *Exchange* movement.

The *initializeParameters* method is void, although some initialization can be done here (for example, initializing a seed for a random number generation routine).

The *generateInitialSolution* method simply selects random points to be included in a solution, unless an initial solution is provided, in which case the initialization does nothing.

3.2 Scatter Search for the *p*-median

PMedian_SS
– **subsetI** : int – **subsetJ** : int
+ **getSubsetI()** : int + **getSubsetJ()** : int # **setSubsetI**(i : int) : void # **setSubsetJ**(j : int) : void # **initializeParameters()** : void # **generateInitialPopulation()** : void # **generateRefSet()** : void # **selectSubset()** : void # **combineSolutions(subset : Population)** : Population # **improveSolutions(subset : Population)** : Population # **updateRefSet()** : void

The Scatter Search for the *p*-median problem uses the standard parameter setting and rules explained above. The key idea to apply the scatter search principles to an optimization problem is the distance between solutions used to evaluate the dispersion among them. This distance for the *p*-median problem is defined using the same objective function. Let $f_Y(X)$ be the objective function for the set of users in Y:

$$f_Y(X) = \sum_{v \in Y} \min_{u \in X} Dist(u, v)$$

The distance between two solutions X and Y is given by $Dist(X, Y) = f_Y(X) + f_X(Y)$.

We now summarize an implementation of the components of the Scatter Search for the *p*-median problem, explaining the key methods mentioned above.

1. **generateInitialPopulation.** A simple way to generate a population consists in randomly creating solutions. We select p times a new point from U that is successively included in the set X. Given the previously fixed size $PopSize$ of random solutions, the population Pop is obtained by applying the local search to each random solution as the improvement method. See Figure 16.
2. **generateRefSet.** To generate a reference set from the population we first include in $RefSet$ the $RefSetSize_1$ best solutions. Then we iteratively include in the $RefSet$ the farthest solution from the solutions already in $RefSet$, repeating this procedure $RefSetSize_2$ times. We then obtain the reference set $RefSet$ with size $RefSetSize = RefSetSize_1 + RefSetSize_2$. The code of this method is given in Figure 17. The *getFurthermostSolution* method used in the code should return the furthermost solution of a given population from those in the reference set, using the *distance* concept defined above.
3. **selectSubset.** The selection of a subset to apply the combination consists in considering all the subsets of fixed size r (usually $r = 2$). Figure 18 contains the code of this method. This method maintains its main indexes as class attributes for allowing a *StopCriterion* class to determine if the subset generation loop has finished.
4. **combineSolutions.** The combination of each two solutions consists in the following. In the first place this method selects the points common to both solutions. Let X be the set of these points. For every point $u \in U \setminus X$ let
$$L(u) = \{v \in U : Dist(u, v) \leq \beta Dist_{max}\}$$
where
$$Dist_{max} = \max_{u,v \in U} Dist(u, v).$$
Choose the point $u^* \in U$ such that
$$Dist(X, u^*) = \max_{u \in U} Dist(X, u)$$
and select at random a point $v \in L(u^*)$ that is included in X. This step is iteratively applied until $|X| = p$. The code of this method is shown in Figure 19. There, the method calls several procedures like *getCommonPoints*, that should return a solution with the points that are part of all the solutions; *getDiffPoints*, that does exactly the opposite, returning the points that do not appear in all the solutions; *getFurthermostPoint*, that returns the point with the largest distance to a set of points, selected from a set of possibles; finally, *selectNearNeighbour* returns a random point from a set of possibles that are considered "near" a given one (here, the limit of "near" is controlled by the parameter β).
5. **improveSolutions.** Given a solution, the **improveSolutions** performs the local search on a population of solutions using the interchange moves.

6. **updateRefSet.** Let $ImpSolSet$ be the set of all the solutions reached by the **improveSolutions**. Apply **generateRefSet** to the set $RefSet \cup ImpSolSet$.

4 Conclusions

The ability of OOP to develop encapsulated, extensible software makes this paradigm one of the most suitable for programming metaheuristics. Most metaheuristics appearing in the literature share some common aspects in their design (a main loop, attributes like the number of iterations, etc), which makes them good candidates for establishing a class hierarchy.

The hierarchy we have proposed tries to differentiate between *point-based* and *population-based* metaheuristics, but maintaining at the same time an intuitive parallelism between their respective methods, which makes them easier to understand. We have also proposed an implementation of a metaheuristic of each class, *VNS* for *point-based*, and *Scatter Search* for *population-based*. Although it is a very reduced set of examples, having in mind the number of metaheuristics currently created, we hope that they give a significant insight of how other metaheuristics could be implemented with this approach. The purpose is to make the reader able to program new metaheuristics without getting stuck in programming details, just caring of coding the relevant parts of the algorithm.

Another property of OOP is the encapsulation and mobility it provides to objects. For example, as we mentioned in the *MhT_VNS* class, we can produce an object to perform a local search in the *improveSolution* method, but we could perfectly use a metaheuristic object to do that task. This example illustrates how *objects* may help improving versatility and usability in metaheuristic's software design.

The reader may have also noted that, although the code provided has a $C++$-like style, we have tried not to stick to a specific programming language, since our purpose was the design of the class hierarchy, which is code-independent. The classes and their respective attributes and methods where designed to use commonly extended features and data types, in order to allow portability between languages.

As an additional comment the hierarchy proposed in this work has been successfully implemented in practice in the context of the *Weka Project* ([34], [36]). *Weka* is a collection of machine learning algorithms oriented to data mining tasks, that is implemented in Java, and that provides a graphical interface for dealing with the algorithms it contains. The project defines the interface any *Data Mining* class should have to be able to interoperate with the Weka environment, and all the classes included in the project conforming to the interface are accessible to the final user. This environment allows a user to graphically interact with these algorithms, including metaheuristics. This illustrates the power of encapsulated, top-down design, allowing a problem-oriented group of classes (like *Metaheuristic*) to be transparently integrated in a graphical environment.

Acknowledgments

This research has been partially supported by the Spanish Ministry of Science and Technology through the project TIC2002-04242-C03-01; 70% of which are FEDER funds.

The research of the co-author M. García Torres has been partially supported by a CajaCanarias grant.

References

1. J. Belisle. OMG Standards for Object-Oriented Programming. AIXpert, pp. 38-42, Aug. 1993.
2. C. Blum and A. Roli. Metaheuristics in combinatorial optimization: Overview and conceptual comparison *ACM Computing Surveys*, 35-3:268-308, 2003.
3. M. Dorigo and T. Stutzle. The ant colony optimization metaheuristic: Algorithms applications, and advances. In F. Glover and G. Kochenberger, editors, *Handbook on MetaHeuristics*. 2003.
4. Z. Drezner. *Facility location: A survey of applications and methods*, Springer, 1995.
5. F. García-López, B. Melián-Batista, J.A. Moreno-Pérez, J.M. Moreno-Vega. The Parallel Variable Neighborhood Search for the p-Median Problem *Journal of Heuristics* 8 (2002) 375–388.
6. F. García-López, B. Melián-Batista, J.A. Moreno-Pérez, J.M. Moreno-Vega. Parallelization of the scatter search for the p-median problem. *Parallel Computing* 29 (2003) 575–589.
7. F. Glover. Heuristics for Integer Programming using Surrogate Constraints, *Decision Sciences* 8, (1977) 156–166.
8. F. Glover. Future paths for integer programming and links to artificial intelligence. *Computers and Operations Research*, 5:533–549, 1986.
9. F. Glover. Tabu Search for Nonlinear and Parametric Optimization (with Links to Genetic Algorithms). *Discrete Applied Mathematics* 49 (1994) 231–155.
10. F. Glover. A template for scatter search and path relinking. In J.-K. Hao and E. Lutton, editors, *Artificial Evolution*, volume 1363, pages 13–54. Springer-Verlag, 1998.
11. F. Glover Scatter Search and Path Relinking. in D. Corne, M. Dorigo, F. Glover (Eds.) *New Ideas in Optimisation*, Wiley, (1999).
12. F. Glover and G. Kochenberger (eds.). *Handbook of Metaheuristics*. Kluwer, 2003.
13. F. Glover and M. Laguna. *Tabu Search*. Kluwer, 1997.
14. F. Glover, M. Laguna, R. Martí. Fundamentals of Scatter Search and Path Relinking *Control and Cybernetics*, 39, (2000) 653-684.

15. F. Glover, M. Laguna, R. Martí. Scatter Search. in *Theory and Applications of Evolutionary Computation: Recent Trends*, A. Ghosh, S. Tsutsui (Eds.) Springer-Verlag, (2003).
16. P. Hansen, N. Mladenović. Variable Neighborhood Search for the p-Median. *Location Science* 5 (1997) 207–226.
17. P. Hansen and N. Mladenović. An Introduction to Variable Neighborhood Search. in: S. Voss et al. eds., Metaheuristics, Advances and Trends in Local Search Paradigms for Optimization (Kluwer, 1999) 433-458.
18. P. Hansen and N. Mladenović, Variable Neighborhood Search: Principles and applications, *European Journal of Operational Research* 130:449-467, 2001.
19. P. Hansen and N. Mladenović. Developments in variable neighbourhood search. In C. Ribeiro and P. Hansen, editors, *Essays and Surveys in Metaheuristics*, pages 415–439. 2002.
20. P. Hansen and N. Mladenović. Variable neighborhood search. In P.M. Pardalos and M.G.C. Resende, editors, *Handbook of Applied Optimization*, pages 221–234. Oxford University Press, 2002.
21. P. Hansen, N. Mladenović. Variable Neighborhood Search. In F. Glover and G. Kochenberger (eds.), *Handbook of Metaheuristics* Chapter 6, 2003.
22. D. Henderson, S.H. Jacobson and A.W. Johnson. Theory and Practice of Simulated Annealing. In F. Glover and G. Kochenberger, editors, *Handbook on MetaHeuristics*, chapter 10. 2003.
23. O. Kariv, S.L. Hakimi. An algorithmic approach to network location problems; part 2. The p-medians. *SIAM Journal on Applied Mathematics*, 37(1969), 539-560.
24. M. Laguna and R. Martí. *Scatter Search: Metodology and Implementations in C.* Kluwer Academic Press, (2003).
25. J.A. Lozano and P. Larrañaga. *Estimation of Distribution Algorithms. A New Tool for Evolutionary Computation.* Kluwer Academic, 2002.
26. P. Mirchandani and R. Francis, (eds.). *Discrete location theory.* Wiley-Interscience, 1990.
27. J.-Y. Potvin, K. Smith. Artificial Neural Networks for Combinatorial Optimization. In F. Glover and G. Kochenberger, editors, *Handbook on MetaHeuristics*, chapter 15. 2003.
28. C.R. Reeves. Genetic algorithms. In F. Glover and G. Kochenberger, editors, *Handbook on MetaHeuristics*, chapter 3. 2003.
29. C.R. Reeves. *Modern Heuristic Techniques for Combinatorial Problems.* Blackwell Scientific Press, 1993.
30. M. Resende, C. Ribeiro. Greedy Randomized Adaptive Search Procedures. In F. Glover and G. Kochenberger, editors, *Handbook on MetaHeuristics*, chapter 8. 2003.
31. C.C. Ribeiro and P. Hansen, editors. *Essays and Surveys in Metaheuristics*, volume 15. Kluwer, 2002.
32. E.A. Silver, R. Victor, V. Vidal, and D. De Werra. A tutorial on heuristic methods. *European Journal of Operational Research*, 5:153–162, 1980.
33. C. Voudouris and E.P.K. Tsang. Guided local search. In F. Glover and G. Kochenberger, editors, *Handbook on MetaHeuristics*, chapter 7. 2003.
34. Weka project webpage.
http://www.cs.waikato.ac.nz/ ml/weka/
35. Wikipedia on-line definition for OOP.
http://en.wikipedia.org/wiki/Object-oriented_programming

36. I.H. Witten, E. Frank Data Mining: Practical Machine Learning Tools and Techniques with Java Implementations Morgan Kaufmann, 1999
37. M. Yagiura and T. Ibaraki. On metaheuristic algorithms for combinatorial optimization problems. *Systems and Computers in Japan*, 32(3):33–55, 2001.
38. M. Yagiura and T. Ibaraki. Local search. In P.M. Pardalos and M.G.C. Resende, editors, *Handbook of Applied Optimization*, pages 104–123. Oxford University Press, 2002.
39. S.H. Zanakis, J.R. Evans, and A.A. Vazacopoulos. Heuristic methods and applications: a categorized survey. *European Journal of Operational Research*, 43:88–110, 1989.

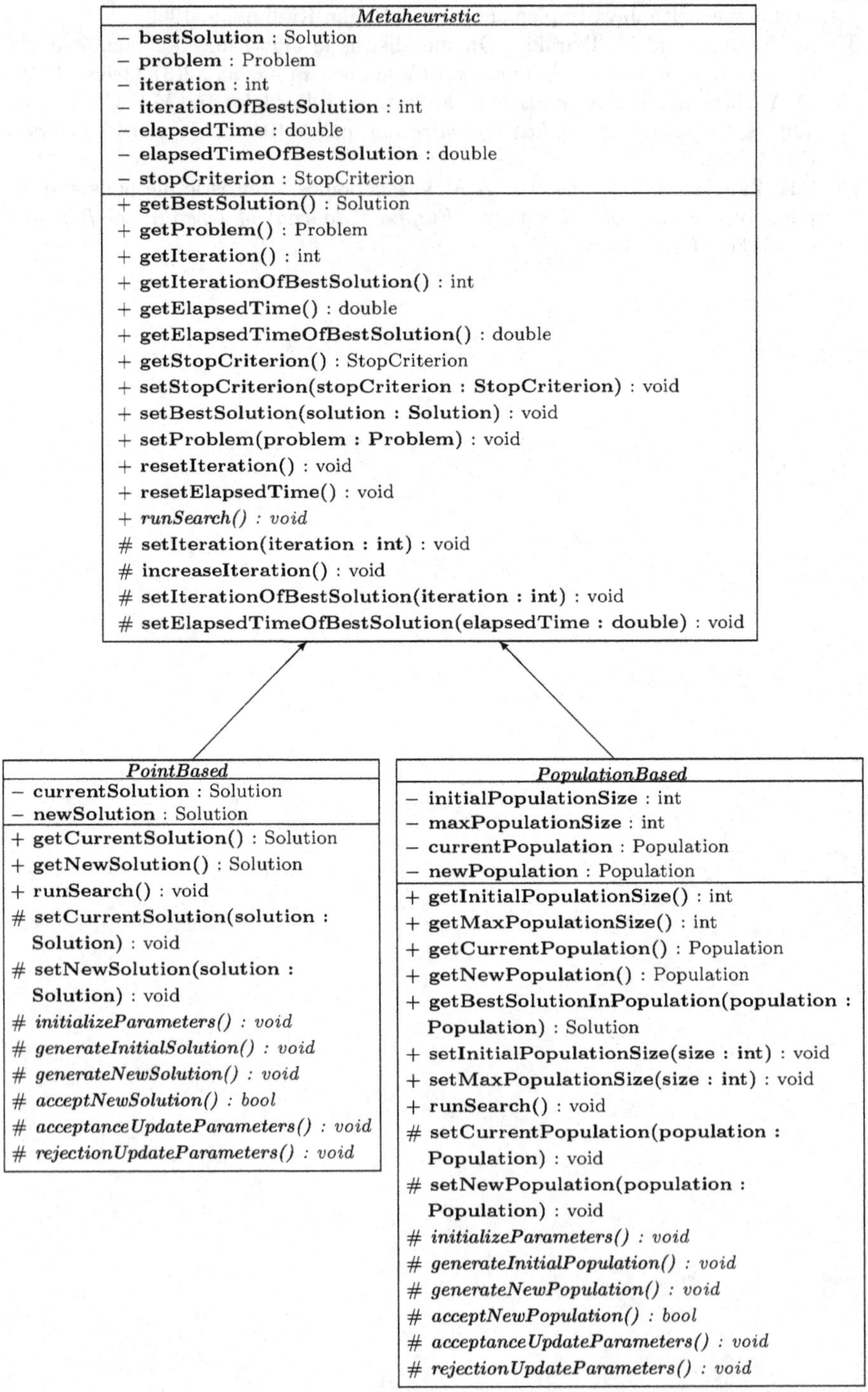

Fig. 3. Class Hierarchy for the proposed metaheuristics

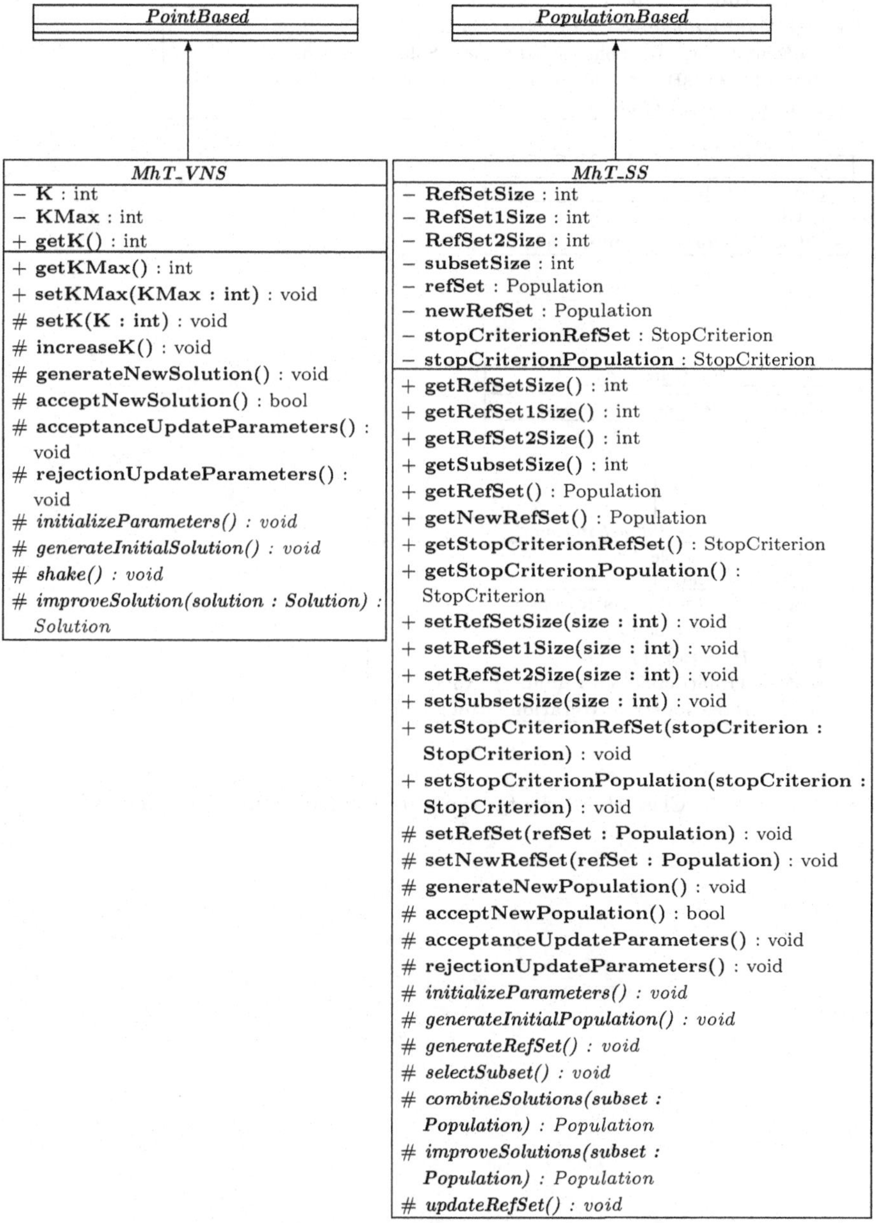

Fig. 4. Class Hierarchy for the proposed metaheuristics (cont.)

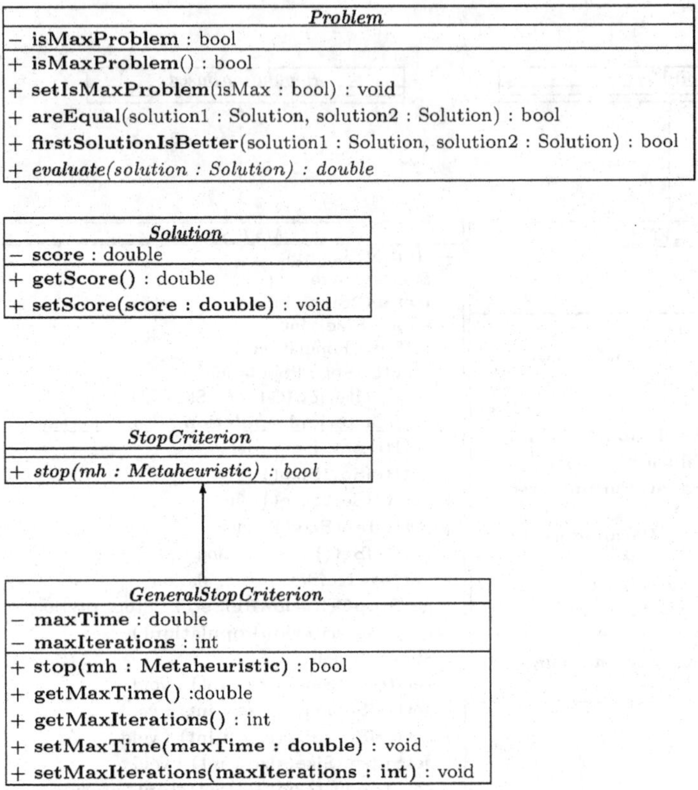

Fig. 5. Class Hierarchy for the proposed metaheuristics (cont.)

PointBased::runSearch

```
void PointBased::runSearch()
{
resetEllapsedTime();
resetIterations();
generateInitialSolution();
do {
    generateNewSolution();
    if (acceptNewSolution()) {
        acceptanceUpdateParameters();
        setCurrentSolution (getNewSolution());
        if (getProblem().firstSolutionIsBetter
            (getCurrentSolution(),getBestSolution()))
        {
            setTimeOfBestSolution(time());
            setIterationOfBestSolution(getIteration());
            setBestSolution(getCurrentSolution());
        }
    } else {
        rejectionUpdateParameters();
    }
    increaseIteration();
} while (!this.getStopCriterion().stop());
}
```

Fig. 6. PointBased::runSearch code

PopulationBased::runSearch

```
void PopulationBased::runSearch()
{
Solution bestInPopulation;

resetEllapsedTime();
resetIterations();
generateInitialPopulation();
do {
    generateNewPopulation();
    if (acceptNewPopulation()) {
        acceptanceUpdateParameters();
        setCurrentPopulation (getNewPopulation());
        bestInPopulation = getBestSolutionInPopulation
            (getCurrentPopulation());
        if (getProblem().firstSolutionIsBetter
            (bestInPopulation, getBestSolution()))
        {
            setTimeOfBestSolution(time());
            setIterationOfBestSolution(getIteration());
            setBestSolution(bestInPopulation);
        }
    } else {
        rejectionUpdateParameters();
    }
    increaseIteration();
} while (!this.getStopCriterion().stop());
}
```

Fig. 7. PopulationBased::runSearch code

MhT_VNS::generateNewSolution

```
void MhT_VNS::generateNewSolution()
{
    shake();
    improveSolution();
}
```

Fig. 8. MhT_VNS::generateNewSolution code

MhT_VNS::acceptNewSolution

```
void MhT_VNS::acceptNewSolution()
{
    return getProblem().firstSolutionIsBetter
        (getNewSolution(), getCurrentSolution());
}
```

Fig. 9. MhT_VNS::acceptNewSolution code

MhT_VNS::acceptanceUpdateParameters

```
void MhT_VNS::acceptanceUpdateParameters()
{
    setK( 1 );
}
```

Fig. 10. MhT_VNS::acceptanceUpdateParameters code

MhT_VNS::rejectionUpdateParameters

```
void MhT_VNS::rejectionUpdateParameters()
{
    increaseK();
}
```

Fig. 11. MhT_VNS::rejectionUpdateParameters code

MhT_SS::generateNewPopulation

```
void MhT_SS::generateNewPopulation()
{
    Population subset;
    do {
        setNewRefSet(new Population());
        do {
            subset = selectSubset();
            subset = combineSolutions(subset);
            subset = improveSolutions(subset);
            setNewRefSet(getNewRefSet().add(subset));
        } while(!getStopCriterionRefSet().stop());
        updateRefSet();
    } while(!getStopCriterionPopulation().stop());
}
```

Fig. 12. MhT_SS::generateNewPopulation code

Local Search

```
void local_search(sol cur_sol)
{
 init_sol = cur_sol ;
 while improved(cur_sol,init_sol))) {
 for (i=p;i<n;i++)
   for (j=0;j<p;j++) {
      exchange(new_sol,cur_sol,i,j) ;
      if improved(new_sol,cur_sol)
        cur_sol = new_sol
      } /* for */
 } /* while */
} /* local_search */
```

Fig. 13. Local Search Pseudocode

Exchange

```
void exchange(Solution new_sol, Solution cur_sol,
              int i, int j)
{
 facilities = cur_sol.getFacilities();
 aux = facilities[i] = facilities[j];
 facilities[i] = facilities[j];
 facilities[j] = aux;
 new_sol.setFacilities(facilities);
}
```

Fig. 14. Exchange Pseudocode

PMedian_VNS::shake

```
void PMedian_VNS::shake(sol cur_sol)
{
 init_sol = cur_sol ;
 for (r=0;r<k;r++) {
   i = rnd % p ;
   j = p + rnd % (n-p) ;
   exchange(cur_sol,new_sol,i,j) ;
   cur_sol = new_sol ;
 } /* for */
} /* shake */
```

Fig. 15. Shake Pseudocode

PMedian_SS::generateInitialPopulation

```
void PMedian_SS::generateInitialPopulation()
{
 for (i=0;i<getInitialPopulationSize();i++) {
   cur_sol = generateRndSolution();
   cur_sol = improveSolution(curr_sol);
   getCurrentPopulation().add(cur_sol);
 }
}
```

Fig. 16. generateInitialPopulation code

PMedian_SS::generateRefSet

```
void PMedian_SS::generateRefSet()
{
 // evaluate the solutions
 for (i=0;i<getCurrentPopulation().size();i++) {
   cur_sol = (curr_sol);
   getProblem().evaluate(getCurrentPopulation().get(i));
 }

 // order the solutions by their score
 sort(getCurrentPopulation());

 // add refSetSize1 best solutions to the refSet
 for (i=0;i<getRefSetSize1();i++) {
   getRefSet().add(getCurrentPopulation().get(i));
 }

 // add refSetSize2 distant solutions to the refSet
 for (i=0;i<getRefSetSize2();i++) {
   cur_sol = getFurthermostSol(population,refSet);
   refSet.add(cur_sol);
 }
}
```

Fig. 17. generateRefSet code

PMedian_SS::selectSubset

```
Population PMedian_SS::selectSubset()
{
 i = getSubsetI();
 j = getSubsetJ();
 if (NaN(i) || NaN(j)) {
   i = 0;
   j = 1;
 } else if (j == getRefSetSize()-1) {
   i++;
   j = i + 1;
 }
 setSubsetI(i);
 setSubsetJ(j);
 return list(getCurrentPopulation.get(i),
     (getCurrentPopulation.get(j));
}
```

Fig. 18. selectSubset code

PMedian_SS::combineSolutions

```
Population PMedian_SS::combineSolutions(Population solutions)
{
 newSolution = getCommonPoints(solutions);
 possibleSolutions = getDiffPoints(solutions);
 while (newSolution.getFacilities().size() <
        problem.getP()) {
   point = getFurthermostPoint(newSolution,
       possibleSolutions);
   point = selectNearNeighbour(possibleSolutions,
       point,beta);
   newSolution.getFacilities().add(point);
 }
 return list(newSolution);
}
```

Fig. 19. combineSolutions code

ooMILP – A C++ Callable Object-oriented Library and the Implementation of its Parallel Version using CORBA

Panagiotis Tsiakis and Benjamin Keeping

Process Systems Enterprise Ltd, 107a Hammersmith Bridge Rd, London W6 9DA, United Kingdom {p.tsiakis,b.keeping}@psenterprise.com

Summary. Process systems engineering is one of the many areas in which mixed integer optimisation formulations have been successfully applied. The nature of the problems requires specialised solution strategies and computer packages or callable libraries able to be extended and modified in order to accommodate new solution techniques. Object-oriented programming languages have been identified to offer these features. Process system applications are usually of large scale, and require modelling and solution techniques with high level of customisation. *ooMILP* is a library of C++ callable procedures for the definition, manipulation and solution of large, sparse mixed integer linear programming (MILP) problems without the disadvantages of many existing modelling languages. We first present a general approach to the packaging of numerical solvers as software components, derived from material developed for the CAPE-OPEN project. The presentation is in the context of construction and solution of Mixed Integer Linear Programming (MILP) problems. We then demonstrate how this package, based on the use of CORBA interfaces for synchronous execution within a single process, can be adapted with a minimum of problem-specific changes to provide a distributed solution.

Key words: Object-oriented programming, callable library, optimization, parallel computing, interface, branch-and-bound

1 Introduction

A wide variety of engineering, industrial, and business applications are formulated as Mixed Integer Linear Programming (MILP) problems. The high utilisation of MILPs in all these areas requires flexible integer optimisation techniques and supporting software, able to be customised for different applications. These features are provided by object-oriented languages. These languages are suitable for creating reusable software in an error-free manner.

A modelling language is expected to provide the ability to write a model of the formulation in a manner that can be manipulated in many ways. Currently, given the model written in the modelling language as input, the output is typically a standard format representation, readable by any system. A mixed integer programming system is defined a program that reads in a mixed integer standard form, creates an internal representation of the problem, and then typically attempts to solve it using one of the techniques available to the system. An optimisation technique is required to have three main properties:

1. to be applicable to a wide range of problems,
2. to have efficient implementation in order to handle large-scale problems in reasonable small times,
3. easy formulation of a model.

Available in the market at the moment there are four well-known languages, namely AMPL, GAMS, LINGO and MP-XPRESS, and three major MILP systems-solvers, which are CPLEX, OSL and XPRESS. The latter also have optimisation subroutine libraries. Additionally to the commercial solvers there is a number of available optimisation subroutine libraries but with limited capabilities regarding the size of the problem they can handle.

These modelling languages have the advantages of being generic, interfaced to most of the well-known solvers and are extensively used in industry and academia. But they have their own disadvantages such as limited ability to implement new solution algorithms, difficulty to embed within other applications since they have file-based communication, plus they do not offer the generation of MILPs "on the fly", or they cannot handle more than one MILP simultaneously either in sequence or in parallel while they are within a solving routine. On the other hand, callable libraries can be embedded within other software but are difficult to use for building complex formulations, and most likely they cannot support the solution of more than one MILP.

The lack of computer packages or callable libraries that support more than one solution technique and allow the modification of an existing code or the implementation of a new solution technique, and can handle more than one MILP simultaneously, provided one of the motives for the development of this optimisation package. Another prohibiting factor to the use of these systems is the cost of purchasing and maintaining.

Over recent years parallel-distributed computing environments are developing very rapidly to tackle large problem instances such as occur in real industrial use. With the application of optimization becoming more common within the industry the need for powerful numerical packages has increased. However the use of the optimisation routines (contained in those packages) within their own modelling environments has been very restricted by the lack of standardised interfaces and monolithic view of the current modelling packages.

In addition, parallel distributed computing applications require flexible software tools that allow easy implementation of complex formulations and algorithms.

Identifying the need for common interfaces between software vendors and the necessity of open architecture systems, the CAPE-OPEN MINLP interfaces were proposed, making it possible to access seamlessly, from a process modelling environment, the latest optimisation algorithms available. This will provide more robust and more efficient numerical tools to the end-users. For optimisation algorithm developers, implementing the CAPE-OPEN MINLP interface specification will give them access to a larger market where their tools will be put to use immediately [6].

Global CAPE-OPEN (GCO) has developed a standard and unified software interface for numerical solvers for both Mixed Integer Linear Programs (MILPs) and Mixed Integer Nonlinear Programs (MINLPs). The interfaces proposed take into account matters such as:

1. easy definition of complex MILP formulations
2. unified access to variety of commercial MILP solvers
3. implementation and testing of new MILP solution algorithms on both serial and parallel computer architectures.
4. building software *components* within larger software systems (e.g. for process scheduling, supply chain design, etc.)
5. allowing *multiple* instances of MILP solver simultaneously within the *same* application.

Prior to the specification of Global CAPE-OPEN standards, a number of integer optimisation tools were developed using various implementation techniques and employing different solution methods. MINTO [9] is a software system for the implementation and solution of mixed integer linear programming problems using a branch-and-price algorithm. ABACUS [7] uses an object-oriented framework for the construction and solution of MILP problems using branch-and-cut and branch-and-price algorithms. bc-opt [1] uses branch-and-cut as the main method to optimise MILP problems.

The remainder of this chapter is organized into sections leading to the final implementation of a parallel algorithm. Section 2 gives an overview of the package developed. Our initial effort was focused on producing a system that would allow capitalising on the architecture benefits as they were defined by Global CAPE-OPEN, and build a callable library that can be easy used within complex solution schemes. Section 3 describes the C++ objects as they were developed to provide a modeling environment to mathematically describe MILP problems and its serial pre-CORBA implementation. The implementation of CORBA interfaces in order to provide transparency and remote access to the routines dynamically under any operating system and use distributed computing power is described in Section 4. Section 5 presents a decomposition algorithm suitable for almost any category of MILP problems making use of solving in parallel many smaller problem instances. Section 6 continues with the parallel architecture of the system as it was applied to incorporate the above algorithm. Conclusions and ideas for further work are summarized in Section 7.

2 *ooMILP* Overview

MILPs are optimisation problems with a linear objective function and linear equality and inequality constraints. The variables appearing in the objective function and constraints are generally restricted to lie between specified lower and upper bounds. Furthermore, some of these variables may be restricted to integer values only. The aim of the optimisation is to determine values of the variables that minimise or maximise the objective function while satisfying the constraints and all other restrictions imposed on them.

Many important process engineering problems can be formulated directly as MILPs. These problems include:

1. supply chain optimisation;
2. process planning and scheduling;
3. distribution and transportation planning and scheduling;
4. heat exchanger network synthesis.

Moreover, MILPs appear as important sub-problems in the solution of mixed integer nonlinear programming problems (MINLPs), i.e. optimisation problems with nonlinear objective functions and/or constraints, and both continuous and integer variables.

A simple and fairly standard mathematical description of an MILP can be written as:

$$\begin{aligned}\phi = \min\ & a^{\top}x + b^{\top}y \\ & Ax + By \leq c \\ & x^{l} \leq x \leq x^{u} \\ & y \in \mathbb{Z}^{n},\end{aligned}$$

where $x, x^l, x^u, a, b \in \mathbb{R}^n$, $c \in \mathbb{R}^m$, and A, B are an $m \times n$ real matrices. The vectors x, y represent the unknowns; all other vectors and the matrices A, B are known constants.

We note that:

1. the variables x are characterised by an index $i = 1, \ldots n$ and are bounded between given lower and upper bounds x^l and x^u respectively;
2. the variables y are restricted to take integer values;
3. all constraints are expressed as inequalities of the form $\leq c$ and are indexed over the discrete domain $1, \ldots, m$.

Albeit quite general, the above MILP form is not necessarily easy to construct and/or manipulate. A major reason for this is that the variables and constraints are maintained as unstructured lists or arrays which may contain thousands of elements. On the other hand, most mathematical formulations of practical problems in terms of MILPs are expressed in terms of a relatively small number of distinct sets of variables and constraints.

ooMILP is a library of C++ callable procedures for the definition, manipulation and solution of large, sparse mixed integer linear programming (MILP) problems. In particular, *ooMILP*:

1. facilitates the definition of complex sets of constraints, reducing the required programming effort to a minimum;
2. allows its client programs to create and manipulate simultaneously more than one MILP;
3. decouples the problem to be solved from the solver to be used;
4. provides a common interface to diverse MILP solvers which allows the latter to be used without any changes to client programs (cf. CAPE-OPEN standards for numerical solvers).

3 C++ objects and pre-CORBA serial implementation

The nature of MILP problems forces the use or the combination of more than one technique in solving a problem. The object oriented programming approach offers two major advantages over conventional languages:

1. The functionality of a code can be extended without the availability of the source code.
2. Even if the source code is available for use in conventional languages, it is usually very difficult to extend.

Based on the CAPE-OPEN standards as they were proposed [6] for the interface of Mixed Integer Linear Programming problems our initial implementation involved four C++ objects, namely the MILP, the FlatMILP, the MILPSolverManager and MILPSystem.

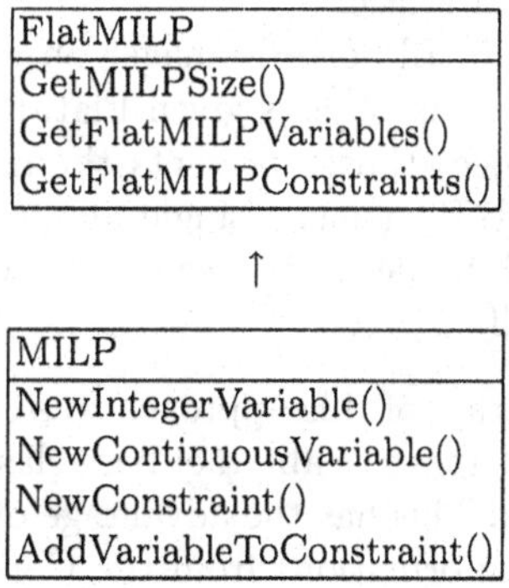

Fig. 1. MILP inheritance structure.

1. MILP provides methods for construction of MILP problems using multi-dimensional variables and constraints. A separate interface is provided for access to the problem by the solver, which treats the problem in a 'flat'

manner (conversion between these two views is the main challenge in implementing the construction tool). For this reason an object hierarchy is used, as shown in Fig. 1.
To illustrate the ease of use of this system, the mathematical form of a typical multidimensional constraint will be as follows. For example, in a process scheduling problem we can use the Resource Task Network (RTN) formulation proposed by Pantelides (1994). The formulation seeks to optimise a process involving NR resources ($r = 1, \ldots, NR$) and NK tasks ($k = 1, \ldots, NK$) over a time horizon discretised into NT time intervals ($t = 1, \ldots, NT$).

$$[RTN] : \max \sum C_r^F \left(R_{r,NT} - R_{r0}\right)$$

subject to:

$$R_{rt} = R_{r,t-1} + \sum_{k \in K_r} \sum_{\theta=0}^{\tau_k} \left(\mu_{kr\theta} N_{k,t-\theta} + \upsilon_{kr\theta} \xi_{k,t-\theta}\right) \quad \forall r, t = 1, \ldots, NT+1$$

$$0 \leq R_{rt} \leq R_r^{\max} \quad \forall r, t = 1, \ldots, NT+1$$

$$V_{kr}^{\min} N_{kt} \leq \xi_{kt} \leq V_{kr}^{\max} N_{kt} \quad \forall k, t = 1, \ldots, NT+1.$$

The corresponding calls to set up the first constraint in $oo\mathcal{MILP}$ are given in Figure 2.
2. The base class FlatMILP is the **only** view of the object required by a normal MILP solver. However, we will later present a parallel solver that requires the ability to construct MILP subproblems, and will thus act as a client of the MILP interface.
3. The main role of the MILPSolverManager object is as a 'factory' which produces the MILPSystem objects given FlatMILP objects.
4. The MILPSystem object thus represents the association of a particular solver and a given MILP. It has a single method, Solve, which returns a success condition and, if successful, leaves the solution to the problem in the MILP object itself.

An initial implementation of this approach was carried out by implementing the interfaces outlined above as abstract base classes, with implementation classes derived from them. This has the advantage of simplifying (and removing possibly proprietary information) from the C++ header file required by a client of the software.[1] The resulting inheritance hierarchy is illustrated in Figure 3.

Thus if the abstract versions of the MILP and FlatMILP are defined in milp.h, and MILPSolverManager and MILPSystem in milpsolver.h, these will be the only headers required by client software using the package.

[1] We assume here that the software is to be used by implementing C++ client software rather than through a more high level tool such as one requiring a declarative language, although such tools could clearly be constructed as clients.

```
for(int r=1 ; r<=NR ; r++) {
  for(int t=1 ; t<=NT ; t++) {
    RTNMilp->AddVariableSliceToConstraintSlice("R",
             Intseq(r,t), IntSeq(r,t), "ResourceBalance",
             IntSeq(r,t),IntSeq(r,t), -1.0);
    RTNMilp->AddVariableSliceToConstraintSlice("R",
             Intseq(r,t-1),IntSeq(r,t-1),"ResourceBalance",
             IntSeq(r,t),IntSeq(r,t), 1.0);
    for(int k=1 ; k<=NK ; k++){
      for(int theta=0 ; theta<=min(tau(k),t-1) ; theta++) {
        RTNMilp->AddVariableSliceToConstraintSlice("N",
                 Intseq(k,t-theta), IntSeq(k,t-theta),
                 "ResourceBalance",IntSeq(r,t),IntSeq(r,t),
                 mu(k,r,theta));
        RTNMilp->AddVariableSliceToConstraintSlice("Xi",
                 Intseq(r,t-theta), IntSeq(r,t-theta),
                 "ResourceBalance",IntSeq(r,t),IntSeq(r,t),
                 nu(k,r,theta));
      }
    }
  }
}
```

Fig. 2. Setting up the first constraint in the RTN example.

3.1 Creation of MILP and MILPSolverManager objects

The MILP and the MILPSolverManager as the main objects require to be defined using the global functions,

```
MILP* newMILP();
```

and

```
MILPSolverManager* newMILPSolverManager(string name);
```

respectively.

Declarations of these functions occur in milp.h and milpsolver.h respectively, but their definitions are implementation specific.

The implementations then simply create the implementation object and return the base class pointer to be used:

```
#include "milp_impl.h"
MILP* newMILP() {
  return new MILP_i();
}
```

Note that the newMILPSolverManager function takes a string argument, intended to specify the required solver implementation.

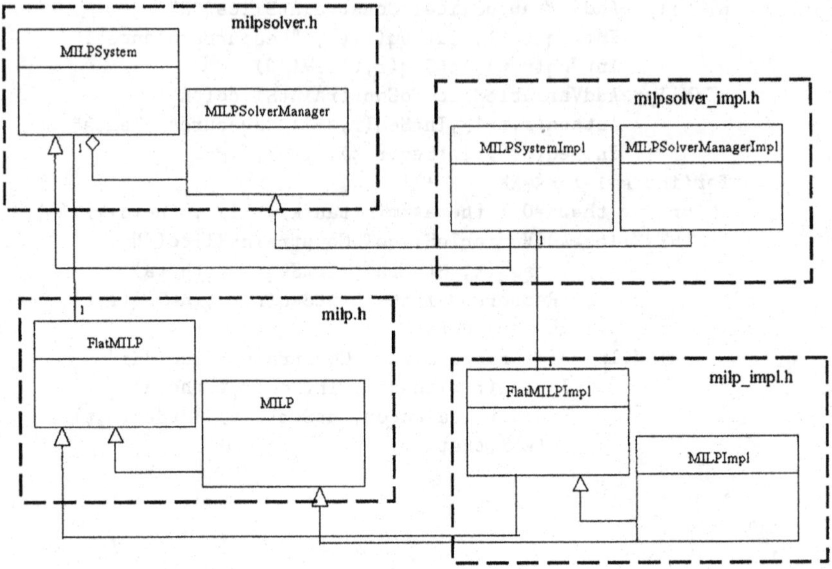

Fig. 3. Total inheritance and structure of the *ooMILP* objects.

```
#include "milpsolver.h"
MILPSolverManager* newMILPSolverManager(string name) {
  return new milpsolvermanager_i(name);
}
```

For our non-CORBA version this had no effect because the solvers had to be loaded in advance and made available — however, we will see that our CORBA-based approach provides a quite natural way of using this name to identify certain solver properties.

3.2 Associated methods and auxiliary objects

As we mentioned in section 2 an MILP comprises a set of linear constraints and a linear objective function, consisting of variable and constraint occurrences. Additionally, the problem data needs to be accessed in various manners, with the problem in structured or flat form, therefore different sets of methods are provided. Variables, constraints and objective function have their own characteristics in this MILP object which are described below and their creation is part of the MILP definition problem.

Methods

Each MILP object offers a number of methods allowing its construction and modification.

1. MILP instantiation method that creates an empty object.
2. MILP construction methods allowing the creation of variables, constraints, objective function and their instances.
3. MILP modification methods allowing to modify existing instances.
4. Structured MILP information access methods allowing to receive information from the structured problem.
5. MILP flat information access and update methods allowing to access the matrix form of the problem and update it.

Variables and Constraints

An MILP is characterised by a number of distinct sets of variables and constraints objects. Each of these sets is either a scalar or an array of an arbitrary number of dimensions.

A variable set (object) is characterised by the following information:

1. the name of the variable set;
2. the type of all variables in this set (continuous or integer);
3. the number of dimensions (may be zero for a scalar variable);
4. the size of each dimension of the array;
5. the upper and lower bounds of each element of the set;
6. the current value of each element of the set.

A constraint set (object) is characterised by the following information:

1. the name of the constraint set;
2. the type of all constraints in this set (equality or inequality);
3. the number of dimensions (may be zero for a scalar constraint);
4. the size of each dimension of the array;
5. the right hand side constant;
6. the variables occurring in these constraints and the corresponding coefficients.

4 Initial CORBA Version

In this section, we will explain the steps taken to package the preceding design into a set of CORBA interfaces. Accomplishing this offers two advantages:

1. **Local/remote transparency** — as we shall see, the ability to use our interfaces either within a single process or across networks is highly valuable.

2. **Componentisation** — provided the same ORB is used for both, a solver component can be provided very cleanly in the form of a dynamically loaded software element (i.e. DLL in a windows context, or shared object in Unix) for execution within the same process as its client.

We will first explore this question of packaging further, before considering how to minimise the impact of CORBA on our code.

4.1 The package object

Additionally to wrapping the original objects discussed in the first section, it proves convenient to provide a package object, so that the service of MILP construction and solution can be provided through a single object reference. The interface is simply:

```
interface IMILPPackage {
  IMILPFactory GetMILPFactory();
  IMILPSolver GetSolver(string name);
};
```

with a corresponding IMILPFactory interface:

```
interface IMILPFactory {
  IMILP CreateMILP();
};
```

Note the string argument to GetSolver, which specifies the implementation to be used. This is likely to imply an attempt to dynamically load the solver software based on this name. For example, if the IMILPPackage is implemented by a server on a UNIX machine, the effect of a call with the string "CPLEX" might be:

1. Load a shared object CPLEX.so into the same process as the server.
2. Call a routine with a standard name within that shared object, which:
   - Creates a CORBA object with the IMILPSolverManager interface
   - Returns its object reference
3. Return the object reference thus created to the client of IMILPPackage.

Since no corresponding need to dynamically select MILP construction tools has yet been identified, there is no such argument to GetMILPFactory.

The overall implementation of the package might be:

1. in the form of a standalone executable, which upon execution generates an object reference which can be given to the client software, e.g. through the CORBA name service.
2. another dynamically loadable object, designed to execute in the client's own process, used through the same mechanism described previously for the individual solvers.

3. directly linked into the client code, since, as with the MILPFactory, there is not currently a need to permit selection between multiple package implementations.

It should be noted that the choice between these three approaches requires only the most minimal differences in implementation code. Essentially, the first requires a main program that creates the object and then serves it in a blocking way, while the other two require an 'init' function which creates the object in the same way but serves it in a non-blocking way and returns its reference. The implementation of both of these is very simple and generic – to the extent that templates can be used to remove the necessity of re-implementing them for every interface.

The first approach listed, i.e. the standalone server, is the most applicable for parallel execution on conventional networked machines, and is discussed further in section 6 of this document.

4.2 Localising CORBA usage

This approach described would give a workable system: however, it is desirable to localise the use of CORBA itself as much as possible within the code for two main reasons:

1. considerable work was done in implementing both MILP construction and solution prior to the introduction of the CORBA approach,
2. this is generally a good principle – the opposite extreme would be to permeate the code with types specific to the CORBA interfaces.

In this section, we will explain in more detail how it was possible for both:

1. client software written to use the original milp.h/milpsolver.h approach, and
2. the implementation side of the interface based on milp_impl.h to remain essentially unchanged.

The first of these was achieved by defining an alternative implementation of our original abstract MILP interface which passes each call to its CORBA equivalent. In particular, new MILP objects are created from an object reference to an existing CORBA MILP (possibly served by a remote process). Thus the newMILP method is implemented by:

1. Obtaining an IMILPPackage reference (e.g. from a file)
2. Calling its GetMILPFactory method
3. Calling the resulting interface's CreateMILP method.
4. Creating a new MILP_c (for CORBA) object from the reference, and returning a pointer to this **local** object.

Similarly, newMILPSolverManager is implemented by:

1. Obtaining the IMILPPackage reference in the same way,

2. Passing the string argument – which thus now has a definite meaning – to the GetSolver method
3. Creating a local MILPSolverManager object.

The interactions and the philosophy are illustrated in Fig. 4. Here we show a MILPSolverManager component (1) wrapping the CPLEX subroutine libraries being used by a client (2) which was written to work with the original (non-CORBA) *ooMILP* interface.

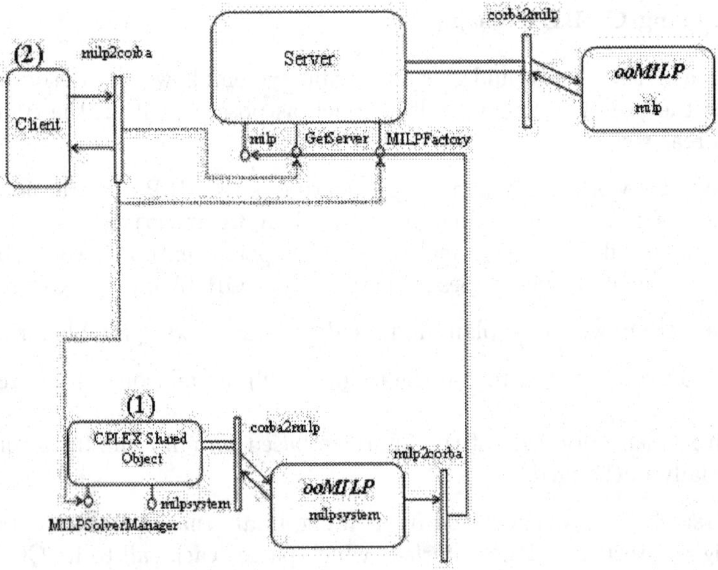

Fig. 4. CORBA wrapping of the initial C++ implementation.

Implementation of the MILPSolverManager's CreateMILPSystem method involves a little subtlety.

The method's argument is a pointer to an abstract MILP object, but in order to use the underlying IMILPSolverManager's CreateMILPSystem method we require an IFlatMILP CORBA object reference. This can be obtained by dynamically down-casting the abstract MILP pointer to the underlying MILP_c object, and thence obtaining the CORBA reference with which this was created. (The MILP_c must therefore make this available through a public method).

The header defining this mapping (for both MILP etc and MILPSolverManager etc.) was named milp2corba.h.

The second part of the mapping process, ensuring that the implementation of MILP construction and solution can remain unchanged, was achieved straightforwardly by wrapping the 'native' interface with a CORBA one: that is, the interfaces were first re-specified in standard CORBA IDL, e.g.:

```
interface IFlatMILP {
   ...
};
```

These IDL files were then compiled into headers and skeleton code. The former is included into both the client and server code, and is used as a base class implementing the server, in much the same way that the original MILP etc were used.

The wrapping can be carried out either through membership:

```
class IFlatMILP_i: public IFlatMILP {
   ...
 private:
  FlatMILP* origflatMILP;
};
```

or **implementation** inheritance:

```
class IFlatMILP_i: public IFlatMILP, private FlatMILP {
   ...
};
```

For consistency with the other wrapping style described earlier, we adopted a membership approach. Note that in either case only the abstract FlatMILP interface is needed. The header file defining this wrapping was designated corba2milp.h.

4.3 Summary

To recap, our initial CORBA implementation provides the capability for client software, given an object reference to an IMILPPackage object, to:

1. Create a MILP object and populate it with variables (both continuous and integer) and constraints involving occurrences of these variables.
2. Get a MILPSolverManager object reference for its choice of solver.
3. Create a MILPSystem object representing the conjunction of this solver with the given problem.
4. Solve the problem — or ask the system to 'solve itself'.
5. Extract and display/analyse the given solution through the original MILP's interface.
6. Make changes — including structural changes — to the MILP if desired, and repeat from step 4.

In the next section we will briefly explain why the need for solver objects to themselves create a number of MILP objects in parallel can arise, before concluding with a demonstration of how cleanly this can be accomplished with our CORBA-based approach.

5 Partially decomposable MILPs

MILPs belong to the category of *combinatorial optimisation* problems since they include decision variables of a discrete nature, in the objective function and/or in some of the problem constraints.

Solution techniques for such problems can be distinguished between *heuristic* and *complete enumeration.* Heuristic techniques seek good solutions at a reasonable computational time but they cannot guarantee either feasibility or optimality; in many cases, they do not even provide a measure of how close to optimality a feasible solution is [10].On the other hand, there are other methods of combinatorial optimisation which are *exact.* These are techniques based on theories of linear programming or graphs, or else use an implicit enumeration approach such a branch-and-bound. Despite recent advances and successes in computational power there are problem instances that cannot be tackled with current solution methods.

In the decomposition methods, the mathematical structure of the models is exploited via variable partitioning, duality theory, and relaxation methods.

Decomposition techniques are applied to cases where strong upper or lower bounds are required in order to accelerate the solution procedure. Such decomposition techniques are:

- Benders Decomposition [4]
- Lagrangian Relaxation [5, 3]
- Cross Decomposition [14]

All methods have extensively been used to solve mixed integer linear programming problems in many disciplines and especially engineering. A drawback of the methods is that their application is problem dependent prohibiting their generalization.

It is not our purpose here to present the theory of decomposable MILPs and the decomposition theory in detail [12, 2]. For our purposes, it is sufficient to note that most (if not all) MILP problems can potentially be decomposed into a number of sub-problems, such that only variables x_p and y_p and constraints C_p occur in sub-problem p ($p = 1, \ldots, NP$), together with the so-called "key" variables x_0, y_0, which occur in all the sub-problems. This structure appears in problems with block diagonal or diagonal constraint structure (see Fig. 5).

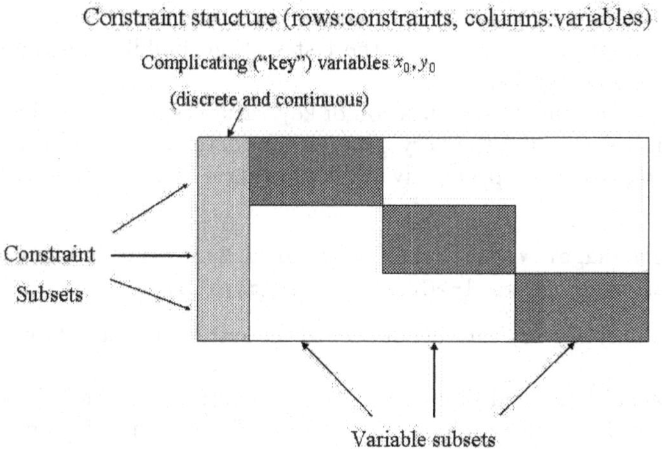

Fig. 5. Constraint structure for decomposable MILPs.

This structure allows one to divide the variables and constraints into disjoint sets (with any finite NP) and they involve variables x_p, y_poccurring in constraints C_p and belong to partition$p$. Taking advantage of this property and the existence of the variables x_0, y_0 and constraints C_0 that belong to all partitions we develop a branch-and-bound method which needs less time to find a solution. However, for the decomposition to be of practical use in problem solution, we must identify a small number of (probably integer) variables x_0, y_0which permit the remaining variables to be partitioned into NP sets, the largest of which is significantly smaller than the original problem. Fortunately, this proves possible for a large class of problems. Of particular significance for this paper is the fact that the selection of key variables must currently be explicitly specified to the solver, although in principle its determination could be automated using a partitioning algorithm.

Thus we envisage a specialised solver which will receive additional information from the client software identifying the key variables before attempting the solution. The algorithm [12] then involves construction of NP subproblems corresponding to x_p, y_p and C_p. These can then be solved in parallel – potentially a number of times during the solution of the original MILP – with different right hand sides (these arise from trying different values of the key variables).

6 Parallel solution software architecture

We will now see how a MILPSolverManager component can be written to implement the approach described in the last section, building on the serial architecture described earlier.

Firstly, a mechanism for specification of key variables is needed. This can readily be handled through the very general parameter specification mechanism used for our solvers. Specifically, after creating a MILPSystem with the calls

```
MILPSolverManager *msol = new MILPSolverManager("PDMILP");
MILPSystem *msys = msol->CreateSystem(mymilp);
```

We will thus supply the key variable indices to the system object before asking it to "solve itself".

As mentioned at the end of section 4, the semantics of the MILPSystem are that structural changes can be made to its MILP between calls to its Solve method. This means that relatively little can be done by the constructor of the MILPSystem, as the structure of the MILP cannot be relied upon to remain fixed during its lifetime. In particular, the fairly routine graph-theoretical analysis required to partition the MILP given its key variables can only be carried out when Solve is called. The problem is decomposed automatically using a partitioning algorithm which takes as input the "key" variables. The decomposition branch-and-bound method is employed to solve the mixed integer linear programming problem, taking advantage of the properties of the decomposable problem (see Fig. 6).

The following is a simple structure for the Solve routine as it applies:

1. Perform partitioning analysis, producing NP pMILPs.
2. Create a MILP object for each partial pMILP.
3. Populate it with variables and constraints.
4. Create a MILPSystem object for each pMILP object.
5. Begin solution of original MILP.
6. When certain conditions apply, solve the pMILPs –
   a) Set RHSs of the pMILPs.

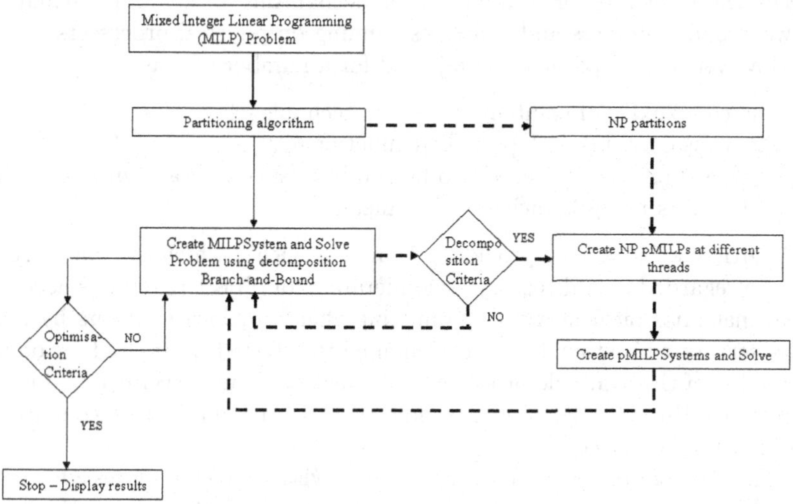

Fig. 6. Parallelised solution algorithm called within the `Solve` method.

   b) Create NP threads – in each, call the Solve method of a MILPSystem, and terminate on return.
   c) Wait until all threads have completed.
   d) Gather overall answer from these solutions, and continue solution of original MILP using this as an upper bound, if we have a solution.
7. Repeat step 6 when next appropriate.

In the next section, we will present the infrastructure used to ensure that steps 2 and 4 of the above algorithm create problem/solver pairs on NP distinct processors (where available).

6.1 Providing the infrastructure for parallel computation

One possible drawback of basing the solution architecture on CORBA is the fact that CORBA provides no built-in mechanisms for allocation of work to available parallel processors. In fact, a solution based on combining CORBA with MPI was considered. MPI [8] does provide the concept of a "virtual machine", where every node is identified by an integer index between 0 and NN-1, where NN is the total number of nodes. Each node can determine both the total number of nodes and its own identity: it can then use such identifiers to transmit messages to other nodes, either singly or in groups. MPI implementations then provide their own mechanisms for defining the mapping between node numbers and processes running on physical processors.

However, this approach was rejected for a number of reasons:

1. The complexity of combining the two technologies
2. The considerable overlap in their functionality
3. The need for MPI processes to be run in a 'special' way, with both client and servers being launched simultaneously

CORBA does in fact provide a "name service", which for our purposes may be regarded as a directory-style hierarchy of object references accessible to all machines on a network, permitting clients to locate servers by name: these servers will either have been manually initiated or created automated from a list of the available processors. Thereafter client-server communication is initiated through the object reference, requiring no further contribution from the name service.

A mechanism based on this service was devised, which requires a minimum of problem-specific coding, and which fits better with the envisaged usage of our code: i.e.:

1. servers being left running, one-per-machine, on a network, at any given time processing work for zero or more clients.
2. clients being run normally, making use of the existing servers.

This mechanism simply consists of a special server executable, MILPPackageServer, being started on each of the chosen processors. Its implementation will be:

1. Create a single IMILPPackage object (see section 4.1).
2. Register the reference of this object with the CORBA name service under the category "MILPPackage".
3. Block indefinitely, waiting for CORBA interactions or an interrupt signal.
4. When such a signal is received:
   - Remove the package reference from the name service
   - Complete processing of current object interactions if possible
   - Terminate

In its simplest form, the client code for steps 2-4 of the algorithm of the previous section will then be:

1. Obtain list of all IMILPPackage servers from name service, and create NP MILP objects in a 'round-robin' manner.
2. Populate the MILPs. Note that this involves network communication – through the local/remote-transparent CORBA mechanism – with each of the servers involved.
3. For each pMILP, use the **same** package object that yielded its IMILPFactory reference to obtain an IMILPSolver reference, and create a MILPSystem by passing the pMILP to this solver. This will mean that the

communication carried out between MILPSystem and MILP during solution of the pMILP will be intra-process.

The mechanisms for creation of MILPs at step 2, and of MILPSystems at step 4, will be as outlined in section 4.2.

As a refinement on this simple approach, a separate interface such as the following can be defined:

```
interface IRateProcessor {
  double GetSpeed();
  double GetLoad();
};
```

which provides indications of both the speed of this particular processor (of use for a heterogeneous network) and a dynamic measure of the existing load on that processor.

This interface can then be "mixed in" to the IMILPPackage interface through inheritance. A general routine can then be written with the following signature:

```
list<Object> GetBestNServers(string type, int N);
```

which given a type (in this case "MILPPackage") and the number of required servers (in this case NP, the number of partitions) does the following:

1. Obtains the list of object references to all available servers of the appropriate type.
2. Applies the CORBA "narrow" mechanism to each object reference in order to interact with the underlying remote object through its IRateProcessor interface (this is the mechanism by which genericity is ensured).
3. Determines the speed $s$and load $l$of each processor and computes an estimate of the relative solution rate it would achieve if given a MILP to solve: the formula (neglecting priority effects) will be approximately: $r = \frac{s}{l+1}$.
4. Returns the object references of the N servers deemed likely to give the best performance (including repeat values if fewer than $N$are available, or if there are sufficient disparities in the rates computed by the previous step (it may be difficult to estimate the effect of giving more than one job to a given server which currently has none, particularly as the processes creating the existing load may be running at different priorities from the server process).

The client code can then free its existing server objects each time solution is completed, and call this routine whenever pMILP solution commences, in order to ensure that resources are allocated as efficiently as possible.

We have implemented a solver component based on the above algorithm, which we refer to as "dBB" (for decomposition Branch-and-Bound).

Fig. 7 (an extended version of Fig. 4) presents the interactions involved when using this MILPSolverManager (1). The dBB software's milpsystem (3) makes use of the Server (4) to create the additional MILPs for each partition. These are then solved using the CPLEX MILPSolverManager (1).

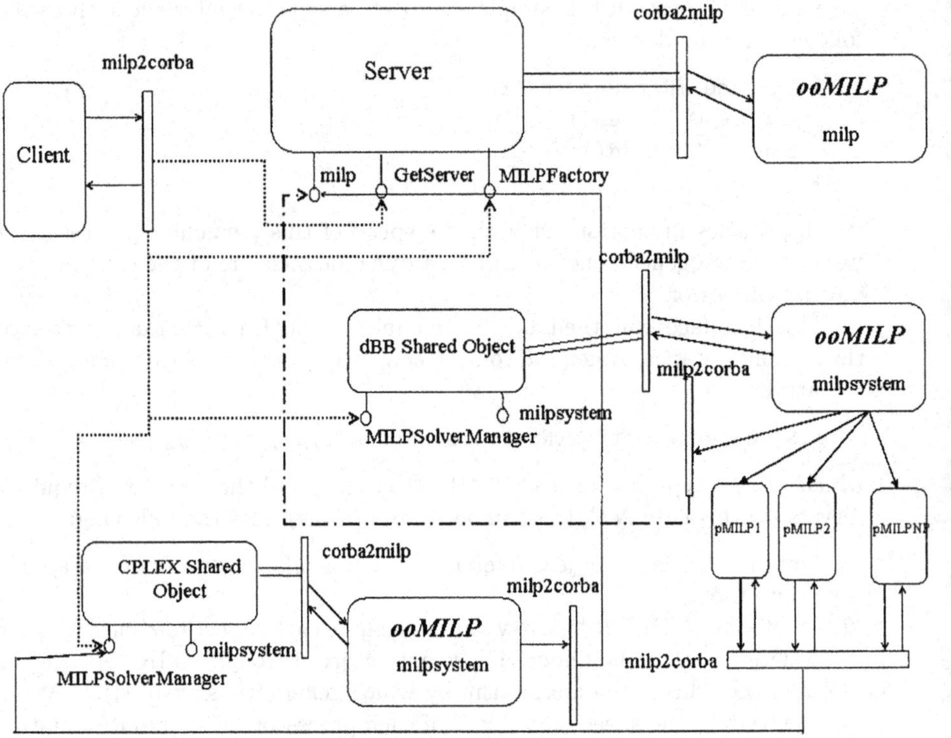

Fig. 7. Interactions involved in using the parallel solver.

6.2 Notes on multithreading

It is worthwhile to record decisions taken in the design of the client-side mechanism to achieve parallelism. CORBA provides a mechanism for "one-way" calling, which at first appeared attractive. This would mean that a single-threaded client could initiate solution of each MILPSystem with non-blocking calls, collecting results through a call-back mechanism. However, the CORBA standard makes it clear that the semantics of one-way calling are "best effort" only, meaning that there is no certainty that a particular call will be received by the server concerned. A particular CORBA implementation, or "ORB", might offer a better guarantee in this respect, but reliance on this would be a loss in terms of portability. Another concern with the use of the mechanism is that an unnecessary difference between the serial and parallel client implementations might arise.

Thus all CORBA calls used in our implementation are of the blocking kind, that is control is only returned to the calling thread when the receiving process has completed execution of each method.

It will be noted that the initialisation stage (steps 2, 3 and 4 of the algorithm presented at the beginning of this section) were carried out from a single thread, where as multiple threads were created for the solution stage (step 6). This is **not** because the time taken for initialisation is expected to be trivial compared with solution, but rather because the initialisation stage inevitably requires a high proportion of client CPU input. Indeed, it might be considered that knowledge of the characteristics of the MILPSystem implementation was relied on in this design – i.e. the fact that only the Solve method is computationally intensive, with only information storage being performed in the construction stages.

The actual mechanism used at step 6 to create the distinct threads is a very straightforward one based on "omni_thread", a library class supplied as a 'bonus' part of our preferred CORBA implementation, omniORB from AT&T Laboratories. We use a single method of this class, create, which takes two arguments: a call-back routine and a pointer to be passed as an argument to that routine. It then creates a thread which calls this routine, and terminates when it returns.

Using this, the thread creation process at substep b of step 6 has the form:

```
omni_thread::create(SolveAMILP, (void*) &(thesystems[i]));
```

where SolveAMILP is a function which calls the Solve routine of its MILPSystem argument, and decrements a counter on completion in order to carry out substep c, i.e. suspending the main thread of execution until all the Solve calls have returned.

6.3 An example from supply chain design problems

An example of the application of the dBB algorithm to the solution of a supply chain design problem under product demand uncertainty [13] is presented.

The problem considers the design of a mutli-echelon, multiproduct production and distribution network operating under a number of possible demand scenarios. The network consists of a number of existing multiproduct, multipurpose plants at fixed locations, a number of warehouses and distribution centres of unknown locations (to be selected by a set of candidatelocations , determined by the optimisation) and a number of customer zones at fixed locations. Except the network structure the optimisation has to determine operational issues, such as production amounts, material flows, warehouse and distribution centres capacities, together with transportation modes through the network.

The formulation for modelling the supply chain for the steady-state design under uncertainty [13] and for the case of dynamic modelling with or without uncertainty fall directly in this category of problems. More specifically, the network design variables and some continuous variables form the set of *key variables* which can decompose the original problem in as many partial MILPs as the number of demand scenarios. Similarly, the dynamic formulation can be decomposed into partial problems corresponding to individual time periods.

We use two instances of the problem with the size presented in Table 1. The problems are for the steady-state case under demand uncertainty. In the first case we consider a three demand scenarios case while in the second a four demand scenarios case.

Demand scenarios	3	4
Constraints	54228	72159
Continuous variables	34176	45561
Binary variables	13989	18525
Continuous key variables	21	21
Binary key variables	381	381
Non-zero elements	192996	257152
Partitions	3	4

Table 1. Problem size statistics.

The results obtained attempting to solve the problem using the two methods are presented in Table 2.

Demand scenarios	3		4	
Method	dBB	CPLEX	dBB	CPLEX
Solution found	1,959,630	1,967,552	2,656,980	2,779,952
Nodes Examined	563	1,381	670	121,000
CPU time (s)	3,716	5,423	2,844	244,321

Table 2. Solution statistics.

The advantage of the decomposition algorithm is obvious comparing with standard branch-and-bound as used within CPLEX. Overall, dBB examines fewer nodes and in less time thus reducing the computational effort. Moreover in the case of four scenarios our solution approach finds the optimal solution where CPLEX only finds a sub-optimal. Although CPLEX was used for this example, the system can use any MILP solver with a callable library and standard interface. The competitive advantage of the method against standard branch-and-bound approaches will be maintained because of the concept of exploiting the structure of the problem and generating better bounds.

7 Conclusions

ooMILP is a general object-orientated interface for MILP definition, construction and solution. It provides a number of methods for this reason requiring minimal programming effort for the construction of the MILP problem formulation by the client, offering at the same time high flexibility.

In addition, it provides other applications ("clients") with unified access to all solvers. Solvers are independent of the milp problem. This adds the advantage of being a convenient platform for implementation and testing of solution algorithms.

This work presented a general approach of packaging numerical solvers for mixed integer linear programming problems based on the CAPE-OPEN standards. The advantages of tools based on open-architecture are clear and the ease of implementation state-of-the-art solution algorithms is inherent.

Also utilising CORBA we provided solution in implementing a parallel algorithm that can use either multi-processor or network computers to reduce the solution time required to a general type of problem widely found in many applications. Additionally, the decomposition Branch-and-Bound (dBB) is generally implemented to be applied to any type of problem that has the properties described.

References

1. Cordier, C., H. Marchand, R. Laundry and L.A. Wolsey, "bc-opt: a Branch-and-Cut Code for Mixed Integer Programs", *Mathematical Programming*, **86**, 335-353, 1999.
2. Dimitriadis, A.D., "Algorithms for the Solution of Large-Scale Scheduling Problems", PhD thesis, Imperial College, University of London, 2000.
3. Fisher, M.L., "The Lagrangean Relaxation Method for Solving Integer Programming Problems", *Management Science*, **27**, 1-18, 1981.
4. Geoffrion, A.M., "Generalized Benders Decomposition", *Journal of Optimization Theory and Applications*, **4**, 237-260, 1972.
5. Geoffrion, A.M., "Lagrangean Relaxation for Integer Programming", *Mathematical Programming Study 2*, **2**, 82-114, 1974.
6. Global CAPE-OPEN, "Mixed Integer Linear/Nonlinear Programming Interface Specifications", `http://www.co-lan.org`, 2002.
7. Junger, M. and S. Thienel, "Introduction to ABACUS - a Branch-and-Cut System", *Operation Research Letters*, **22**(2-3), 83-95, 1998.
8. Message Passing Interface Forum, "MPI: A Message-Passing Interface standard", *International Journal of Supercomputer Applications*, **8**(3/4), 165-414, 1994.
9. Nemhauser, G.L., M.W.P. Savelsbergh and G.C. Sigismondi, "MINTO, a Mixed INTeger Optimizer", *Operation Research Letters*, **15**(1), 47-58, 1994.
10. Reeves, C.R., Modern Heuristic Techniques for Combinatorial Problems, Blackwell Scientific Publications, London, 1993.

11. Tsiakis, P., B.R. Keeping and C.C. Pantelides, "*ooMILP*: A C++ Callable Object-Orientated Library for the Definition and Solution of Large, Sparse Mixed Integer Linear Programming (MILP) Problems", Research Report, Centre for Process Systems Engineering, Imperial College of Science Technology and Medicine, London, 1999.
12. Tsiakis, P., A.D. Dimitriadis, N. Shah, C.C. Pantelides, "Solution of Nearly Decomposable MILP [ND-MILP] Problems", AIChE Annual Meeting, Los Angeles, USA, 2000.
13. Tsiakis, P., N. Shah and C.C. Pantelides, "Design of Multi-Echelon Supply Chain Networks under Demand Uncertainty", *Industrial Engineering and Chemistry Research*, **44**(16), 3585-3604, 2001.
14. VanRoy, T.J., "A Cross Decomposition Algorithm for Capacitated Facility Location", *Operations Research*, **34**, 145-263, 1986.

Part III: Applications

Global Order-Value Optimization by means of a Multistart Harmonic Oscillator Tunneling Strategy

R. Andreani, J.M. Martinez, M. Salvatierra, and F. Yano

Department of Applied Mathematics, IMECC-UNICAMP, University of Campinas, CP 6065, 13081-970 Campinas SP, Brazil
{andreani,martinez,otita,yano}@ime.unicamp.br

Summary. The OVO (Order-Value Optimization) problem consists in the minimization of the Order-Value function $f(x)$, defined by $f(x) = f_{i_p(x)}(x)$, where $f_{i_1(x)}(x) \leq \ldots \leq f_{i_m(x)}(x)$. The functions $f_1, \ldots, f_m$ are defined on $\Omega \subset \mathbb{R}^n$ and p is an integer between 1 and m. When x is a vector of portfolio positions and $f_i(x)$ is the predicted loss under the scenario i, the Order-Value function is the discrete Value-at-Risk (VaR) function, which is largely used in risk evaluations. The OVO problem is continuous but nonsmooth and, usually, has many local minimizers. A local method with guaranteed convergence to points that satisfy an optimality condition was recently introduced by Andreani, Dunder and Martínez. The local method must be complemented with a global minimization strategy in order to be effective when m is large. A global optimization method is defined where local minimizations are improved by a tunneling strategy based on the harmonic oscillator initial value problem. It will be proved that the solution of this initial value problem is a smooth and dense trajectory if Ω is a box. An application of OVO to the problem of finding hidden patterns in data sets that contain many errors is described. Challenging numerical experiments are presented.

Key words: Order-Value optimization, local methods, harmonic oscillator, tunneling, hidden patterns.

1 Introduction

Given m continuous functions $f_1, \ldots, f_m$, defined in a domain $\Omega \subset \mathbb{R}^n$ and an integer $p \in \{1, \ldots, m\}$, the $p-$Order-Value (OVO) function f is given by

$$f(x) = f_{i_p(x)}(x)$$

for all $x \in \Omega$, where $i_p(x)$ is an index function such that

$$f_{i_1(x)}(x) \leq f_{i_2(x)}(x) \leq \ldots \leq f_{i_p(x)}(x) \leq \ldots \leq f_{i_m(x)}(x).$$

The OVO function is continuous [2]. However, even if the functions f_i are differentiable, the function f may not be smooth. The OVO problem consists in the minimization of the Order-Value function:

$$\text{Minimize } f(x) \quad \text{subject to} \quad x \in \Omega. \tag{1}$$

The definition of the OVO problem was motivated by two main applications.

1. Assume that Ω is a space of decisions and, for each $x \in \Omega$, $f_i(x)$ represents the cost of decision x under the scenario i. The Minimax decision corresponds to choose x in such a way that the maximum possible cost is minimized. This is a very pessimistic alternative and decision-makers may prefer to discard the worst possibilities in order to proceed in a more realistic way. For example, the decision maker may want to discard the 10 % more pessimistic scenarios. This corresponds to minimize the $p-$Order-Value function with $p \approx 0.9 \times m$. When $f_i(x)$ represents the predicted loss for the set x of portfolio positions under the scenario i, the function $f(x)$ is the Value-at-Risk function, which is largely used in the finance industry. See [15].
2. Assume that we have a parameter-estimation problem where the space of parameters is Ω and $f_i(x)$ is the error corresponding to the observation i when the parameter x is adopted. The Minimax estimation problem corresponds to minimize the maximum error. As it is well-known this estimate is very sensitive to the presence of outliers [14]. Sometimes, we want to eliminate (say) the 15% larger errors because they can represent wrong observations. This leads to minimize the $p-$Order-Value function with $p \approx 0.85 \times m$. The OVO strategy is adequate to avoid the influence of systematic errors.

In [2] the continuity and differentiability properties of the Order-Value function was proved, nonsmoothness was discussed, local optimality conditions were introduced and, according to them, a local algorithm was defined. This algorithm is guaranteed to converge only to points that satisfy the optimality conditions, which are not necessarily global minimizers. A different approach was used in [1]. In this paper a nonlinear-programming reformulation of the OVO problem was defined. So, a general nonlinear programming solver can be used for solving it but, again, global solutions may be very difficult to find. Nonlinear programming methods that take advantage of the structure of the OVO reformulation were presented in [22, 23] but it is too soon for an evaluation of the effectiveness of these reformulations for solving practical problems.

The objective of the present work is to insert the local algorithm in a global heuristic and to apply the resulting method to the problem of finding hidden patterns. It is interesting to observe that, both in the application to decision problems and in the application to robust estimation of parameters, the OVO problem used corresponds to large values of p (generally close to m) whereas in the hidden-pattern problem the small values of p are the interesting ones. This is because we assume, in the latter problems, that all except a small number of data are corrupted.

This paper is organized as follows. The local algorithm is presented in Section 2. In Section 3 we introduce Lissajous motions, which are the basis of the global heuristic. In Section 4 we describe the global optimization algorithm. The hidden-pattern problem is discussed in Section 5. Numerical experiments are shown in Section 6 and in Section 7 we give some conclusions and discuss the lines for future research.

2 Local algorithm

In this section we present the local algorithm which will be used in the calculations. Before, let us define for all $\varepsilon > 0$ and $x \in \Omega$:

$$I_\varepsilon(x) = \{j \in \{1, \dots, m\} \mid f(x) - \varepsilon \leq f_j(x) \leq f(x) + \varepsilon\}.$$

Algorithm 2.1

Let $x_0 \in \Omega$ be an arbitrary initial point. Let $\theta \in (0,1)$, $\Delta > 0$, $\varepsilon > 0$, $0 < \sigma_{min} < \sigma_{max} < 1$, $\eta \in (0,1]$.

Given $x_k \in \Omega$ the steps of the k-th iteration are:

Step 1. (Solving the subproblem)

Define

$$M_k(d) = \max_{j \in I_\varepsilon(x_k)} \nabla f_j(x_k)^T d.$$

Consider the subproblem

$$\text{Minimize } M_k(d) \text{ subject to } x_k + d \in \Omega,\ \|d\|_\infty \leq \Delta. \tag{2}$$

Let $\bar{d}_k$ be a solution of (2). Let d_k be such that $x_k + d_k \in \Omega$, $\|d_k\| \leq \Delta$ and

$$M_k(d_k) \leq \eta M_k(\bar{d}_k). \tag{3}$$

If $M_k(d_k) = 0$ stop.

Step 2. (Steplength calculation)

Set $\alpha \leftarrow 1$.

If

$$f(x_k + \alpha d_k) \leq f(x_k) + \theta \alpha M_k(d_k) \tag{4}$$

set $\alpha_k = \alpha$, $x_{k+1} = x_k + \alpha_k d_k$ and finish the iteration. Otherwise, choose $\alpha_{new} \in [\sigma_{min}\alpha, \sigma_{max}\alpha]$, set $\alpha \leftarrow \alpha_{new}$ and repeat the test (4).

Observe that, when Ω is convex, (2) is equivalent to the convex optimization problem

$$\text{Minimize } w$$

$$\nabla f_j(x_k)^T d \le w, \ \forall\, j \in I_\varepsilon(x_k),$$
$$x_k + d \in \Omega, \ \|d\|_\infty \le \Delta.$$

Assume that $\Omega \subset \mathbb{R}^n$ is closed and convex, and $f_1, \ldots, f_m$ have continuous partial derivatives in an open set that contains Ω.

For all $x, y \in \Omega$, $j = 1, \ldots, m$, we assume that

$$\|\nabla f_j(x)\|_\infty \le c,$$

and

$$\|\nabla f_j(y) - \nabla f_j(x)\|_\infty \le L\|y - x\|_\infty.$$

Definition. We say that x is ε-optimal (or critical) if

$$\mathcal{D} \equiv \{d \in \mathbb{R}^n \mid x + d \in \Omega \text{ and } \nabla f_j(x)^T d < 0, \ \forall\, j \in I_\varepsilon(x)\} = \emptyset.$$

The following theorem was proved in [2].

Theorem 2.1. *Assume that $x_k \in \Omega$ is the $k-$th iterate of Algorithm 2.1. Then:*

1. *The algorithm stops at x_k if, and only if, x_k is a critical point. If the algorithm does not stop at x_k, then the $k-$th iteration is well-defined and finishes at Step 2 with the computation of x_{k+1}.*
2. *Suppose that $x_* \in \Omega$ is a limit point of a sequence generated by Algorithm 2.1. Then x_* is critical.*

3 Lissajous motions

Assume that $\Omega \subset \mathbb{R}^n$ is a bounded box with nonempty interior. That is:

$$\Omega = \{x \in \mathbb{R}^n \mid \ell \le x \le u\}.$$

The Harmonic Oscillator Initial-Value problem is the system of n independent harmonic oscillators given by:

$$\frac{d^2}{dt^2} x_i(t) + \theta_i^2 x_i(t) = 0, \ i = 1, \ldots, n \tag{5}$$

where $\theta_i^2 = \frac{[k_e]_i}{g}$, g is the mass of the body and the $[k_e]_i$'s are the elasticity constants.

The solutions of (5) are:

$$\alpha(t) = (\cos(\theta_1 t + \varphi_1), \ldots, \cos(\theta_n t + \varphi_n)), \tag{6}$$

where $\varphi_1, \ldots, \varphi_n$ are constants. The trajectory defined by each solution (6) is called a *Lissajous curve*. See [12] p. 36. In Figures 1 and 2 we show examples of these curves for $n = 2$ and $n = 3$, respectively.

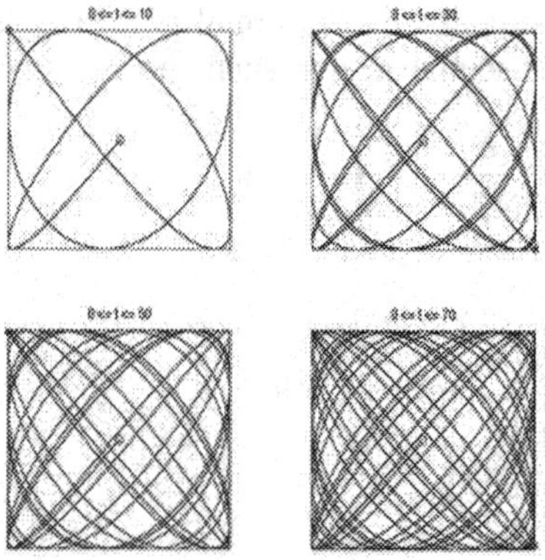

Figure 1

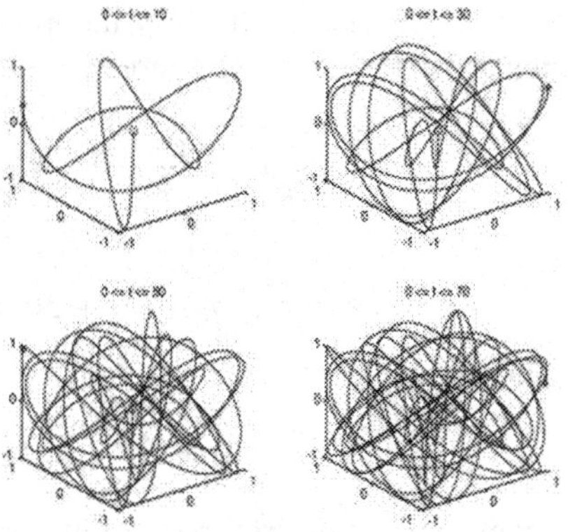

Figure 2

Given $x_0 \in \Omega$ and choosing appropriately $\varphi_1, \ldots, \varphi_n$ we can find a Lissajous curve such that $x(0) = x_0$. Clearly, Lissajous curves are smooth. In this section we give a simple proof that the image of a Lissajous curve is dense in $[-1, 1]^n$.

Definition. We say that $\theta_1, \ldots, \theta_n \in \mathbb{R}$ are linearly independent over $\mathbb{Q}$ if

$$\sum_{i=1}^{n} r_i \theta_i = 0 \quad \text{and} \quad r_1, \ldots, r_n \in \mathbb{Q}$$

only if $r_1 = \cdots = r_n = 0$.

Theorem 3.1. **(Kronecker's Approximation Theorem)**
Let $h_1, h_2, \ldots, h_n \in \mathbb{R}$ be linearly independent over $\mathbb{Q}$, $\xi_1, \ldots, \xi_n \in \mathbb{R}$ and $\varepsilon > 0$. Then, there exists $t \in \mathbb{R}$ and $k_i \in \mathbb{Z}$ such that

$$|h_i t - \xi_i - k_i| < \varepsilon, \ \forall \ i = 1, \ldots, n.$$

Proof. See [13] pp. 431-437. ■

Theorem 3.2. *Let $\alpha : \mathbb{R} \longrightarrow [-1, 1]^n$ be the Lissajous curve given by (6), where $\theta_1, \ldots, \theta_n$ are linearly independent over $\mathbb{Q}$. Then, the image of $\alpha(t)$ is dense in $[-1, 1]^n$.*

Proof. Let $\varepsilon > 0$ be arbitrarily small and $x \in [-1, 1]^n$.

Let $\lambda \in \mathbb{R}^n$ be such that

$$\cos(\lambda_i) = x_i, \ \forall \ i = 1, \ldots, n. \tag{7}$$

By the uniform continuity of the cosine function, there exists $\delta > 0$ such that

$$|t_1 - t_2| \leq \delta \ \Rightarrow \ |\cos(t_1) - \cos(t_2)| \leq \varepsilon, \ \forall \ t_1, t_2 \in \mathbb{R}. \tag{8}$$

Define

$$h_i = \frac{\theta_i}{2\pi}, \ \ \xi_i = \frac{\lambda_i - \varphi_i}{2\pi}, \quad i = 1, \ldots, n.$$

Since $\theta_1, \dots, \theta_n$ are linearly independent, $h_1, \dots, h_n$ are also linearly independent. By Theorem 3.1 there exists $t \in \mathbb{R}$ and $k_i \in \mathbb{Z}$ such that

$$\left| \frac{\theta_i}{2\pi} t - \frac{\lambda_i - \varphi_i}{2\pi} - k_i \right| < \frac{\delta}{2\pi}, \ i = 1, \dots, n.$$

So,

$$|\theta_i t + \varphi_i - (\lambda_i + 2k_i \pi)| < \delta, \ i = 1, \dots, n.$$

Then, by (8) and the periodicity of the cosine function,

$$|\cos(\theta_i t + \varphi_i) - \cos(\lambda_i)| \le \varepsilon.$$

From (7), the desired result follows. ■

Let Φ be the obvious linear diffeomorphism between $[-1,1]^n$ and Ω. By Theorem 3.2, $\{\Phi(\alpha(t)), t \in \mathbb{R}\}$ is dense in Ω. Moreover, if $\beta : (-1,1) \to \mathbb{R}$ is one-to-one and continuous, we have that the set

$$\{\Phi[\alpha(\beta(t))] \mid t \in (-1,1)\}$$

is also dense in Ω. Let us define $F : (-1,1) \to \mathbb{R}$ by

$$F(t) = f[\Phi[\alpha(\beta(t))]]. \tag{9}$$

The problem of minimizing f on Ω is equivalent to the problem of minimizing F on $(-1,1)$ in the sense given by the following theorem.

Theorem 3.3. *Let x_* be a global minimizer of f on Ω and $\varepsilon > 0$. Then, there exists $t \in (-1,1)$ such that*

$$F(t) < f(x_*) + \varepsilon.$$

Proof. By the continuity of f, there exists $\delta > 0$ such that $f(x) < f(x_*) + \varepsilon$ whenever $\|x - x_*\| < \delta$. Since $\{\Phi[\alpha(\beta(t))] \mid t \in (-1,1)\}$ is dense in Ω, there exists $t \in (-1,1)$ such that $\|\Phi[\alpha(\beta(t))] - x_*\| < \delta$. Then, $f[\Phi[\alpha(\beta(t))] < f(x_*) + \varepsilon$ as we wanted to prove. ■

4 Global algorithm

The local Algorithm 2.1 can be very effective in many cases for finding global minimizers of the OVO problem. However, when m is large, the number of local minimizers increases dramatically. Moreover, critical points are not necessarily local minimizers and, therefore, the number of possible limit points of Algorithm 2.1 that are not global solutions is enormous.

Our strategy for solving the OVO problem consists of using the local algorithm for finding a critical point x_* and, then, trying to "escape" from this critical point using a Lissajous curve that passes through it. The linearly independent parameters $\theta_1, \dots, \theta_n$ that define the Lissajous curves are chosen to be the square roots of the n first prime numbers. So, $\{\theta_1, \theta_2, \theta_3, \ldots\} = \{\sqrt{2}, \sqrt{3}, \sqrt{5}, \ldots\}$. Therefore, the escaping strategy is reduced to a one-dimensional tunneling procedure. See [19, 25, 26, 28], among others.

After many trials and errors, we defined the global algorithm described below. Besides the local minimization and the tunneling phases we introduced a multistart procedure for generating different initial points, defining criteria for discarding poor initial points and establishing an upper limit of one second for each call to the tunneling phase.

Algorithm 4.1

Let $k_{max} > 0$ be an algorithmic parameter. Initialize $k \leftarrow 1$, $f_{min} \leftarrow \infty$, $\mathcal{C} \leftarrow \emptyset$, $\mathcal{A} \leftarrow \emptyset$, $\delta = 0.1 \times \min_{1 \le i \le n} \{u_i - l_i\}$.

Step 1. *Random choice*

Choose a random (uniformly distributed) initial point $x_{I_k} \in \Omega$. If $k = 1$ update

$$\mathcal{A} \leftarrow \mathcal{A} \cup \{x_{I_k}\},$$

and go to Step 5.

Step 2. *Functional discarding test*

Define

$$f_{max} = \max\{f(x) \mid x \in \mathcal{A}\}.$$

The probability *Prob* of discarding x_{I_k}, is defined in the following way:

- If $f(x_{I_k}) \le f_{min}$, $Prob \leftarrow 0$.
- If $f(x_{I_k}) \ge f_{max}$, $Prob \leftarrow 0.8$.
- If $f_{min} < f(x_{I_k}) < f_{max}$, $Prob \leftarrow 0.8 \left(\dfrac{f(x_{I_k}) - f_{min}}{f_{max} - f_{min}} \right)$.

Discard the initial point x_{I_k} with probability *Prob*. If x_{I_k} was discarded, return to Step 1.

Step 3. *Neighborhood discarding test*

Define

$$d_{min} = \min\{\|x - x_{I_k}\|_\infty \mid x \in \mathcal{C}\}$$

Update *Prob*, the new probability of discarding x_{I_k}, in the following way:

- If $d_{min} \le \delta$, $Prob \leftarrow 0.8$.
- If $d_{min} > \delta$, $Prob \leftarrow 0$.

Discard the initial point x_{I_k} with probability *Prob*. If x_{I_k} was discarded, return to Step 1.

Step 4. *10-Iterations discarding test*

Perform 10 iterations of the local method (Algorithm 2.1) obtaining the iterate $x_{10,k}$. Define

$$f_{10} = f(x_{10,k})$$

and

$$f_{aux} = f(x_{I_k}) - 0.1(f(x_{I_k}) - f_{min}).$$

Update the probability *Prob* in the following way:

- If $f_{10} \le f_{min}$, $Prob \leftarrow 0$.
- If $f_{10} \ge f_{aux}$, $Prob \leftarrow 0.8$.
- If $f_{min} < f_{10} < f_{aux}$, $Prob \leftarrow 0.8 \left(\frac{f_{10} - f_{min}}{f_{aux} - f_{min}} \right)$.

Discard x_{I_k} with probability *Prob*. If x_{I_k} was discarded, return to Step 1. Otherwise, update

$$\mathcal{A} \leftarrow \mathcal{A} \cup \{x_{I_k}\},$$

and go to Step 5.
Step 5. *Local minimization*

Taking x_{I_k} as initial point, execute Algorithm 2.1 obtaining a critical point $x_{*,k}$. Update the set of critical points:

$$\mathcal{C} \leftarrow \mathcal{C} \cup \{x_{*,k}\}.$$

Update the best functional value:

$$f_{min} \leftarrow \min\{f_{min}, f(x_{*,k})\},$$

$$x_{min} \leftarrow x_{*,k} \text{ if } f_{min} = f(x_{*,k}).$$

Set $time \leftarrow 0$ and go to Step 6.
Step 6. *Tunneling*

Using the Lissajous curve that passes through $x_{*,k}$ and the definition (9), try to obtain $t \in (-1, 1)$ such that $F(t) < f(x_{*,k})$. If such a point is obtained, update

$$x_{I_k} \leftarrow \Phi[\alpha(\beta(t))]$$

and go to Step 5. At each step of the tunneling process, update the computer time parameter *time*. If *time* exceeds one second and $k < k_{max}$ set $k \leftarrow k + 1$ and go to Step 1.

The random choice at Step 1 of Algorithm 4.1 trivially guarantees that a global minimizer is found with probability 1. This is stated, for completeness, in the following theorem.

Theorem 4.1 *Let x^* be a global minimizer of (1) and let $\varepsilon > 0$ be arbitrarily small. Assume that $k_{max} = \infty$. Then, with probability 1, there exists $k \in \{1, 2, \ldots\}$ such that the point x_{min} computed at Step 5 of the algorithm satisfies $f(x_{min}) \le f(x_*) + \varepsilon$.*

Proof. Since f is continuous there exists a ball $\mathcal{B}$ centered in x^* such that

$$f(x) \le f(x_*) + \varepsilon \ \forall \ x \in \mathcal{B}.$$

Since the initial point at Step 1 is chosen randomly and according to the uniform distribution, the probability of choosing $x_{I_k} \in \mathcal{B}$ at a particular iteration k is strictly positive. By the structure of Steps 2–4, the probability of discarding x_{I_k} is strictly smaller than 1.

Therefore, at a fixed iteration k, a point belonging to the ball $\mathcal{B}$ is chosen as the initial point for Algorithm 2.1 with positive probability. Since Algorithm 2.1 does not increase the objective function value, it follows that, given k, a point $x_{min} \in \mathcal{B}$ is computed at Step 5 with positive probability. The desired result follows since $k_{max} = \infty$. ■

5 Hidden patterns

The search of hidden patterns is one of the most challenging issues in modern data mining. Many papers address the problem of hidden-patterns discovery in different areas, as Ecology [9, 24], Web-log Analysis [16], Public Health Administration [17], Spatio-temporal Dynamics of Wave Modes [8], Art [35], Psychoanalytic Literary Criticism [27], Psychiatry [11], Social History [29], Demography [6, 7], City Systems [18] and many others.

Consider the clouds of points given in Figures 3a and 3b. At a first sight, the two clouds of points look qualitatively similar. However, in Figure 3a there are exactly five points that lie on the same parabola whereas in Figure 3b such set of points does not exist.

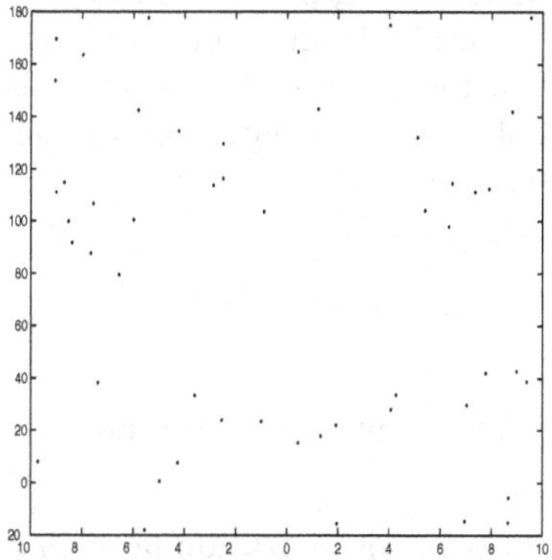
Figure 3a

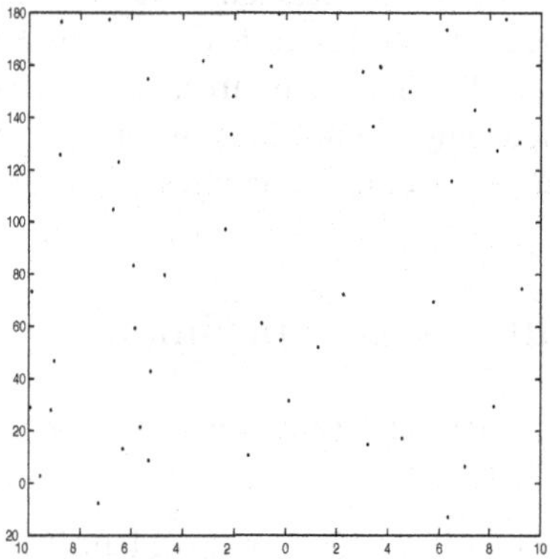
Figure 3b

In Figure 4 we show again the points of Figure 3a together with the parabola that fits those special points. In situations like this, we say that Figure 3a hides the pattern of a parabola (or, simply, hides a parabola), whereas Figure 3b does not.

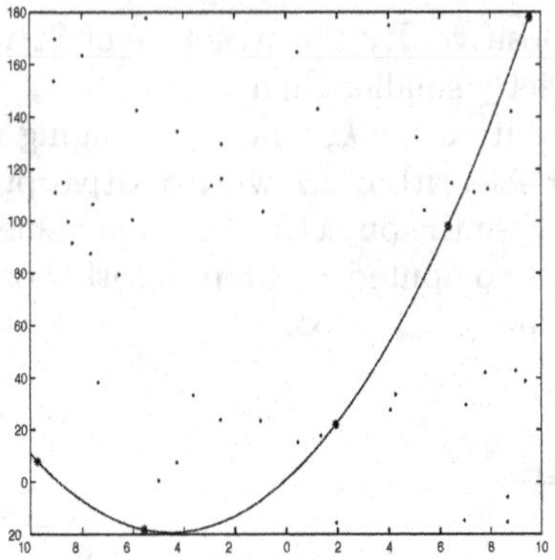

Figure 4

In the following section we show how the OVO problem and Algorithm 4.1 may be used to find hidden patterns of this type. Several examples presented in the following section are simple and small enough to be solved by exhaustive enumerative methods. However, if the number of data or the number of unknown parameters is moderately increased the use of combinatorial procedures is prohibitive. This is the reason why we did not consider enumerative algorithms in our analysis.

It must be mentioned that in many data-mining papers and applications, patterns are hidden in the sense that they are difficult to find, at least using standard fitting procedures. In our case patterns are hidden because they are revealed only taking into account a small amount of data, the reason being that most available information is severely corrupted.

6 Numerical experiments

Some practical features concerning the implementation of the algorithm are given below:

- As mentioned before, problem (2) is a convex optimization problem. Moreover, in our applications the constraints Ω will be linear, therefore (2) is a Linear Programming problem. For solving it we use the IMSL Library routine DDLPRS.
- The subproblems were solved exactly. This means that we used $\eta = 0$.
- The algorithmic parameters used were:

$$\theta = 0.5, \ \Delta = 1, \ \varepsilon = 10^{-3}, \ \sigma_{\min} = 0.1, \ \sigma_{\max} = 0.9.$$

- In the backtracking process (4) we took $\alpha_{new} = 0.5\alpha$.
- All the numerical experiments were run on a Pentium 4, 2.4 Ghz, 1Gb RAM in double precision FORTRAN.

6.1 Finding hidden polynomials

Assume that $\{(t_1, y_1), \ldots, (t_m, y_m)\} \subset \mathbb{R}^2$ is a set of data and we know that "most of them are wrong". Nevertheless, a few of these points contain valuable uncorrupted information, represented by a low-degree polynomial $x_1 t^{n-1} + \cdots + x_{n-1}t + x_n$. Least-squares fitting of the form $y_i \approx x_1 t_i^{n-1} + \ldots + x_{n-1}t_i + x_n$ leads to disastrous results due to the overwhelming influence of outliers.

	Number of minimizations		Best solution obtained			
p	k_{max}	Successful Tunnelings	x	$f(x)$	Calls to Algorithm 2.1	Successful Step
3	10	2	-0.34 1.65 3.97	1.40E-16	4	step 1
4	150	190	1.00 -5.00 2.00	2.68E-13	233	step 1
5	250	395	1.00 -5.00 2.00	2.94E-13	590	step 1
6	200	284	1.51 0.90 6.23	0.2687	307	step 6
7	200	397	1.66 1.39 2.59	1.3313	431	step 6
8	200	376	1.54 0.84 5.14	2.3728	384	step 6

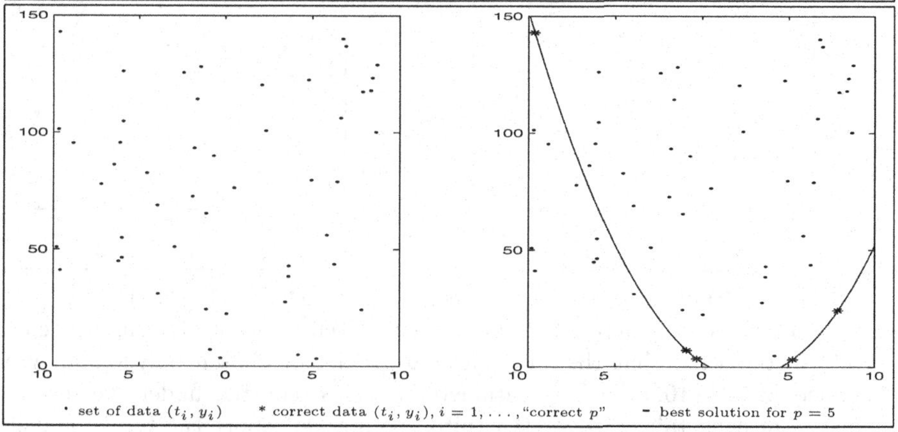

Table 1: $m = 50$, "correct p"$= 5$, $x^* = (1, -5, 2)$

The OVO approach for finding the hidden polynomial consists in defining, for each $i = 1, \ldots, m$, the error function

$$f_i(x) = (\sum_{j=1}^{n} x_j t_i^{n-j} - y_i)^2.$$

Given $p \in \{1, \ldots, m\}$, this set of functions defines an OVO problem (1) for which Algorithm 4.1 may be employed. The idea is to solve this problem for different values of p. If p is close to m we expect a large value of the OVO function at the solution found, showing that there are wrong data among the points that correspond to $f_{i_1}, \ldots, f_{i_p}$. When p is decreased, the OVO function at the solution tends to decrease as well. We expect that, when we take "the correct p", the OVO function would decrease abruptly, taking a value close to zero.

The results are summarized in Tables 1 to 6 and the corresponding pictures.

	Number of minimizations		Best solution obtained			
p	k_{max}	Successful Tunnelings	x	$f(x)$	Calls to Algorithm 2.1	Successful Step
8	50	99	-1.00 -6.00 4.00	7.65E-13	66	step 6
9	50	102	-1.00 -6.00 4.00	2.28E-12	72	step 6
10	50	93	-1.00 -6.00 4.00	4.94E-12	45	step 6
11	200	397	-0.95 -6.94 8.08	1.00	406	step 6
12	200	368	-1.04 -5.90 5.45	1.42	75	step 6
13	200	371	-0.94 -6.19 1.95	3.14	133	step 1

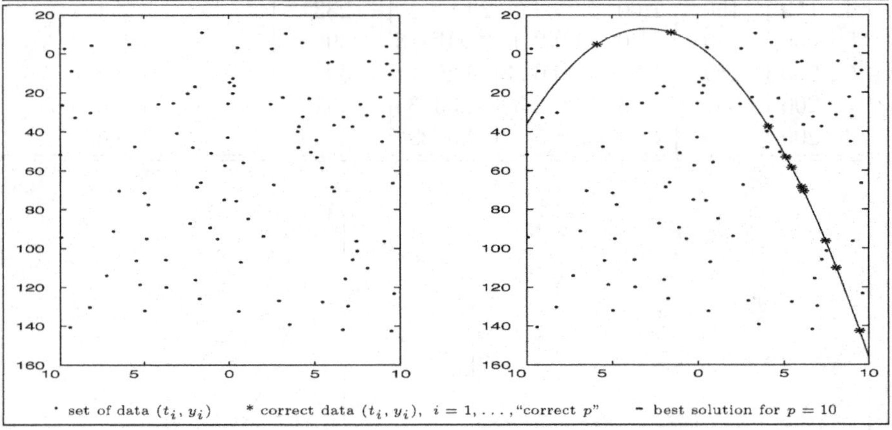

Table 2: $m = 100$, "correct p"$= 10$, $x^* = (-1, -6, 4)$

In all cases we generated m random data. Ten per cent of them are "correct" in the sense that they fit exactly a previously chosen polynomial. We defined $\Omega = [-10, 10]^n$. The parameter k_{max} is reported under the second column of each table. Under the third column we report the total number of "better points" obtained by the tunneling Lissajous procedure. The last column indicates the step of Algorithm 4.1 that gave rise to the initial point that produced the solution. The penultimate column indicates the number of calls to Algorithm 2.1 that were necessary to find the best point obtained.

6.2 Finding hidden circles

The experiments are entirely analogous to the ones reported for polynomials. In this case we need to estimate three parameters $x_1, x_2, x_3 \in \Omega$ where

$$\Omega = \{x \in \mathbb{R}^3 \mid -10 \leq x_1, x_2 \leq 10,\ 0 \leq x_3 \leq 10\}.$$

	Number of minimizations		Best solution obtained			
p	k_{max}	Successful Tunnelings	x	$f(x)$	Calls to Algorithm 2.1	Successful Step
13	50	91	-1.00 1.00 -6.00	9.92E-14	33	step 6
14	50	95	-1.00 1.00 -6.00	5.18E-14	36	step 6
15	100	177	-1.00 1.00 -6.00	2.45E-14	221	step 6
16	200	350	-1.00 1.00 -5.96	0.0272	14	step 6
17	200	358	-1.00 1.00 -6.18	0.0339	179	step 6
18	200	332	-1.00 0.97 -6.19	0.0865	288	step 6

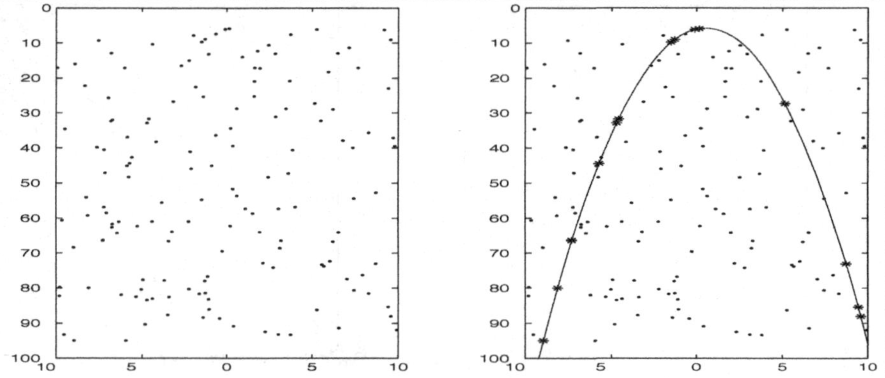

Table 3: $m = 150$, "correct p"$= 15$, $x^* = (-1, 1, -6)$

The center of the unknown circle is (x_1, x_2) and its radius is x_3. The functions f_i are:

$$f_i(x) = [(t_i - x_1)^2 + (y_i - x_2)^2 - x_3^2]^2.$$

The results, following the same conventions as before, are summarized in Tables 7, 8 and 9 and the corresponding pictures.

6.3 Finding hidden "bananas"

The hidden pattern is a curve in the ty-plane, of the form

$$(y - x_2 - (t - x_1)^2)^2 + (1 - (t - x_1))^2 = x_3,$$

where x_1, x_2 and x_3 are the parameters that we need to estimate. We defined

$$\Omega = \{x \in \mathbb{R}^3 \mid -10 \leq x_1, x_2 \leq 10, \ 0 \leq x_3 \leq 20\}.$$

	Number of minimizations		Best solution obtained			
p	k_{max}	Successful Tunnelings	x	$f(x)$	Calls to Algorithm 2.1	Successful Step
3	10	4	1.21 3.34 -5.84 1.36	8.41E-20	4	step 6
4	50	74	1.24 3.55 -5.90 -0.58	1.14E-16	84	step 1
5	50	85	1.00 5.00 3.00 -7.00	1.26E-14	101	step 6
6	200	466	1.05 5.10 -0.95 -10.0	140.50	309	step 6
7	200	397	0.97 4.76 5.77 -10.0	679.53	119	step 6
8	200	450	0.48 -0.52 -7.83 -10.0	3655.75	627	step 6

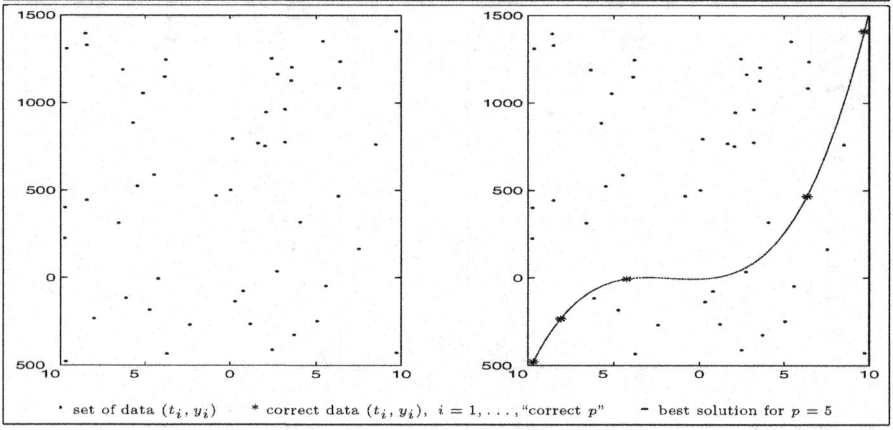

Table 4: $m = 50$, "correct p"$= 5$, $x^* = (1, 5, 3, -7)$

And, for each $i = 1, \ldots, m$ the error function is:

$$f_i(x) = [(y_i - x_2 - (t_i - x_1)^2)^2 + (1 - (t_i - x_1))^2 - x_3]^2.$$

The results, following the same conventions as before, are summarized in Tables 10, 11 and 12 and the corresponding pictures.

6.4 Finding hidden ellipses

In this section the hidden patterns are ellipses:

$$\frac{(t_i - x_1)^2}{x_3^2} + \frac{(y_i - x_2)^2}{x_4^2} = 1.$$

In this case we need to estimate four parameters x_1, x_2, x_3 and $x_4 \in \Omega$ where

$$\Omega = \{x \in \mathbb{R}^4 \mid -10 \leq x_1, x_2 \leq 10, \ 0 \leq x_3, x_4 \leq 10\}.$$

	Number of minimizations		Best solution obtained			
p	k_{max}	Successful Tunnelings	x	$f(x)$	Calls to Algorithm 2.1	Successful Step
8	100	181	-1.00 3.00 3.00 9.00	3.46E-17	144	step 6
9	50	100	-1.00 3.00 3.00 9.00	9.15E-15	93	step 6
10	50	80	-1.00 3.00 3.00 9.00	4.51E-14	23	step 6
11	200	407	-1.00 3.00 2.97 8.86	0.0272	369	step 6
12	200	414	-0.98 3.21 0.80 1.73	118.89	115	step 6
13	200	429	-0.98 3.46 2.17 -5.14	207.12	527	step 6

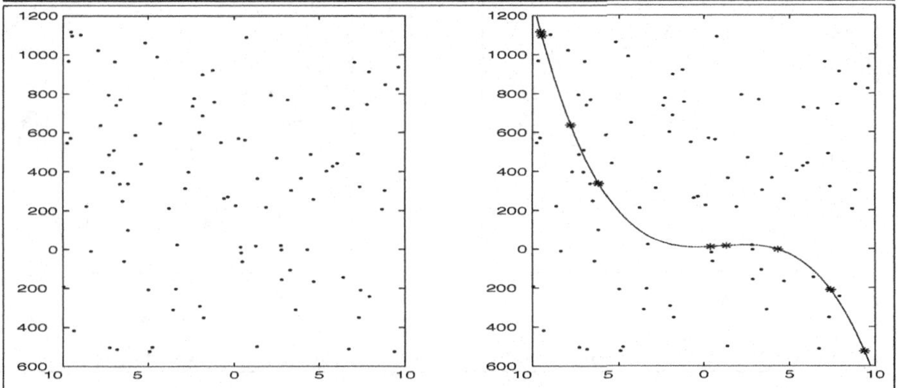

Table 5: $m = 100$, "correct p"$= 10$, $x^* = (-1, 3, 3, 9)$

Therefore, the error functions are:

$$f_i(x) = \left[\frac{(t_i - x_1)^2}{x_3^2} + \frac{(y_i - x_2)^2}{x_4^2} - 1\right]^2.$$

The results, following the same conventions as before, are summarized in Tables 13, 14 and 15 and the corresponding pictures.

We used the third problem of Table 15 for comparing the efficiency of Algorithm 4.1 with respect to a straightforward multistart strategy. The result reported in Table 15 was obtained using 25 minutes of CPU time. However, the pure random multistart method, which uses the local algorithm without discarding and tunneling, did not obtain a solution of similar quality after 12 hours of computation.

6.5 Finding a hidden polynomial-trigonometric function

Consider the points in Figure 5a. We wish to fit a polynomial-trigonometric function of the form

$$y(x, t) = x_1 + x_2 t + x_3 t^2 + x_4 t^3 + x_5 \cos(x_6 t + x_7) + x_8 \sin(x_9 t + x_{10}).$$

Consequently, the error functions are:

$$f_i(x) = [y_i - y(x, t_i)]^2.$$

Taking $p = 50$ we obtained using Algorithm 4.1 and after 2 hours of CPU time the function given in Figure 5b. Observe that since we need at least 10 points to fit this function and $C_{10}^{100} \cong 1.7 \times 10^{13}$, the use of enumerative schemes is completely impossible.

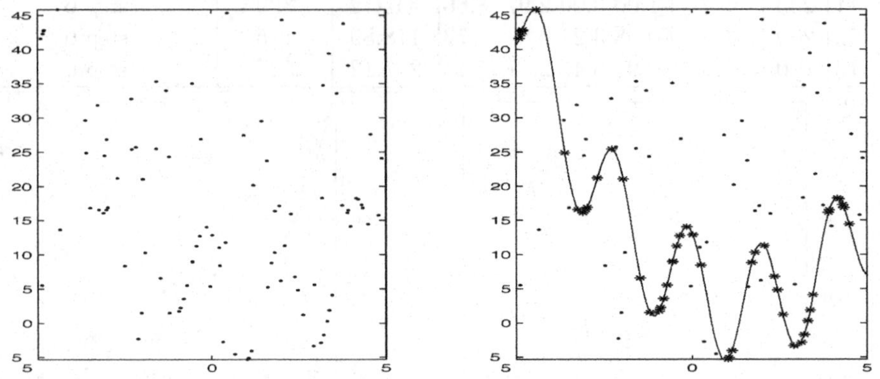

Figure 5a Figure 5b

· set of data (t_i, y_i) * correct data (t_i, y_i), $i = 1, \ldots,$ "correct p" - best solution for $p = 50$

Figure 5: $m = 100$, "correct p"$= 50$

7 Conclusions

Order-Value Optimization is a serious global minimization problem with important applications, many of which remain to be discovered. In this paper we emphasized the application to finding hidden patterns in the presence of massive corrupted data. The local algorithm introduced in [2] is not efficient enough to cope the problem of multiple critical points, therefore the definition of a globalization strategy was necessary. In order to preserve the local efficiency of Algorithm 2.1, our global strategy incorporates multiple starts with discarding strategies and a one-dimensional tunneling procedure for escaping from critical points, based on Lissajous harmonic oscillator dense curves.

Although we do not have rigorous theoretical arguments to support the point of view that this strategy is the best possible for the global OVO problem (and probably it is not) a rather extensive numerical experimentation (some of which is reported here) suggests that we are not far from discovering a satisfactory methodology for many practical problems. In particular, the use of Lissajous curves has been a pleasant experience. The idea of transforming n-dimensional minimization problems into one-dimensional ones by means of dense curves is not new but the theoretical question about which is "the best" curve to fill the n-dimensional box does not seem to be explicitly

	Number of minimizations		Best solution obtained			
p	k_{max}	Successful Tunnelings	x	$f(x)$	Calls to Algorithm 2.1	Successful Step
13	50	80	-1.00 3.00 7.00 0.00	3.24E-16	914	step 6
14	50	82	-1.00 3.00 7.00 0.00	8.30E-12	69	step 6
15	100	210	-1.00 3.00 7.00 0.00	1.45E-12	213	step 6
16	200	413	-1.01 3.06 7.79 -1.75	12.04	34	step 6
17	200	447	-1.01 2.96 8.57 0.66	37.10	61	step 6
18	200	409	-0.98 3.10 4.89 -2.21	55.24	225	step 6

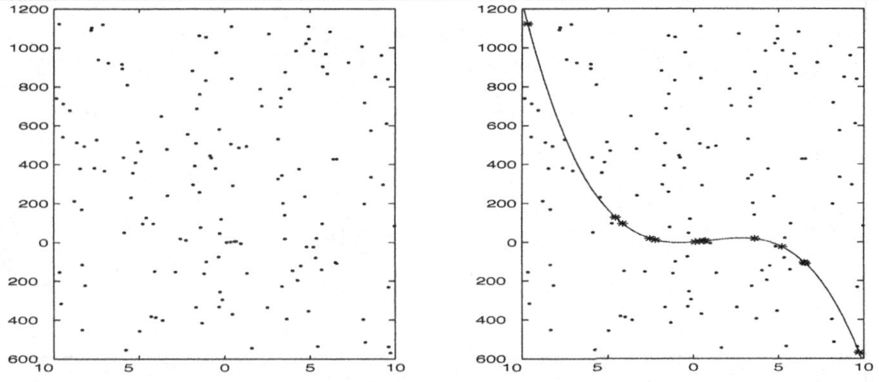

Table 6: $m = 150$, "correct p"$= 15$, $x^* = (-1, 3, 7, 0)$

formulated. Even the criteria that should define "the best" filling curve are not completely clear. Although we feel that Lissajous curves satisfy many of these (non-formulated) criteria, we would like to point out the relevance of future theoretical research on this subject.

Valuable research on solving multi-dimensional multi-extremal optimization problem employing Peano-space-filling curves [4, 5, 10, 21, 30, 31, 32, 33, 34] may complement our Lissajous-based approach. A common drawback of space-filling algorithms is that closeness of points in the multidimensional space does not correspond to closeness in the corresponding one-dimensional interval. Strongin [33] introduced an attractive scheme which allows one to reflect, in the reduced one-dimensional problem, some information on the nearness of points in the multidimensional domain. His ideas should be adapted to our scheme in future research. We also need to improve the escaping procedure, which in the present implementation is rather naive.

Finally, we would like to mention that the application of the hidden-pattern technology introduced in this paper to the Common Reflection Surface (CRS) problem [3, 20] has been suggested by Lúcio T. Santos and other members of the Computational Geophysics Group at the University of Campinas. Advances on this research will be reported in the near future.

	Number of minimizations		Best solution obtained			
p	k_{max}	Successful Tunnelings	x	$f(x)$	Calls to Algorithm 2.1	Successful Step
3	50	3	-4.99 -4.81 5.68	3.48E-19	4	step 1
4	2800	3002	5.00 -3.00 7.00	8.77E-18	5670	step 6
5	200	263	5.00 -3.00 7.00	2.93E-16	378	step 1
6	200	302	5.66 3.04 4.32	0.01	363	step 6
7	200	288	5.71 3.09 4.29	0.08	341	step 6
8	200	315	0.01 1.50 5.63	0.18	366	step 6

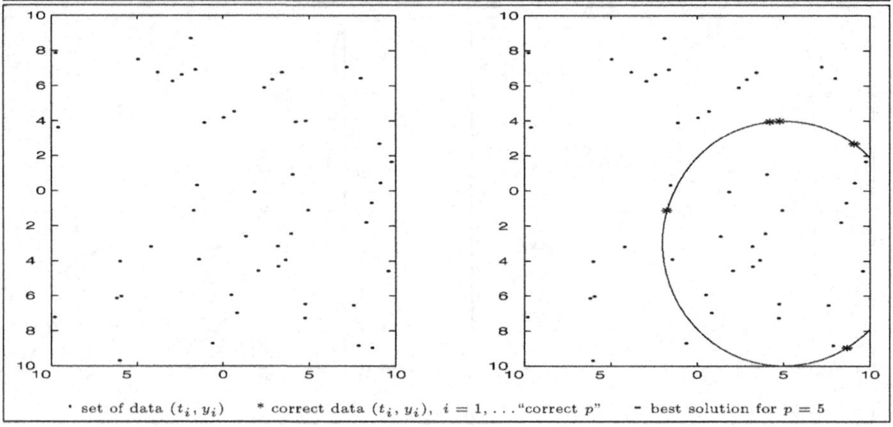

Table 7: $m = 50$, "correct p"$= 5$, $x^* = (5, -3, 7)$

A particularly interesting field of future research concerns the use of Order-Value Optimization for training neural networks in the presence of a large number of corrupted data.

Acknowledgments

We are indebted to Lúcio T. Santos for his careful reading of the first draft of this paper and to Leo Liberti for suggesting many improvements.

R. Andreani was supported by PRONEX-Optimization 76.79.1008-00, FAPESP (Grant 01-04597-4) and CNPq (Grant 301115/96-6), J.M. Martinez was supported by PRONEX-Optimization 76.79.1008-00, FAPESP (Grant 01-04597-4) and CNPq, M. Salvatierra was supported by PRONEX-Optimization 76.79.1008-00 and FAPESP (Grant 02-00171-5) and F. Yano was supported by PRONEX-Optimization 76.79.1008-00 and FAPESP (Grant 02-14203-6).

	Number of minimizations		Best solution obtained			
p	k_{max}	Successful Tunnelings	x	$f(x)$	Calls to Algorithm 2.1	Successful Step
8	50	89	1.00 -2.00 5.00	1.21E-15	78	step 6
9	300	543	1.00 -2.00 5.00	1.35E-17	781	step 6
10	50	88	1.00 -2.00 5.00	5.11E-17	109	step 6
11	200	382	0.92 -1.99 5.00	0.6348	355	step 6
12	200	407	1.00 -2.00 4.88	1.4563	527	step 6
13	200	371	1.00 -2.00 4.87	1.6275	185	step 6

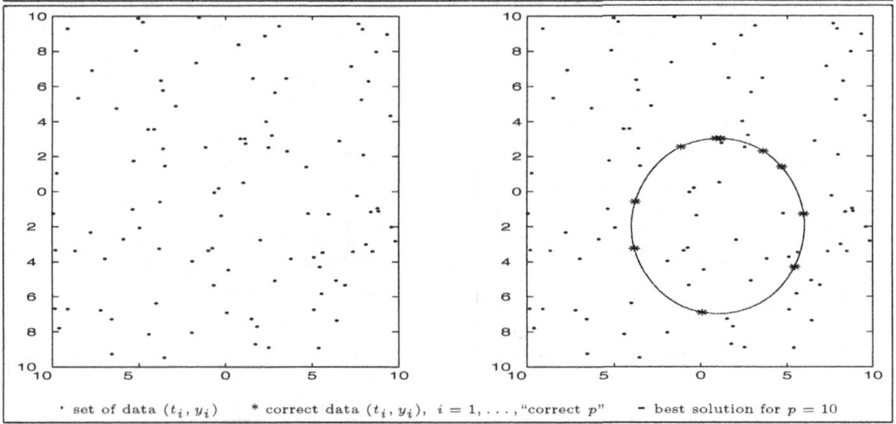

Table 8: $m = 100$, "correct p"$= 10$, $x^* = (1, -2, 5)$

References

1. R. Andreani, C. Dunder, and J. M. Martínez. Nonlinear-Programming reformulation of the Order-Value Optimization problem. Technical Report MCDO 22/05/01, Department of Applied Mathematics, University of Campinas, 2001. To appear in *Mathematical Methods of Operations Research* (2005).
2. R. Andreani, C. Dunder, and J. M. Martínez. Order-Value Optimization: formulation and solution by means of a primal Cauchy method. *Mathematical Methods of Operations Research*, 58:387–399, 2003.
3. R. Biloti, L. T. Santos, and M. Tygel. Multi-parametric traveltime inversion. *Stud. geophysica et geodaetica*, 46:177–192, 2002.
4. A. R. Butz. Space filling curves and mathematical programming. *Inform. Control*, 12(4):314–330, 1968.
5. A. R. Butz. Alternative algorithm for Hilbert's space filling curve. *IEEE Trans. Computers*, C-20(4):424–426, 1971.
6. E. Cam and J. Y. Monnat. Stratification based on reproductive state reveals contrasting patterns of age-related variation in demographic parameters in the kittiwake. *OIKOS*, 90:560–574, 2000.
7. L. L. Cavalli-Sforza. Genes, peoples, and languages. *Proceedings of the National Academy of Sciences of the United States of America*, 94:7719–7724, 1997.
8. A. Dinklage and C. Wilke. Spatio-temporal dynamics of hidden wave modes in a dc glow discharge plasma. *Physics Letters A*, 277:331–338, 2000.

	Number of minimizations		Best solution obtained			
p	k_{max}	Successful Tunnelings	x	$f(x)$	Calls to Algorithm 2.1	Successful Step
13	300	567	-3.00 1.00 7.00	1.52E-15	815	step 6
14	100	174	-3.00 1.00 7.00	1.99E-13	256	step 6
15	100	173	-3.00 1.00 7.00	5.98E-16	241	step 6
16	200	360	-3.00 1.04 7.00	0.3760	287	step 6
17	200	349	-3.07 1.68 7.69	3.3274	228	step 6
18	200	363	-3.05 1.69 7.69	3.6929	105	step 6

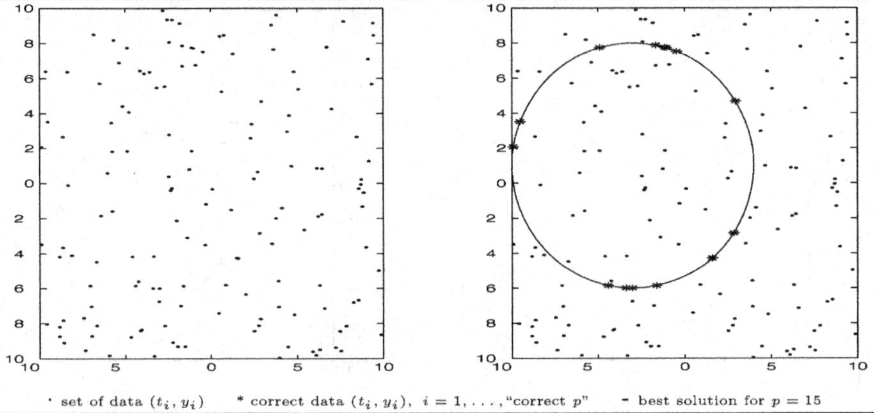

Table 9: $m = 150$, "correct p"= 15, $x^* = (-3, 1, 7)$

9. X. Espadaler and C. Gómez. The species body-size distribution in Iberian ants is parameter independent. *Vie et Milieu*, 52:103–107, 2002.
10. V. P. Gergel, L. G. Strongin, and R. G. Strongin. The vicinity method in pattern recognition. *Engineering Cybernetics.* (Transl. from Izv. Acad. *Nauk* USSR, Techn, Kibernetika 4:14-22, 1987).
11. R. Gronner and L. Manheim. Hidden patterns. *Arch. Gen. Psychiat.*, 17:248, 1967.
12. J. Haag. *Oscillatory Motions.* Wadsworth, Belmont, California, 1962.
13. E. Hewitt and K. A. Ross. *Abstract Harmonic Analysis I.* Springer-Verlag, Berlin, 1963.
14. P. J. Huber. *Robust Statistics.* Wiley, New York, 1981.
15. P. Jorion. *Value at Risk: the new benchmark for managing financial risk.* Mc Graw-Hill, New York, 2nd edition, 2001.
16. K. P. Joshi, A. Joshi, and Y. Yesha. On using a warehouse to analyze web logs. *Distributed and Parallel Databases*, 13:161–180, 2003.
17. H. C. Koh and S. K. Leong. Data mining applications in the context of casemix. *Annals Academy of Medicine Singapore*, 30:41–49, 2001.
18. J. Kropp. A neural network approach to the analysis of city systems. *Applied Geography*, 18:83–96, 1998.
19. A. V. Levy and A. Montalvo. The tunneling algorithm for the global minimization of functions. *SIAM J. Sci. Stat. Comp.*, 6:15–29, 1985.

	Number of minimizations		Best solution obtained			
p	k_{max}	Successful Tunnelings	x	$f(x)$	Calls to Algorithm 2.1	Successful Step
3	50	9	4.70 5.73 5.82	3.52E-18	53	step 1
4	1000	1258	1.00 -4.00 7.00	1.38E-16	1828	step 6
5	1000	1899	1.00 -4.00 7.00	4.80E-17	2141	step 6
6	1000	2114	1.86 -5.49 12.9	0.0857	2835	step 6
7	1000	2284	1.89 -5.48 13.0	0.2796	2081	step 6
8	1000	2459	1.87 -5.52 13.2	0.4496	551	step 6

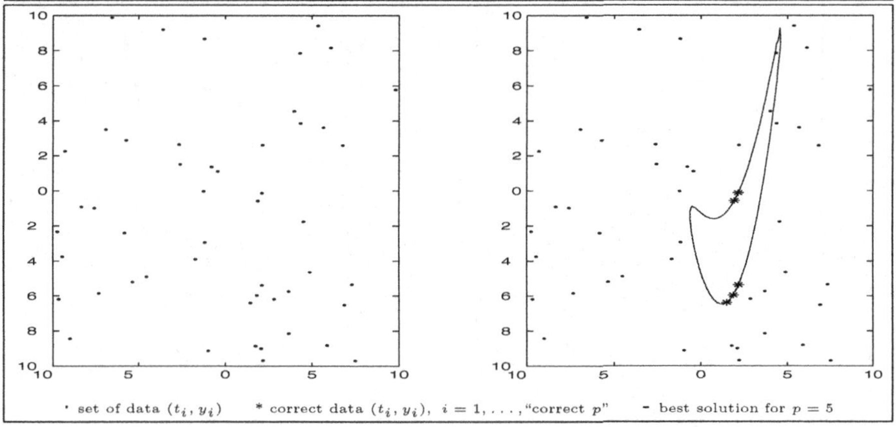

Table 10: $m = 50$, "correct p"$= 5$, $x^* = (1, -3, 7)$

20. J. Mann, R. Jager, T. Muller, G. Hocht, and P. Hubral. Common-reflection-surface stack - a real data example. *Journal of Applied Geophysics*, 42, 3-4:301–318, 1999.
21. D. L. Markin and R. G. Strongin. A method for solving multi-extremal problems with non-convex constraints, that uses a priori information about estimates of the optimum. *U.S.S.R. Comput. Maths. Math. Phys.*, 27(1):33–39, 1987. Pergamon Press 1988 (*Zh. vychisl. Mat. mat. Fiz.* 27(1):52-62, 1987).
22. J. M. Martínez. Inexact restoration: advances and perspectives. Invited talk at *Workshop on Control and Optimization.* Erice, Italy, July, 2001.
23. J. M. Martínez and E. A. Pilotta. Inexact restoration methods for nonlinear programming: advances and perspectives. To appear in *Optimization and Control with applications*, edited by L. Q. Qi, K. L. Teo and X. Q. Yang. Kluwer Academic Publishers.
24. M. May and L. Ragia. Spatial subgroup discovery applied to the analysis of vegetation data. *Lecture Notes in Artificial Intelligence*, 2569:49–61, 2002.
25. D. V. Nichita, S. Gómez, and E. Luna-Ortiz. Multiphase equilibria calculation by direct minimization of Gibbs free energy with a global optimization method. *Comput. Chem. Eng.*, 26:1703–1724, 2002.
26. D. V. Nichita, S. Gómez, and E. Luna-Ortiz. Multiphase equilibria calculation by direct minimization of Gibbs free energy using the tunneling global optimization method. *J. Can. Petrol. Technol.*, 43:13–16, 2004.

	Number of minimizations		Best solution obtained			
p	k_{max}	Successful Tunnelings	x	$f(x)$	Calls to Algorithm 2.1	Successful Step
8	500	1018	-1.00 -7.00 9.00	1.24E-05	403	step 6
9	500	1013	-1.00 -7.00 8.99	6.66E-05	40	step 6
10	500	1152	-1.00 -7.00 9.00	2.28E-08	760	step 6
11	500	1102	-1.00 -7.00 9.00	0.0001	695	step 6
12	500	1068	-1.00 -6.95 9.01	0.0646	458	step 6
13	500	1103	-1.00 -6.89 8.98	0.3416	1263	step 1

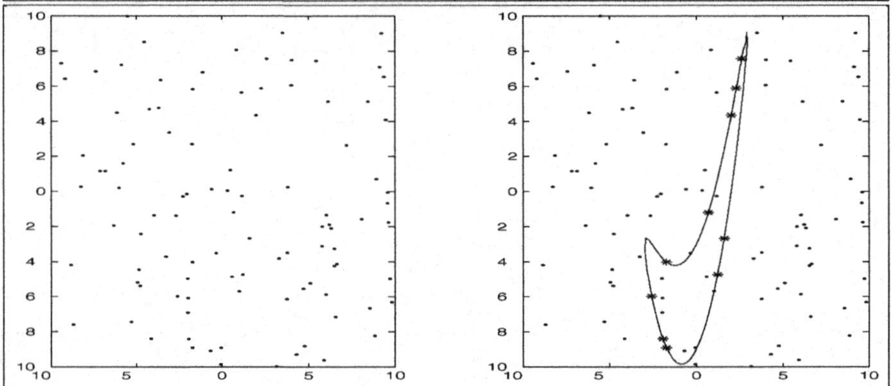

Table 11: $m = 100$, "correct p"$= 10$, $x^* = (-1, -7, 9)$

27. N. H. Pronko and L. Manheim. Hidden patterns - Studies in psychoanalitic literary criticism. *Psychol. Rec.*, 17:283–&, 1967.
28. D. Romero, C. Barron, and S. Gómez. The optimal geometry of Lennard-Jones clusters: 148-309. *Comput. Phys. Commun.*, 123:87–96, 1999.
29. P. Street. The logic and limits of "plant loyalty": Black workers, white labor, and corporate racial paternalism in Chicago's stockyards, 1916-1940. *Journal of Social History*, 29:659–&, 1996.
30. R. G. Strongin. On the convergence of an algorithm for finding a global extremum. *Engineering Cybernetics*, 11:549–555, 1973.
31. R. G. Strongin. *Numerical Methods for Multi-extremal Problems.* Nauka, Moscow, 1978. (in Russian).
32. R. G. Strongin. Numerical methods for multi-extremal nonlinear programming problems with nonconvex constraints. In V. F. Demyanov and D. Pallaschke, eds., *Lecture Notes in Economics and Mathematical Systems*, 225:278–282, 1984. Proceedings 1984. Springer-Verlag. IIASA, Laxenburg/Austria 1985.
33. R. G. Strongin. Algorithms for multi-extremal mathematical programming problems employing the set of joint space-filling curves. *Journal of Global Optimization*, 2:357–378, 1992.
34. R. G. Strongin and D. L. Markin. Minimization of multi-extremal functions with nonconvex constraints. *Cybernetics*, 22(4):486–493, 1986. Translated from Russian. Consultant Bureau. New York.

	Number of minimizations		Best solution obtained			
p	k_{max}	Successful Tunnelings	x	$f(x)$	Calls to Algorithm 2.1	Successful Step
13	100	216	0.00 -6.00 15.0	1.45E-15	107	step 6
14	100	207	-0.00 -6.00 15.0	1.31E-16	41	step 6
15	100	221	0.00 -6.00 15.0	2.49E-16	70	step 6
16	200	427	-0.00 -6.00 14.7	0.0675	169	step 6
17	200	481	-0.01 -6.01 15.5	0.6014	458	step 6
18	200	450	-0.02 -6.07 15.2	0.8290	515	step 6

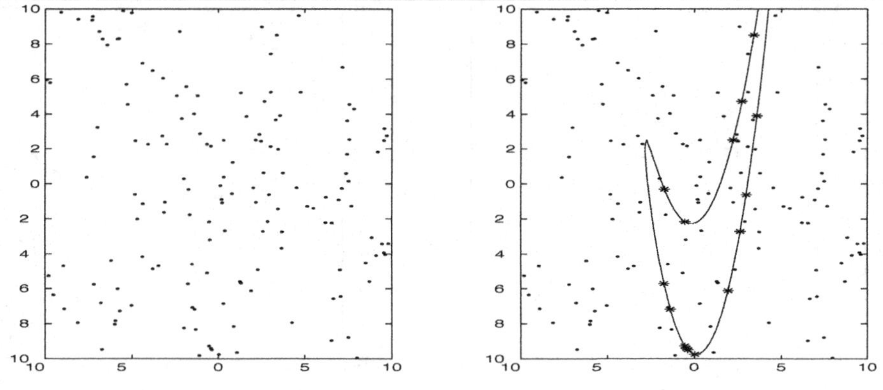

Table 12: $m = 150$, "correct p"$= 15$ $x^* = (0, -6, 15)$

35. J. W. Weaver. Hidden patterns in Joyce 'Portrait of the artist as a young man'. *South Atlantic Bulletin*, 41:63–63, 1976.

	Number of minimizations		Best solution obtained			
p	k_{max}	Successful Tunnelings	x	$f(x)$	Calls to Algorithm 2.1	Successful Step
8	1819	4181	0.00 4.00 9.00 5.00	1.73E-13	5063	step 6
9	500	1151	0.00 4.00 9.00 5.00	3.18E-13	1587	step 6
10	500	1091	0.00 4.00 9.00 5.00	2.39E-13	993	step 6
11	500	1218	-0.02 4.01 9.12 4.94	0.0006	1601	step 6
12	500	1226	-0.08 4.03 9.12 4.93	0.0008	1229	step 6
13	500	1183	0.31 4.22 8.14 6.06	0.0033	303	step 6

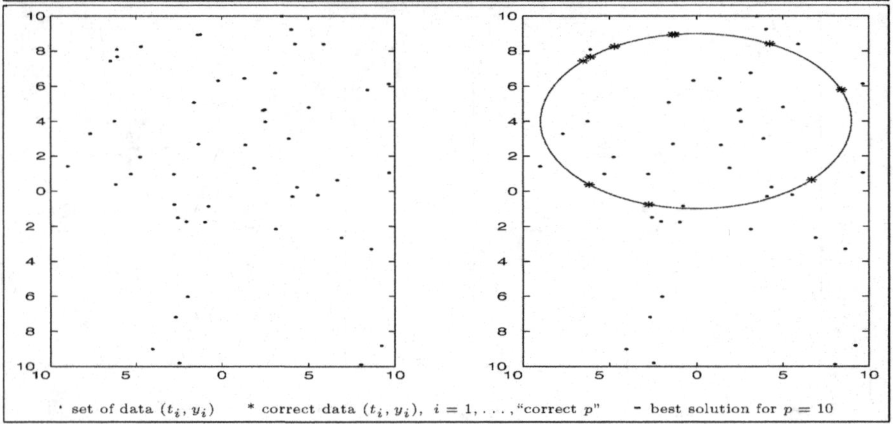

Table 13: $m = 50$, "correct p"$= 10$, $x^* = (0, 4, 9, 5)$

	Number of minimizations		Best solution obtained			
p	k_{max}	Successful Tunnelings	x	$f(x)$	Calls to Algorithm 2.1	Successful Step
18	1000	3008	0.01 -2.00 9.00 6.99	7.07E-6	1085	step 6
19	1000	2954	0.00 -2.00 8.99 7.00	6.00E-6	3057	step 1
20	6000	17667	-0.00 -2.00 9.00 7.00	2.58E-6	23201	step 6
21	1000	2966	0.01 -1.98 8.98 6.99	7.76E-5	3914	step 6
22	1000	2950	0.03 -1.99 8.93 7.04	0.0002	229	step 6
23	1000	2859	0.00 -1.99 8.90 7.04	0.0004	1728	step 6

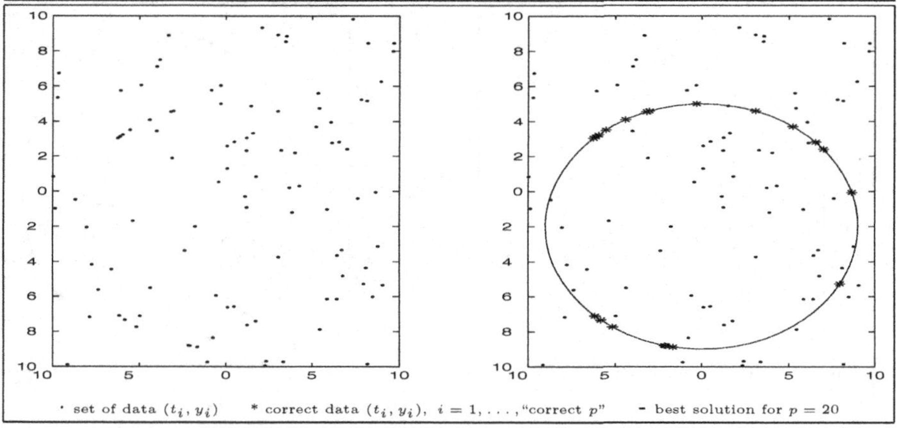

Table 14: $m = 100$, "correct p"$= 20$, $x^* = (0, -2, 9, 7)$

	Number of minimizations		Best solution obtained			
p	k_{max}	Successful Tunnelings	x	$f(x)$	Calls to Algorithm 2.1	Successful Step
28	1000	3494	0.97 -0.00 7.99 9.00	7.43E-5	1660	step 6
29	1000	3428	1.00 -0.02 8.00 9.00	2.35E-5	1959	step 6
30	1000	3311	1.00 -0.00 8.00 8.99	3.85E-6	1492	step 6
31	1000	3388	1.02 0.01 7.96 8.99	0.0002	333	step 6
32	1000	3376	1.04 0.01 8.01 9.00	0.0003	4149	step 6
33	1000	3338	1.00 0.02 8.03 9.02	0.0004	1761	step 6

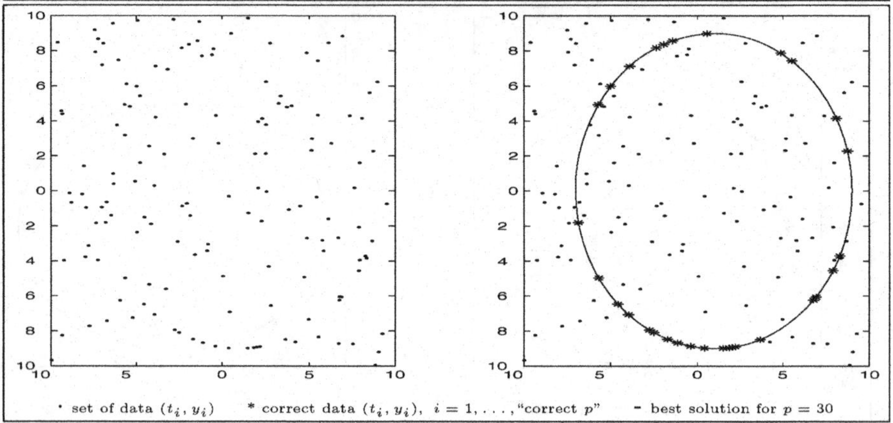

Table 15: $m = 150$, "correct p"$= 30$, $x^* = (1, 0, 8, 9)$

On generating Instances for the Molecular Distance Geometry Problem

Carlile Lavor

Department of Applied Mathematics (IMECC-UNICAMP), State University of Campinas, CP 6065, 13081-970, Campinas-SP, Brazil carlile@ime.unicamp.br

Summary. The molecular distance geometry problem can be stated as the determination of the three-dimensional structure of a molecule using a set of distances between pairs of atoms. It can be formulated as a global minimization problem, where the main difficulty is the exponential increasing of local minimizers with the size of the molecule. The aim of this study is to generate new instances for the molecular distance geometry problem that can be used in order to test algorithms designed to solve it.

Key words: Molecular distance geometry problem, instance generation, NMR spectroscopy.

1 Introduction

The molecular distance geometry problem (MDGP) can be defined as the problem of finding Cartesian coordinates $x_1, ..., x_N \in \mathbb{R}^3$ of the atoms of a molecule such that

$$||x_i - x_j|| = d_{i,j} \qquad ([i,j] \in S), \tag{1}$$

where S is the set of pairs of atoms $[i,j]$ whose Euclidean distances $d_{i,j}$ are known. If all distances are given, the problem can be solved in linear time [3]. Otherwise, the problem is NP-hard [8].

The distances $d_{i,j}$, in (1), can be obtained, for example, with nuclear magnetic resonance (NMR) data and with knowledge on bond lengths and bond angles of a molecule. Usually, NMR data only provide distances between certain close-range hydrogen atoms [1].

The MDGP can be formulated as a global minimization problem, where the objective function can be given by

$$f(x_1, ..., x_N) = \sum_{[i,j]\in S} \left(||x_i - x_j||^2 - d_{i,j}^2\right)^2. \tag{2}$$

It may easily be seen that $x_1, ..., x_N \in \mathbb{R}^3$ solve the problem if and only if $f(x_1, ..., x_N) = 0$. For more details and other approaches for the MDGP see, for example, [3, 2, 4, 5, 6].

In [5], Moré and Wu consider the MDGP by applying the global continuation approach to obtain a global minimizer of the function (2). To test the method, they used instances based on a molecule positioned in a three-dimensional lattice.

The aim of this study is to generate new instances for the MDGP which are based on the parameters of a molecular potential energy function defined in [7].

In Section 2, we briefly describe Moré-Wu instances. Section 3 gives the model in which the new instances are based, explains how the instances are generated and presents some examples of the new instances. Section 4 ends with some conclusions.

2 Moré-Wu instances

The instances used in [5] are based on a molecule with s^3 atoms ($s = 1, 2, 3, ...$) positioned in the three-dimensional lattice defined by

$$\{(i_1, i_2, i_3) \in \mathbb{R}^3 : 0 \leq i_k \leq s - 1,\ k = 1, 2, 3\}.$$

There are two sets of instances. The first one has distances for both near and relatively distant atoms, while the second one only has distances for near atoms.

In the first set of instances, an order is defined for the atoms of the lattice by letting atom i be the atom at position (i_1, i_2, i_3), where

$$i = 1 + i_1 + s i_2 + s^2 i_3,$$

and the set S, in (1), is defined by

$$S = \{[i, j] : |i - j| \leq s^2\}.$$

For example, for a molecule with 8 atoms ($s = 2$), the sequence of atoms is

$$\begin{aligned}
x_1 &= (0, 0, 0),\\
x_2 &= (1, 0, 0),\\
x_3 &= (0, 1, 0),\\
x_4 &= (1, 1, 0),\\
x_5 &= (0, 0, 1),\\
x_6 &= (1, 0, 1),\\
x_7 &= (0, 1, 1),\\
x_8 &= (1, 1, 1),
\end{aligned}$$

and the set S is given by

$$\begin{aligned} S = \{&[1,2],[1,3],[1,4],[1,5],\\ &[2,1],[2,3],[2,4],[2,5],[2,6],\\ &[3,1],[3,2],[3,4],[3,5],[3,6],[3,7],\\ &[4,1],[4,2],[4,3],[4,5],[4,6],[4,7],[4,8],\\ &[5,1],[5,2],[5,3],[5,4],[5,6],[5,7],[5,8],\\ &[6,2],[6,3],[6,4],[6,5],[6,7],[6,8],\\ &[7,3],[7,4],[7,5],[7,6],[7,8]\}. \end{aligned}$$

In the second set of instances, the set S is defined by

$$S = \{[i,j] : ||x_i - x_j|| \leq \sqrt{r}\},$$

where r is a parameter that defines the cutoff value. For example, for $r = 16$, we have

$$\begin{aligned} S = \{&[1,2],[1,3],[1,4],[1,5],[1,6],[1,7],[1,8],\\ &[2,1],[2,3],[2,4],[2,5],[2,6],[2,7],[2,8],\\ &[3,1],[3,2],[3,4],[3,5],[3,6],[3,7],[3,8],\\ &[4,1],[4,2],[4,3],[4,5],[4,6],[4,7],[4,8],\\ &[5,1],[5,2],[5,3],[5,4],[5,6],[5,7],[5,8],\\ &[6,1],[6,2],[6,3],[6,4],[6,5],[6,7],[6,8],\\ &[7,1],[7,2],[7,3],[7,4],[7,5],[7,6],[7,8]\}. \end{aligned}$$

3 New instances

3.1 The background model

The new instances for the MDGP are based on the model proposed by Philips et al. [7]. This model considers a molecule as being a chain of N atoms with Cartesian coordinates given by $x_1, ..., x_N \in \mathbb{R}^3$. For every pair of consecutive atoms i, j, let r_{ij} be the bond length which is the Euclidean distance between them. For every three consecutive atoms i, j, k, let θ_{ik} be the bond angle corresponding to the relative position of the third atom with respect to the line containing the previous two. Likewise, for every four consecutive atoms i, j, k, l, let ω_{il} be the angle, called the torsion angle, between the normals through the planes determined by the atoms i, j, k and j, k, l. The three-dimensional structure of a molecule is determined by minimizing the sum of the following terms:

$$\begin{aligned} f_d &= \sum_{[i,j]\in M_1} c^r_{ij}(r_{ij} - r^0_{ij})^2, \\ f_a &= \sum_{[i,j]\in M_2} c^\theta_{ij}(\theta_{ij} - \theta^0_{ij})^2, \\ f_\omega &= \sum_{[i,j]\in M_3} c^\omega_{ij}(1 + \cos(n_{ij}\omega_{ij} - \omega^0_{ij})), \\ f_v &= \sum_{[i,j]\in M_4} \left(\frac{A_{ij}}{r^{12}_{ij}} - \frac{B_{ij}}{r^6_{ij}}\right). \end{aligned} \tag{3}$$

The terms f_d, f_a, f_ω are the potentials corresponding to bond lengths, bond angles, and torsion angles, respectively. The factor c^r_{ij} is the bond stretching force constant, c^θ_{ij} is the angle bending force constant, and c^ω_{ij} is the torsion force constant. The factors r^0_{ij} and θ^0_{ij} represent the equilibrium values for bond length and bond angle, respectively. The constant n_{ij} defines the number of minima involved and the constant ω^0_{ij} is the phase angle that defines the position of the minima. The term f_v is the Lennard-Jones potential, where A_{ij} and B_{ij} are constants defined by each pair $[i,j]$ and r_{ij} is the Euclidean distance between atoms i and j. The sets M_1, M_2, M_3, M_4 are the sets of pairs of atoms separated by one covalent bond, two covalent bonds, three covalent bonds, and more than two covalent bonds, respectively.

In most conformation calculations, all bond lengths and bond angles are assumed to be fixed at their equilibrium values r^0_{ij} and θ^0_{ij}, respectively. Thus, the first three atoms in the chain can be fixed: the first one is fixed at $x_1 = (0,0,0)$, the second one is positioned at $x_2 = (-r_{12}, 0, 0)$, and the third one is fixed at $x_3 = (r_{23}\cos(\theta_{13}) - r_{12}, r_{23}\sin(\theta_{13}), 0)$. The fourth atom is determined by the torsion angle ω_{14}; the fifth atom is determined by the torsion angles ω_{14} and ω_{25}; the sixth atom is determined by the torsion angles ω_{14}, ω_{25}, and ω_{36}; and so on.

In that model, the bond lengths and bond angles are set to $r_{ij} = 1.526$ Å (for all $[i,j] \in M_1$) and $\theta_{ij} = 109.5°$ (for all $[i,j] \in M_2$), respectively. Also, $c^\omega_{ij} = 1$, $n_{ij} = 3$, and $\omega^0_{ij} = 0$ (for all $[i,j] \in M_3$), providing three "preferred" torsion angles at $60°, 180°$, and $300°$. Using these parameters, we can generate distances between pairs of atoms and obtain instances for the MDGP.

3.2 Generation of instances

Considering bond lengths r_{ij} and bond angles θ_{ij} fixed ($r_{ij} = 1.526$ Å, $\forall [i,j] \in M_1$; $\theta_{ij} = 109.5°$, $\forall [i,j] \in M_2$), the three-dimensional structure of a molecule can be completely determined by its torsion angles. In [7], the computational results were obtained for problems with 4 through 30 atoms. For a molecule with 18 atoms, for example, 15 torsion angles were obtained:

$$\begin{gathered} 181°, 176°, 293°, 292°, 165°, 294°, 193°, \\ 166°, 61°, 197°, 65°, 66°, 193°, 67°, 181°. \end{gathered} \tag{4}$$

Note that these values can be viewed as perturbations of the "preferred" torsion angles $60°, 180°$, and $300°$, cited above. Thus, based on the model described in the previous subsection, we can generate torsion angles for a molecule, for example, selecting one value ω from the set

$$\{60°, 180°, 300°\} \tag{5}$$

and another one from the set

$$\{\omega + i : i = -15°, ..., 15°\}. \tag{6}$$

Both of these selections are random.

To generate distances in order to define the set S, in (1), we first obtain Cartesian coordinates for each atom of the chain $(x_{n_1}, x_{n_2}, x_{n_3})$, using the following matrices [7] (note that the first three atoms of the chain are fixed):

$$\begin{bmatrix} x_{n_1} \\ x_{n_2} \\ x_{n_3} \\ 1 \end{bmatrix} = B_1 B_2 ... B_n \begin{bmatrix} 0 \\ 0 \\ 0 \\ 1 \end{bmatrix} \quad (n = 1, ..., N),$$

where

$$B_1 = \begin{bmatrix} 1&0&0&0 \\ 0&1&0&0 \\ 0&0&1&0 \\ 0&0&0&1 \end{bmatrix}, \quad B_2 = \begin{bmatrix} -1&0&0&-r_{12} \\ 0&1&0&0 \\ 0&0&-1&0 \\ 0&0&0&1 \end{bmatrix},$$

$$B_3 = \begin{bmatrix} -\cos\theta_{13} & -\sin\theta_{13} & 0 & -r_{23}\cos\theta_{13} \\ \sin\theta_{13} & -\cos\theta_{13} & 0 & r_{23}\sin\theta_{13} \\ 0 & 0 & 1 & 0 \\ 0 & 0 & 0 & 1 \end{bmatrix},$$

and $B_i =$

$$\begin{bmatrix} -\cos\theta_{(i-2)i} & -\sin\theta_{(i-2)i} & 0 & -r_{(i-1)i}\cos\theta_{(i-2)i} \\ \sin\theta_{(i-2)i}\cos\omega_{(i-3)i} & -\cos\theta_{(i-2)i}\cos\omega_{(i-3)i} & -\sin\omega_{(i-3)i} & r_{(i-1)i}\sin\theta_{(i-2)i}\cos\omega_{(i-3)i} \\ \sin\theta_{(i-2)i}\sin\omega_{(i-3)i} & -\cos\theta_{(i-2)i}\sin\omega_{(i-3)i} & \cos\omega_{(i-3)i} & r_{(i-1)i}\sin\theta_{(i-2)i}\sin\omega_{(i-3)i} \\ 0 & 0 & 0 & 1 \end{bmatrix},$$

for $i = 4, ..., N$ ($r_{ij} = 1.526$ Å, $\forall\ [i,j] \in M_1$; $\theta_{ij} = 109.5°$, $\forall\ [i,j] \in M_2$; and ω_{ij}, $\forall\ [i,j] \in M_3$, is obtained according to the rule explained above).

Recall that the problem is to determine the Cartesian coordinates for atoms of a molecule using only a subset of all distances between them. With a prescribed cutoff value d, we can generate an instance for the MDGP defining the set S by

$$S = \{[i,j] : ||x_i - x_j|| \leq d\}.$$

3.3 Examples

Usually, the distance data obtained from NMR experiments are less than or equal to 5 Å [1]. For each example in this subsection, we will select distances with a cutoff value equal to 4 Å.

Using torsion angles defined by (4), from a total of 153 pairs of atoms, 67 pairs were selected (to represent the set S, we consider $[i,j]$ and $[j,i]$ associated to the same pair of atoms):

$$
\begin{aligned}
S = \{ \; & [1,2],[1,3],[1,4],[1,14],[1,15],[1,16],[1,17],[1,18],\\
& [2,1],[2,3],[2,4],[2,5],[2,15],[2,16],[2,17],[2,18],\\
& [3,1],[3,2],[3,4],[3,5],[3,6],[3,7],[3,15],[3,16],[3,17],[3,18],\\
& [4,1],[4,2],[4,3],[4,5],[4,6],[4,7],[4,17],[4,18],\\
& [5,2],[5,3],[5,4],[5,6],[5,7],[5,8],\\
& [6,3],[6,4],[6,5],[6,7],[6,8],[6,9],\\
& [7,3],[7,4],[7,5],[7,6],[7,8],[7,9],[7,10],\\
& [8,5],[8,6],[8,7],[8,9],[8,10],[8,11],\\
& [9,6],[9,7],[9,8],[9,10],[9,11],[9,12],[9,15],[9,17],\\
& [10,7],[10,8],[10,9],[10,11],[10,12],[10,13],\\
& [11,8],[11,9],[11,10],[11,12],[11,13],[11,14],[11,15],\\
& [12,9],[12,10],[12,11],[12,13],[12,14],[12,15],\\
& [13,10],[13,11],[13,12],[13,14],[13,15],[13,16],\\
& [14,1],[14,11],[14,12],[14,13],[14,15],[14,16],[14,17],\\
& [15,1],[15,2],[15,3],[15,9],[15,11],[15,12],[15,13],[15,14],[15,16],[15,17],[15,18],\\
& [16,1],[16,2],[16,3],[16,13],[16,14],[16,15],[16,17],[16,18],\\
& [17,1],[17,2],[17,3],[17,4],[17,9],[17,14],[17,15],[17,16],[17,18],\\
& [18,1],[18,2],[18,3],[18,4],[18,15],[18,16],[18,17] \quad \}.
\end{aligned}
$$

We can obtain a better representation of selected pairs of atoms by defining a matrix $A \in \{0,1\}^{N \times N}$ such that

$$
A_{ij} = \begin{cases} 1, & \text{if } [i,j] \in S \\ 0, & \text{if } [i,j] \notin S \end{cases} .
$$

We define $A_{ii} = i$ to indicate, at the i-th row, which atoms are close to atom i according to the prescribed cutoff value. For example, the matrix A associated to the set S above is:

$$A = \begin{bmatrix}
1&1&1&1&0&0&0&0&0&0&0&0&0&1&1&1&1&1\\
1&2&1&1&1&0&0&0&0&0&0&0&0&0&1&1&1&1\\
1&1&3&1&1&1&1&0&0&0&0&0&0&0&1&1&1&1\\
1&1&1&4&1&1&1&0&0&0&0&0&0&0&0&0&1&1\\
0&1&1&1&5&1&1&1&0&0&0&0&0&0&0&0&0&0\\
0&0&1&1&1&6&1&1&1&0&0&0&0&0&0&0&0&0\\
0&0&1&1&1&1&7&1&1&1&0&0&0&0&0&0&0&0\\
0&0&0&0&1&1&1&8&1&1&1&0&0&0&0&0&0&0\\
0&0&0&0&0&1&1&1&9&1&1&1&0&0&1&0&1&0\\
0&0&0&0&0&0&1&1&1&10&1&1&1&0&0&0&0&0\\
0&0&0&0&0&0&0&1&1&1&11&1&1&1&1&0&0&0\\
0&0&0&0&0&0&0&0&1&1&1&12&1&1&1&0&0&0\\
0&0&0&0&0&0&0&0&0&1&1&1&13&1&1&1&0&0\\
1&0&0&0&0&0&0&0&0&0&1&1&1&14&1&1&1&0\\
1&1&1&0&0&0&0&0&1&0&1&1&1&1&15&1&1&1\\
1&1&1&0&0&0&0&0&0&0&0&0&1&1&1&16&1&1\\
1&1&1&1&0&0&0&0&1&0&0&0&0&1&1&1&17&1\\
1&1&1&1&0&0&0&0&0&0&0&0&0&0&1&1&1&18
\end{bmatrix}.$$

Note that, using this matrix, it is easy to see that in addition to the 4-th atom being close to the atoms $1, 2, 3$ and $5, 6, 7$, it is also close to the last two atoms of the chain. The three-dimensional structure of the molecule defined by the torsion angles (4) is given in Figure 1.

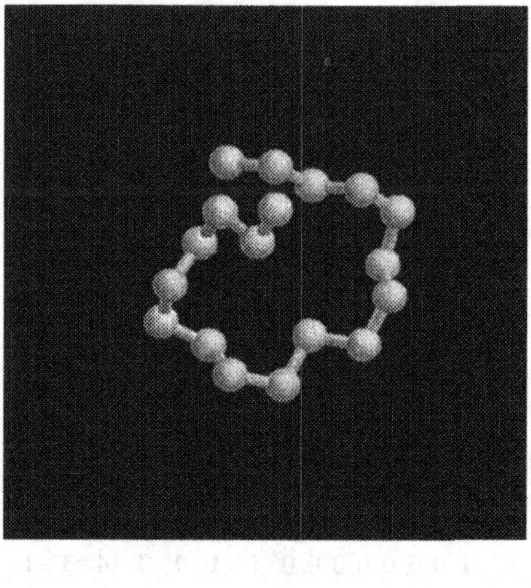

Fig. 1. Molecule associated to the matrix A.

Now we briefly describe three new examples, also with 18 atoms. Using the schema given in subsection 3.2, 10 instances were randomly generated. We consider one with 78 selected pairs of atoms (intermediate number of 10 runs), another with 56 selected pairs of atoms (minimum number of 10 runs), and the last one with 113 selected pairs of atoms (maximum number of 10 runs). Below, we present the matrix representations A_{int}, $A_{\min}$, and $A_{\max}$, for these three cases. For the first example, we also give the three-dimensional structure of the corresponding molecule (Figure 2).

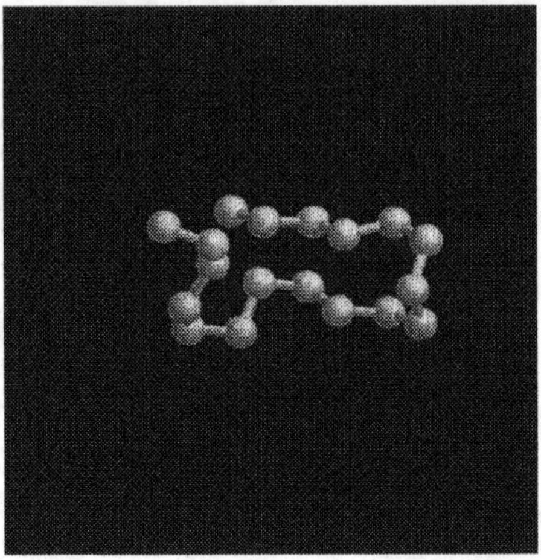

Fig. 2. Molecule associated to the matrix A_{int}.

Instance with 78 selected pairs:

$$A_{int} = \begin{bmatrix}
1&1&1&1&0&0&0&0&0&0&1&1&1&1&0&1&0&0\\
1&2&1&1&1&0&0&0&0&1&1&1&1&1&1&1&1&0\\
1&1&3&1&1&1&0&0&1&1&1&1&1&0&0&1&0&0\\
1&1&1&4&1&1&1&1&1&1&1&1&0&0&0&0&0&0\\
0&1&1&1&5&1&1&1&1&1&1&0&0&0&0&0&0&0\\
0&0&1&1&1&6&1&1&1&1&0&0&0&0&0&0&0&0\\
0&0&0&1&1&1&7&1&1&1&0&0&0&0&0&0&0&0\\
0&0&0&1&1&1&1&8&1&1&1&0&0&0&0&0&0&0\\
0&0&1&1&1&1&1&1&9&1&1&1&0&0&0&0&0&0\\
0&1&1&1&1&1&1&1&1&10&1&1&1&0&0&0&0&0\\
1&1&1&1&1&0&0&1&1&1&11&1&1&1&0&1&0&0\\
1&1&1&1&0&0&0&0&1&1&1&12&1&1&1&1&0&0\\
1&1&1&0&0&0&0&0&0&1&1&1&13&1&1&1&0&0\\
1&1&0&0&0&0&0&0&0&0&1&1&1&14&1&1&1&0\\
0&1&0&0&0&0&0&0&0&0&0&1&1&1&15&1&1&1\\
1&1&1&0&0&0&0&0&0&0&1&1&1&1&1&16&1&1\\
0&1&0&0&0&0&0&0&0&0&0&0&0&1&1&1&17&1\\
0&0&0&0&0&0&0&0&0&0&0&0&0&0&1&1&1&18
\end{bmatrix}.$$

Instance with 56 selected pairs:

$$A_{\min} = \begin{bmatrix}
1&1&1&1&0&0&0&0&0&0&0&0&0&0&0&0&0&0\\
1&2&1&1&1&0&0&0&0&0&0&0&0&0&0&0&0&0\\
1&1&3&1&1&1&0&0&0&0&0&0&0&0&0&0&0&0\\
1&1&1&4&1&1&1&0&0&0&0&0&0&0&0&0&0&0\\
0&1&1&1&5&1&1&1&1&0&0&0&0&0&0&0&0&0\\
0&0&1&1&1&6&1&1&1&1&1&0&0&0&0&0&0&0\\
0&0&0&1&1&1&7&1&1&1&1&1&0&0&0&0&0&0\\
0&0&0&0&1&1&1&8&1&1&1&0&0&0&0&0&0&0\\
0&0&0&0&1&1&1&1&9&1&1&1&0&0&0&0&0&0\\
0&0&0&0&0&1&1&1&1&10&1&1&1&0&0&0&0&0\\
0&0&0&0&0&1&1&1&1&1&11&1&1&1&0&0&0&0\\
0&0&0&0&0&0&1&0&1&1&1&12&1&1&1&0&1&0\\
0&0&0&0&0&0&0&0&0&1&1&1&13&1&1&1&1&1\\
0&0&0&0&0&0&0&0&0&0&1&1&1&14&1&1&1&0\\
0&0&0&0&0&0&0&0&0&0&0&1&1&1&15&1&1&1\\
0&0&0&0&0&0&0&0&0&0&0&0&1&1&1&16&1&1\\
0&0&0&0&0&0&0&0&0&0&0&1&1&1&1&1&17&1\\
0&0&0&0&0&0&0&0&0&0&0&0&1&0&1&1&1&18
\end{bmatrix}.$$

Instance with 113 selected pairs:

$$A_{\max} = \begin{bmatrix}
1&1&1&1&1&1&1&1&1&0&0&0&1&1&1&1&1&1\\
1&2&1&1&1&1&1&0&0&0&0&0&0&0&1&1&1&1\\
1&1&3&1&1&1&0&0&0&0&0&0&1&0&0&1&1&1\\
1&1&1&4&1&1&1&0&0&0&0&0&1&0&1&1&1&1\\
1&1&1&1&5&1&1&1&1&0&0&1&1&1&1&1&1&1\\
1&1&1&1&1&6&1&1&1&0&0&0&1&1&1&1&1&1\\
1&1&0&1&1&1&7&1&1&1&0&1&1&1&1&1&1&1\\
1&0&0&0&1&1&1&8&1&1&1&1&1&1&1&1&1&1\\
1&0&0&0&1&1&1&1&9&1&1&1&1&1&1&1&1&0\\
0&0&0&0&0&0&1&1&1&10&1&1&1&1&1&0&0&0\\
0&0&0&0&0&0&0&1&1&1&11&1&1&1&0&0&0&0\\
0&0&0&0&1&0&1&1&1&1&1&12&1&1&1&0&1&0\\
1&0&1&1&1&1&1&1&1&1&1&1&13&1&1&1&1&1\\
1&0&0&0&1&1&1&1&1&1&1&1&1&14&1&1&1&1\\
1&1&0&1&1&1&1&1&1&1&0&1&1&1&15&1&1&1\\
1&1&1&1&1&1&1&1&1&0&0&0&1&1&1&16&1&1\\
1&1&1&1&1&1&1&1&1&0&0&1&1&1&1&1&17&1\\
1&1&1&1&1&1&1&1&0&0&0&0&1&1&1&1&1&18
\end{bmatrix}.$$

4 Conclusion

This study presented a new way to generate instances for the molecular distance geometry problem which is based on the parameters of a molecular potential energy function given in [7].

In [5], Moré and Wu consider the molecular distance geometry problem by applying the global continuation approach where they used instances based on a molecule positioned in a three-dimensional lattice.

The instances generated by the method proposed here have a more flexible geometric conformation, making the corresponding molecules more "realistic". In the examples of subsection 3.3, we have seen that we can obtain sets of instances with a distinct number of selected pairs of atoms, trying to capture different features in distance data from real problems.

Based on the idea presented here, we can generate many other instances for the molecular distance geometry problem. For example, we can use a potential energy function different from the function (3) and, manipulating its parameters, we can adjust the values of sets (5) and (6) in order to generate instances with distinct characteristics.

Acknowledgments

The author would like to thank FAPERJ and CNPq for their support.

References

1. Creighton, T.E. (1993). *Proteins: Structures and Molecular Properties.* W.H. Freeman and Company, New York.
2. Crippen, G.M. and Havel, T.F. (1988). *Distance Geometry and Molecular Conformation.* John Wiley & Sons, New York.
3. Dong, Q. and Wu, Z. (2002). A linear-time algorithm for solving the molecular distance geometry problem with exact inter-atomic distances. *Journal of Global Optimization*, 22:365-375.
4. Hendrickson, B.A. (1995). The molecule problem: exploiting structure in global optimization. *SIAM Journal on Optimization*, 5:835-857.
5. Moré, J.J. and Wu, Z. (1997). Global continuation for distance geometry problems. *SIAM Journal on Optimization*, 7:814-836.
6. Moré, J.J. and Wu, Z. (1999). Distance geometry optimization for protein structures. *Journal of Global Optimization*, 15:219-234.
7. Phillips, A.T., Rosen, J.B., and Walke, V.H. (1996). Molecular structure determination by convex underestimation of local energy minima. In *DIMACS Series in Discrete Mathematics and Theoretical Computer Science,* Volume 23, pages 181-198, Providence, American Mathematical Society.
8. Saxe, J.B. (1979). *Embeddability of weighted graphs in k-space is strongly NP-hard.* In Proc. 17th Allerton Conference in Communications, Control, and Computing, Monticello, IL, pages 480-489.

Index